Ethical Issues in E–Business:
Models and Frameworks

Daniel E. Palmer
Kent State University, USA

BUSINESS SCIENCE REFERENCE

Hershey · New York

Director of Editorial Content:	Kristin Klinger
Director of Book Publications:	Julia Mosemann
Acquisitions Editor:	Lindsay Johnston
Development Editor:	Mike Killian
Publishing Assistant:	Keith Glazewski
Typesetter:	Keith Glazewski
Production Editor:	Jamie Snavely
Cover Design:	Lisa Tosheff
Printed at:	Yurchak Printing Inc.

Published in the United States of America by
Business Science Reference (an imprint of IGI Global)
701 E. Chocolate Avenue
Hershey PA 17033
Tel: 717-533-8845
Fax: 717-533-8661
E-mail: cust@igi-global.com
Web site: http://www.igi-global.com/reference

Library of Congress Cataloging-in-Publication Data

Ethical issues in e-business : models and frameworks / Daniel E. Palmer, editor.
     p. cm.
  Includes bibliographical references and index.
Summary: "This book provides a comprehensive overview of the most important ethical issues associated with the expanding world of e-business, and offers relevant theoretical frameworks to ethical issues in all significant areas of e-business"--Provided by publisher.   ISBN 978-1-61520-615-5 (hardcover) -- ISBN 978-1-61520-616-2 (ebook) 1. Electronic commerce--Moral and ethical aspects. 2. Business ethics. I. Palmer, Daniel E. II. Title.

HF5548.32.E84 2010
174'.4--dc22

2010000042

British Cataloguing in Publication Data
A Cataloguing in Publication record for this book is available from the British Library.

All work contributed to this book is new, previously-unpublished material. The views expressed in this book are those of the authors, but not necessarily of the publisher.

Editorial Advisory Board

Table of Contents

Section 1
The New Landscape of Business: Markets and Morals in the Age of E-Business

Section 2
Anonymity, Trust, and Loyalty on the Internet

Section 3
Marketing Ethics in E-Business

Section 4
Privacy and Property Rights Online

Section 5
Ethical Issues in Public Policy and Communication

Chapter 14

Detailed Table of Contents

Section 1
The New Landscape of Business: Markets and Morals in the Age of E-Business

Chapter 1

Daniel E. Palmer, Kent State University, USA

Chapter 1 examines the major ways in which e-business is transforming business practices, and the manner in which these transformative changes are raising new ethical issues. The author identifies three key transformative aspects of e-business and seeks to illustrate their ethical significance in terms of their potential to affect stakeholder relations. In doing so, the author demonstrates how the common model of stakeholder theory in business ethics can be adopted to illuminate and direct our thinking about ethical issues in e-business. The chapter concludes by arguing that two key moral norms, the commitment to transparency and respect for persons, can serve as important moral ideals to guide and shape our thinking about ethics in the new frontier of e-business.

Chapter 2

Susan Emens, Kent State University, USA

Chapter 2 investigates the changing landscape of e-business by examining the shift in business models involved in e-business. Utilizing the idea of a paradigm shift in examining these changes, the chapter looks at the way in which business parties, such as buyers and suppliers, relate in business transactions in e-commerce. Here, the chapter uses the example of online reverse auctions to illustrate both the nature of the paradigm change as well as its ethical implications for buyers and sellers. While the chapter argues that e-business practices such as online reverse auctions can have significant benefits for businesses, it shows that they also can affect the trust that exists between business parties. As trust is an essential component of business ethics, the chapter maintains that finding ways to establish trust in e-commerce transactions should be encouraged in the new paradigm of e-business.

Section 2
Anonymity, Trust, and Loyalty on the Internet

Chapter 3 addresses one of the key barriers to fostering trust in e-business: the anonymity of the Internet. E-business transactions are often anonymous in so far as the parties involved have no direct, face-to-face, relationship with the persons they are dealing with. This chapter argues that the problems of trust arise in e-commerce since Internet interactions take place without the kind of basis in personal, face-to-face, interactions that traditionally serve as the source of trust in business. The chapter uses the ethical and social theory of Thomas Hobbes to both illustrate these issues and to offer a potential solution to the problem of anonymity on the Internet. In this respect, the chapter maintains that institutional or social regulations may be necessary in e-commerce in order to appropriately remedy the lack of traditional mechanisms of trust in e-business.

In chapter 4 the author argues that in transcending the limitations of "real world" business transactions, e-business both affords greater market opportunities to businesses, but also significantly exacerbates problems of trust. Because Internet transactions involve no direct personal contact, the chapter maintains that the motivation to be moral can be significantly diminished in e-business transactions. The chapter argues that merely appealing to traditional moral principles will not sufficiently address issues of trust on the Internet, since this does not address the emotional component that typically serves to ground trust in personal relations. As such, the author appeals to the moral theory of sentimentalism as a possible means of resolving issues of trust in e-business. The chapter argues that the sentimentalist approach can provide an account of how to develop trust in e-commerce that will properly motivate persons to engage in ethical forms of e-business.

This chapter seeks to determine how to establish means of fostering trust in Internet transactions from a legal perspective. The decentralized and international nature of the e-commerce makes traditional forms of commercial regulation more difficult to enforce in e-business. Given this difficulty, the chapter suggests that considerations of non-traditional approaches to regulating e-business are warranted. In this regard, the chapter points to fiduciary law as one possible means of facilitating trust in e-business. The fiduciary concept focuses upon the responsibilities that agents have to preserve the interests of those to whom they are entrusted. Fiduciary law, the chapter notes, is geared toward maintaining the integrity of crucial social and economic relations and thus would aid in the facilitation of trust and loyalty in e-commerce.

Section 3
Marketing Ethics in E-Business

Chapter 6

Abe Zakhem, Seton Hall University, USA

In Chapter 6 the author examines the emerging use of Mobile Location Based Services (MLBS) for e-marketing purposes. The chapter argues that the kind of access to and information about consumers that such technologies provide companies generate conditions of vulnerability and dependency for users. Adopting a fiduciary framework, the chapter goes on to suggest that given the nature of this vulnerability and dependency, we are justified in applying fiduciary principles to the relationship between MLBS providers and users. As such, the chapter recommends regulating e-commercial relations utilizing such technologies along fiduciary lines.

Chapter 7

J.J. Sylvia IV, The University of Southern Mississippi, USA

Chapter 7 examines one family of such forms of online research: the use of A/B and multivariate testing for website optimization. Such online modeling techniques are used by companies to test various aspects of website design, with the aim of increasing average user orders and decreasing abandoned shopping carts. While such techniques may seem benign, the chapter maintains that the kind of manipulation they involve is actually quite ethically significant. The chapter examines such practices from the perspective of three different ethical theories. The chapter concludes by examining the manipulative aspects of A/B and multivariate testing and suggesting that more explicit codes of conduct for the usage of such testing need to be developed.

Chapter 8

Erkan Özdemir, Uludag University, Turkey

Chapter 8 uses the traditional notion of the marketing mix to illuminate the ethical issues of online marketing. The traditional idea of the marketing mix categorizes that marketing strategies of companies into four crucial elements: product, price, place, and promotion. In Chapter 8 the author examines each of these four elements in respect to online marketing practices. The author seeks to demonstrate that online marketing practices involve unique issues in regards to each of the four elements, and that in each case ethical issues can be raised as well. By examining the distinctive ethical issues involved in the marketing mix in online marketing practices, the author both illuminates the ethically sensitive areas of online marketing as well as offers a framework for responding to those issues.

This chapter looks at the efforts of companies to market their Corporate Social Responsibility (CSR) efforts to consumers online. The Internet provides companies with the means to do so cheaply and with the ability to reach a greater audience than with more traditional forms of advertising. However, the chapter argues that as a medium of communication, the Internet poses particular challenges in terms of the epistemological constraints it places on users. These issues are especially important, the chapter maintains, when addressing claims of corporate social responsibility online. The chapter argues that while there is nothing intrinsically wrong with seeking to promote a good corporate image online, companies should do so in an ethically responsible manner. In this regard, the chapter concludes by offering practical guidelines for corporations to use in promoting morally good conduct online.

Section 4
Privacy and Property Rights Online

In Chapter 10 the author addresses issues of privacy in e-business. Comparing the traditional tale of Lady Godiva and Peeping Tom with the contemporary case of Facebook's Beacon program, the author maintains that while the underlying concept of privacy has not changed, we must rethink the manner in which we recognize and respect privacy in light of the changes in what can be done with personal information as a result of the widespread adoption of electronic technologies. The chapter argues the privacy is best viewed in terms of individuals' ability to control information about themselves within negotiated zones, and offers an account of privacy zones that can be used to respond to privacy issues in contemporary online practices in a manner that is respectful of all stakeholder interests.

Chapter 11 takes up the issue of strong copyright in e-business. Strong copyright views are those that wish to apply the maximal amount of legal protection for intellectual property and are often appealed to in response to worries about the increasing ability of consumers to access and distribute digital media. The chapter views these moves as potentially disempowering users of digital media, and argues that doing so would undermine the benefits gained by users through the new technological capabilities connected with digital media. In appealing to a number of different moral theories the chapter seeks to show that the appeal to strong copyright is not morally justifiable. Instead, the chapter argues that businesses should adopt innovative business models that are respectful of the new opportunities offered to end users by electronic technologies.

Chapter 12

Chapter 12 argues that online collection and tracking techniques now allow businesses to gain a wealth of information on consumer behavior that provides them with a competitive advantage in predicting consumer preferences and anticipating online behavior. Such a situation creates an asymmetrical balance of knowledge between consumers and businesses, and thus potentially disadvantages consumers in online transactions. This chapter argues that if we view consumer information stemming from online activity as a form of virtual property, then we can establish limits, based on property rights, on how consumer information can be gathered and utilized. The author argues that marketers have an obligation to respect this right in their interactions with online consumers, and that doing so will aid in the achievement of more fair and efficient market transactions in the digital environment.

Section 5
Ethical Issues in Public Policy and Communication

Chapter 13

In Chapter 13 the author examines the question of fairness in regards to taxation of Internet sales. While some people view the lack of direct taxation on Internet sales as a strong market incentive, others view the loss of revenue to states and municipalities as a serious problem. One possible solution offered would be to impose a universal interstate sales tax on all Internet sales. This chapter examines the arguments in favor and against such an Internet tax. From a contractarian perspective, the chapter argues that such an Internet tax would be justified since it more fairly distributes resources in a mutually advantageous manner. The author uses the theories of John Rawls and David Gauthier in articulating his arguments in favor of the Internet sales tax.

Chapter 14

Chapter 14 examines ethical questions related to one of the most common forms of communication used in e-business, e-mail. The author examines the numerous ways in which e-mail differs from other forms of business communication. As an asynchronous, distant, and text-based form of communication, the author demonstrates how e-mail communication can be misused, manipulated, and give rise to misunderstanding. As such, the chapter argues that it is important that companies are sensitive to the potential unethical behaviors that widespread e-mail usage can give rise to in the workplace. After carefully delineating all of the potential ethically problematic uses of e-mail communication, the author considers how corporations might foster ethically appropriate usage of e-mail.

Foreword

An axiom I have repeated often during over two decades of teaching business ethics is that there is great job security for business ethicists given that there is no shortage of new cases to study. If it wasn't enough that ethical foibles by individuals and corporations provided a continuous stream of fodder for ethical analysis, the burgeoning universe of e-commerce has added a whole new technological dimension to business ethics that has left scholars scrambling to catch up. However, e-business does not just represent the next phase in commerce with novel moral issues. The technological advancements are challenging our conceptions of many of the fundamental tenets of liberal theory such as the nature of private property and the definition of privacy. These challenges can have far reaching ethical implications: the tail not only wags the dog, but may redefine the dog.

The issues that arise out of e-business are ethically and theoretically compelling but the increasingly common use of e-business makes the need for moral analysis pragmatically pertinent. For example, e-commerce is one of the most visible and widely used aspects of e-business. From items that only a few years ago consumers would never have dreamed of purchasing online, such as clothes, to purchasing items that only exist virtually, such as buying new hair for an avatar on Second Life, no one can sell a product any more without considering a web presence. With online retail sales approaching $200 billion annually in the U.S., a small but dramatically expanding percentage of overall retail sales, online commerce is clearly not a fringe endeavor. Of course, commerce is only one aspect of e-business. Electronic communications, reporting, and document management are now routine aspects of virtually all business large and small.

Given the current developments in electronic business, the need for the discussions that take place in *Ethical Issues in E-Business: Models and Frameworks* obviate themselves. Although business ethics books are abundant, and some of them have chapters on e-business, and there are plenty of texts on e-business, *Ethical Issues in E-Business* is one of the first book-length treatments of the range of moral issues that arise as a result of these new technological developments in business. What is so exciting about this fresh collection of articles is not only the tremendous breadth of topics addressed but also the international flavor of the authors involved. E-business is a global phenomenon that is redefining international trade in a manner that cannot be easily captured in traditional categories of analysis. The contributors to this volume represent a variety of disciplines and yet write in an accessible style so that one does not need specialized knowledge to grasp the gravity and implications of the issues addressed. And let me emphasize, the seriousness of the discussions that take place in this book by addressing the postmodern moment.

In the 1990's a brief intellectual dialogue took place over the nature and potential of *postmodern business ethics*. The discussion began with Ronald M. Green's contention in "Business Ethics As A Postmodern Phenomenon" that business ethics is postmodern because it "rejects unitary or totalizing explanations of reality" and it values "'de-centering' of perspective and discovery of 'otherness,' 'difference' and marginality as valid modes of approach to experience" (1993). Green's article provoked responses by David M. Rasmussen in "Business Ethics and Postmodernism: A Response" (1993) and Clarence C. Walton in "Business Ethics and Postmodernism: A Dangerous Dalliance" (1993). Neither of the responses was as sanguine about describing business ethics as a postmodern project. Rasmussen and Walton found postmodern analysis lacking in various ways. Andrew Gustafson, although sharing some skepticism about postmodernism, describes how postmodernism can serve as a tool of business ethics inquiry in "Making Sense of Postmodern Ethics" (2000). Although this was an important and fruitful discussion about the relationship of postmodernism to business ethics, these articles focused on ethical analysis as postmodern. *What the authors failed to anticipate was that the field of business was about to enter a postmodern era because of the digital revolution.* On the one hand, *Ethical Issues in E-Business* is a volume that addresses ethical issues that have arisen because of technological advances, but at a deeper level, this text is confronting the postmodern turn in contemporary business. Representing more than just new practices and policies, electronic business is presenting challenges to fundamental categories and values.

In many ways, the shift from traditional forms of business to e-business is characteristically postmodern with the changes creating new epistemological and ethical challenges. A cursory look at a few of the topics covered in *Ethical Issues in E-Business* reveals that because of the complex facets of internet connectivity, business is no longer in the black-and-white confines of Kansas anymore. In "The Ethical Implications of A/B and Multivariate E-Commerce Optimization Testing," J.J. Silvia IV, reveals not only that business ethicists must become familiar with the new language of e-business, but that new information transmission challenges existing categories. Specifically, are websites information and purchasing portals or advertising? Commercial advertising is the source of long standing ethical debate and regulation, but how should web-based communication be assessed and regulated, if at all? Similarly, in "Privacy Revisited: From Lady Godiva's Peeping Tom to Facebook's Beacon Program," Kirsten Martin reexamines issues of privacy given what she describes as the "greased" and "sticky" nature of personal information in this age of technology. Information can get away from protected control and yet it can also be collected and aggregated. What does it mean to place personal information on the web when that information can take on a life of its own? What I have called the postmodern shift in challenging traditional categories, Susan Emens, in "The New Paradigms of Business on the Internet and Its Ethical Implications" describes as a new paradigm. The changes will not be contained in e-business but will spread to all aspects of commerce, and eventually, all of social life.

There may be some who wish to compartmentalize *Ethical Issues in E-Business* as only of interest to business practitioners or business ethicists, however business is merely at the forefront of the melding of technology and self definition. The issues raised here regarding the renegotiation of foundational concepts given a digitally enhanced future have wide ranging implications far beyond narrow business concerns. In 1991, Donna Haraway anticipated the pending technological revolution and its postmodern implications in her discussion of cyborgs, human/machine hybrids: "Cyborg imagery can help express two crucial arguments in this essay: first, the production of universal, totalizing theory is a major mistake that misses most of reality, probably always, but certainly now; and second, taking responsibility for the social relations of science and technology means refusing an anti-science metaphysics, a demonology of technology, and so means embracing the skilful task of reconstructing the boundaries of daily life, in partial connection with others, in communication with all of our parts (p, 181)" *Ethical Issues in E-Business* can also be read as an important discussion in "reconstructing the boundaries of daily life."

REFERENCES

Green, R. M. (1993). Business Ethics As A Postmodern Phenomenon. *Business Ethics Quarterly*, *3*(3), 219-225.

Gustafson, A. (2000). Making Sense of Postmodern Business Ethics. *Business Ethics Quarterly*, *10*(3), 645-658.

Haraway, D. (1991). A Cyborg Manifesto: Science, Technology and Socialist-Feminism in the Late Twentieth Century, In *Simians, Cyborgs and Women: The Reinvention of Nature* (149-181). New York, Routledge.

Rasmussen, D. M. (1993). Business Ethics and Postmodernism: A Response. *Business Ethics Quarterly*, *3*(3), 271-277.

Walton, C. C. (1993). Business Ethics and Postmodernism: A Dangerous Dalliance. *Business Ethics Quarterly*, *3*(3), 285-305.

Maurice Hamington
Metropolitan State College of Denver

Maurice Hamington *is Associate Professor of Women's Studies and Philosophy and Director of the Institute for Women's Studies and Services at Metropolitan State College of Denver. He holds an M.B.A. and a Ph.D. in Religion and Social Ethics from the University of Southern California as well as a Ph.D. in Philosophy from the University of Oregon. His publications include Feminist Interpretations of Jane Addams (Penn State Press, 2010) an edited volume; The Social Philosophy of Jane Addams (University of Illinois Press, 2009); Socializing Care: Feminist Ethics and Public Issues (Rowman & Littlefield, 2006) co-edited with Dorothy C. Miller; Embodied Care: Jane Addams, Maurice Merleau-Ponty and Feminist Ethics (University of Illinois Press, 2004); Revealing Male Bodies (Indiana University Press, 2002) co-edited with Nancy Tuana et al; and, Hail Mary? The Struggle for Ultimate Womanhood in Catholicism (Routledge, 1995). He is currently co-editing a volume on care ethics and business.*

Preface

In the contemporary world of business, the utilization of electronic forms of communication has become an essential aspect of most business models and processes. E-business is, to no small extent, part and parcel of all business today. While many businesses operate exclusively on the Internet, and thus make use entirely of e-business models, what is perhaps even more significant is to the extent to which nearly all businesses have had to adopt elements of e-business in order to remain competitive in the contemporary marketplace. While the more public face of e-business can be found in the various forms of Internet sales and marketing that almost all businesses now utilize to some extent, elements of e-business now aid in the facilitation of every form of business activity, from business to business transactions to operations management and human resources. Internet technologies and associated forms of electronic communication provide businesses a greatly expanded ability to efficiently carry out a wide assortment of tasks and to communicate effectively with an extensive range of stakeholders. It is thus no surprise that e-business is revolutionizing the manner in which business gets done and is a central element of the new globalized economy.

However, while no one would deny the many benefits that Internet technologies have afforded businesses or the expanded opportunities that e-business models are providing for companies, it is not surprising that ethical issues have been raised about various feature of e-business as well. As with any transformative business practice, by changing established models and offering new possibilities in commerce, e-business also has the potential to significantly affect employees, consumers, and other stakeholders in numerous ways. Business ethicists are always concerned with such transformative changes, and seek to clarify how their affects change our understanding of the rights and responsibilities of those involved. Given the range and extent of the transformation being brought about through the adoption of various forms of e-business, it should be readily understandable why the ethics of e-business is such a fertile area of inquiry.

Essentially, the ethics of e-business seeks to assure that those involved in implementing and utilizing forms of e-business do so in ways that are morally justifiable. The great benefits offered by e-business need to be pursued in a manner that is sensitive to the potential ethical implications of various aspects of e-business, if the transformation being brought about by and through e-business is ultimately going to contribute to the common good. This volume is designed with the aim of helping achieve this goal by offering a wide range of essays dealing with the multi-faceted nature of the ethics of e-business. The authors deal with a number of different ethical issues in e-business, and approach their subjects from diverse perspectives, but they share the common aim of calling our attention to the importance of ethics in e-business. The volume thus seeks to provide an overview of the ethics of e-business that is both broad, as the different essays treat many of the most pressing ethical issues in e-business, and deep, in so far as each individual essay provides a sustained and detailed analysis of the particular topic treated. As such, no matter what their particular background or reasons for being interested in ethical issues in e-business, readers should readily find something that will expand their understanding of and appreciation for the ethics of e-business in the essays included.

OBJECTIVES AND AUDIENCE

Aside from the general commitment to providing an extensive and thorough treatment of many of the most important ethical issues in e-business, three other important considerations were central in the planning of this volume. First, the volume was designed to be interdisciplinary in nature. As a branch of applied ethics, business ethics itself is necessarily interdisciplinary in nature, as engaging in ethical analysis of business issues requires a proper understanding of both the relevant ethical concepts as well as of the pertinent economic, social, legal, and organizational characteristics of contemporary business practices. The necessity of multidisciplinary understanding is perhaps even more important when considering e-business, given the complex, diverse and multi-faceted forms and structures of e-business. Properly characterizing the ethical issues raised in e-business thus necessitates calling upon insights from diverse disciplinary approaches. As such, this volume was intended to include representatives from a number of academic fields and areas of specialty. This goal has been met, as the articles contained herein are authored by writers from a number of different disciplines, including philosophy, legal studies, and a number of business specialties.

Second, ethical issues in e-business raise both theoretical and practical considerations. On the theoretical side, it is important to understand the conceptual matters involved in the ethics of e-business in order to both situate them in their proper historical, philosophical, social, and legal contexts and to be able to correctly appreciate the full significance of the issues involved. However, as with all forms of applied ethics, the aim is not merely to arrive at a correct theoretical understanding, as important as this is, but, at the end of the day, to provide guidance for those actually involved on how to act in ethically appropriate ways. That is, the ethics of e-business must also ultimately be geared toward promoting ethically responsible uses of e-business. The essays in this volume are meant to provide a balance between theoretical and practical approaches. Some chapters focus more on the theoretical, while others accent practical applications, but all are cognizant of the need for addressing both theory and practice in responding to ethical issues of e-business. As such, whether readers are more interested in investigating the theoretical constructs relevant to the ethics of e-business or in finding practical guidance on how to implement ethically responsible forms of e-business, they should find the volume useful in their endeavors.

Finally, as e-business effortlessly transcends national boundaries, it is important to realize that ethical issues in e-business are not limited to business in any particular country and, indeed, often have a transnational component. As such, this volume was intended to include perspective on the ethics of e-business originating from around the globe. The goal was to include authors from a number of different countries, in order to provide a more holistic view of the impact and ethical significance of e-business, which, by its very nature, has a global reach. Again, this goal has been met. While the majority of the authors are, perhaps not surprisingly given the origin of the volume, from the United States, the volume also does include chapters written by authors from a number of other areas. In this respect, the volume includes chapters written by authors based in Canada, Europe, and the Middle East. The hope is that the inclusion of chapters written by authors outside the United States will further encourage the inclusion of a broader and more diverse range of voices within business ethics, as well as foster the kind of international cooperation which will certainly be essential to fostering ethical forms of e-business.

Given the range of issues and approaches adopted in the chapters included here, this volume should profitably serve a wide audience. First, scholars of business ethics from various disciplines can use the book as a basic resource on ethical issues in e-business. As the first major scholarly collection of essays solely devoted to the ethics of e-business, the book provides both a comprehensive overview of the state of the field, as well as many references to other scholarly resources on the topics treated for those who wish to follow up further on the issues discussed. Students of business ethics should also find this volume to be a useful resource for studying emerging issues in business ethics, and the text could easily be adapted for classroom use. Indeed, many of the articles include discussions of the hottest current issues in e-business ethics that could easily be used as the basis for case studies in business ethics for classroom discussion. Finally, since the essays in this book include discussions of many issues of practical import for the ethical conduct of e-business, the volume should be useful to any business person who is interested in developing and conducting e-business in an ethical manner as well.

THE CHAPTERS

Section 1: The New Landscape of Business: Markets and Morals in the Age of E-Business

The first section of the book is designed to introduce readers to the ethics of e-business by focusing upon the manner in which e-business is transforming traditional business practices and the ethical significance of these changes. Understanding the shifts in business practices and models brought about through the adoption of forms of e-business is necessary for articulating and responding to the distinctive ethical issues found in e-business. Chapter 1 sets the stage by examining the major ways in which e-business is transforming business practices, and the manner in which these transformative changes are raising new ethical issues. The author identifies three key transformative aspects of e-business and seeks to illustrate their ethical significance in terms of their potential to affect stakeholder relations. In doing so, the author demonstrates how the common model of stakeholder theory in business ethics can be adopted to illuminate and direct our thinking about ethical issues in e-business. The chapter concludes by arguing that two key moral norms, the commitment to transparency and respect for persons, can serve as important moral ideals to guide and shape our thinking about ethics in the new frontier of e-business.

In Chapter 2, Susan Emens continues the discussion of the changing landscape of e-business by examining the shift in business models involved in e-business. Emens examines the way in which the growth of e-business has forced businesses to develop new organizational models and strategies. She utilizes the idea of a paradigm shift in examining these changes. Essentially, Emens maintains that e-business involves the adoption of a new business paradigm that entails changes in the way in which business parties, such as buyers and suppliers, relate in business transactions. Here, the chapter uses the example of online reverse auctions to illustrate both the nature of the paradigm change as well as its ethical implications for buyers and sellers. While the chapter argues that e-business practices such as online reverse auctions can have significant benefits for businesses, it shows they also can affect the trust that exists between business parties. As trust is an essential component of business ethics, the chapter maintains that finding ways to establish trust in e-commerce transactions should be central in the new paradigm of e-business. The chapter concludes by examining several different ways in which trust might be fostered in e-business transactions.

Section 2: Anonymity, Trust, and Loyalty on the Internet

As indicated by Emens, trust is a central element of business relations, and thus key to understanding the ethical issues of e-business. Section 2 takes up questions of trust and related issues in a sustained way. All three chapters in this section see trust as the central issue of the ethics of e-business, and each responds to issues of trust in a unique manner. Chapter 3 addresses one of the key barriers to fostering trust in e-business: the anonymity of the Internet. E-business transactions are often anonymous in so far as the parties involved have no direct, face-to-face, relationship with the persons they are dealing with. Eric Rovie identifies several distinct ethical problems that can arise from a lack of trust in e-commerce. He goes on to suggest that the problems of trust arise in e-commerce since Internet interactions take place without the kind of basis in personal, face-to-face, interactions that traditionally serve as the source of trust in business. Rovie uses the ethical and social theory of Thomas Hobbes to both illustrate these issues and to offer a potential solution to the problem of anonymity on the Internet. In this respect, he argues that institutional or social regulations may be necessary in e-commerce in order to appropriately remedy the lack of traditional mechanisms of trust in e-business. Such remedies, similarly to Hobbes' sovereign, would act to curtail the moral hazards caused by opportunistic free riders in e-commerce.

Much like Rovie, in chapter 4, Andrew Terjesen argues that in transcending the limitations of "real world" business transactions, e-business both affords greater market opportunities to businesses, but also significantly exacerbates problems of trust. Once more, the chapter points to the anonymity of the Internet as the source of distrust. Because Internet transactions involve no direct personal contact, Terjesen maintains that the motivation to be moral can be significantly diminished in e-business transactions. However, unlike Rovie's institutional response, Terjesen moves to an examination of the affective side of the individuals involved. Here, the author argues that merely appealing to traditional moral principles will not sufficiently address issues of trust on the Internet, since this does not address the emotional component that typically serves to ground trust in personal relations. As such, the author appeals to the moral theory of sentimentalism as a possible means of resolving issues of trust in e-business. Terjesen argues that by providing an understanding of how judgments of trust are evoked in the first place, the sentimentalist approach can provide an account of how to develop trust in e-commerce that will properly motivate persons to engage in ethical forms of e-business.

Chapter 5 also views trust as the central ethical issue of e-business. The chapter views the establishment of trust as essential to the continued vitality of e-commerce. The difficulty, given the unique nature of e-business, is how to establish means of fostering trust in Internet transactions. For instance, the decentralized nature of the Internet makes traditional forms of commercial regulation more difficult to enforce in e-business. Given this difficulty, the author believes that considerations of non-traditional approaches to regulating e-business are warranted. In this regard, the chapter points to fiduciary law as one possible means of facilitating trust in e-business. The fiduciary concept focuses upon the responsibilities that agents have to preserve the interests of those to whom they are entrusted. Fiduciary law, the author notes, is geared toward maintaining the integrity of crucial social and economic relations and thus would aid in the facilitation of trust and loyalty in e-commerce. The author goes on to describe the specific ways in which the proper application of fiduciary law could supplement statutory regulation and common law in governing e-commerce.

Section 3: Marketing Ethics in E-Business

In many ways marketing represents the public face of business, and it is thus not surprising that the most visible aspect of e-business is likewise often represented by the new marketing efforts utilized by companies in the age of e-commerce. Certainly e-technologies allow companies the potential to market services and products in numerous innovative ways. However, the ethical implications of these practices also need to be considered, and the chapters in this section take up that task. In Chapter 6, Abe Zakhem examines the emerging use of Mobile Location Based Services (MLBS) for e-marketing purposes. Given that mobile phones are currently the most widely used technology platforms on the planet, Zakhem argues that marketers are well aware of the immense opportunities to them by such technologies. However, Zakhem also believes that the kind of access to and information about consumers that such technologies provide companies generate conditions of vulnerability and dependency for users. Adopting a fiduciary framework much like that explored by Rotman in the previous chapter, Zakhem goes on to suggest that given the nature of this vulnerability and dependency, we are justified in applying fiduciary principles to the relationship between MLBS providers and users. As such, Zakhem recommends regulating e-commercial relations utilizing such technologies along fiduciary lines.

Online marketers have used a number of techniques for researching consumer behavior as well, and Chapter 7 examines one family of such forms of online research: the use of A/B and multivariate testing for website optimization. J. J. Sylvia shows how companies use such online modeling techniques to test various aspects of website design, with the aim of increasing average user orders and decreasing abandoned shopping carts. While such techniques may seem benign, Sylvia maintains that the kind of manipulation they involve is actually quite ethically significant. Examining the practices from the perspective of three different ethical theories, Sylvia argues that while some uses of such testing may be ethically acceptable, other forms are not ethically justifiable. The chapter concludes by examining the manipulative aspects of A/B and multivariate testing and suggesting that more explicit codes of conduct for the usage of such testing need to be developed.

Chapter 8 uses the traditional notion of the marketing mix to illuminate the ethical issues of online marketing. The traditional idea of the marketing mix categorizes that marketing strategies of companies into four crucial elements: product, price, place, and promotion. In Chapter 8, the author examines each of these four elements in respect to online marketing practices. The author seeks to demonstrate that online marketing practices involve unique issues in regards to each of the four elements, and that in each case ethical issues can be raised as well. By examining the distinctive ethical issues involved in the marketing mix in online marketing practices, the author both illuminates the ethically sensitive areas of online marketing as well as offers a framework for responding to those issues. The chapter concludes with a consideration of how businesses can create a corporate climate conducive to ethically responsible forms of online marketing.

The last chapter in this section takes up the issue of marketing corporate image online. Mary Lyn Stoll notes that many companies have an interest in marketing their Corporate Social Responsibility (CSR) efforts to consumers. The Internet provides such companies with the means to do so cheaply and with the ability to reach a greater audience than with more traditional forms of advertising. However, the author argues that as a medium of communication, the Internet poses particular challenges in terms of the epistemological constraints it places on users. These issues are especially important, the chapter argues, when addressing claims of corporate social responsibility online. Stoll examines the nature of online communication and the benefits and problems involved with marketing good corporate conduct online given the particular challenges of online communication. She argues that while there is nothing intrinsically wrong with seeking to promote a good corporate image online, companies should do so in an ethically responsible manner. In this regard, she concludes by offering practical guidelines for corporations to use in promoting morally good conduct online.

Section 4: Privacy and Property Rights Online

Two of the most prominent concepts often invoked in discussions of ethical issues in e-business are those of privacy and property. This is no doubt due to the fact that various Internet technologies have seriously altered the manner in which information is gathered and distributed in e-business. Such changes raise questions about how to approach issues of privacy and property rights in an online environment. In Chapter 10, Kristen Martin addresses issues of privacy in e-business. Comparing the traditional tale of Lady Godiva and Peeping Tom with the contemporary case of Facebook's Beacon program, Martin maintains that while the underlying concept of privacy has not changed, we must rethink the manner in which we recognize and respect privacy in light of the changes in what can be done with personal information as a result of the widespread adoption of electronic technologies. Martin argues the privacy is best viewed in terms of individuals' ability to control information about themselves within negotiated zones, and offers an account of privacy zones that can be used to respond to privacy issues in contemporary online practices in a manner that is respectful of all stakeholder interests.

Moving from privacy rights to property rights, Chapter 11 takes up the issue of strong copyright in e-business. Strong copyright views are those that wish to apply the maximal amount of legal protection for intellectual property and are often appealed to in response to worries about the increasing ability of consumers to access and distribute digital media. The chapter views these moves as potentially disempowering users of digital media, and argues that doing so would undermine the benefits gained by users through the new technological capabilities connected with digital media. In appealing to a number of different moral theories, such as Lockean, Kantian, and Utilitarian theories, the chapter seeks to show that the appeal to strong copyright is not morally justifiable. Instead, the chapter argues that businesses should adopt innovative business models that are respectful of the new opportunities offered to end users by electronic technologies. Such a proactive response would seek to secure mutually beneficial models of business that foster enhanced technological usage rather than stifle user capabilities.

Chapter 12 also takes up issues of property rights in e-business, but from the consumer side. Matt Hettche argues that while electronic commerce has generally been seen as a tool for consumer empowerment, it is also well recognized that it affords companies the ability to gather vast data on consumer behavior. Online collection and tracking techniques now allow businesses to gain a wealth of information on consumer behavior that provides them with a competitive advantage in predicting consumer preferences and anticipating online behavior. Such a situation creates an asymmetrical balance of knowledge between consumers and businesses, and thus potentially disadvantages consumers in online transactions. Hettche argues that if we view consumer information stemming from online activity as a form of virtual property, then we can establish limits, based on property rights, on how consumer information can be gathered and utilized. The chapter appeals to the traditional Lockean defense of property rights, and seeks to show to show how this justification can be extended to cover virtual property rights in the age of e-business. Hettche argues that marketers have an obligation to respect this right in their interactions with online consumers, and that doing so will aid in the achievement of more fair and efficient market transactions in the digital environment.

Section 5: Ethical Issues in Public Policy and Communication

The final section involves discussions of several areas of pragmatic concern in the ethics of e-business. One of the most vexing public policy issues of Internet commerce revolves around questions of taxation. Currently, most businesses are not required to collect sales taxes on e-commerce transactions in the United States, as long as the sales originate from out of state. In Chapter 13, Brian Coleman examines the question of fairness in regards to taxation of Internet sales. While some people view the lack of direct taxation on Internet sales as a strong market incentive, others view the loss of revenue to states and municipalities as a serious problem. One possible solution offered would be to impose a universal interstate sales tax on all Internet sales. Coleman examines the arguments in favor and against such an Internet tax. From a contractarian perspective, Coleman argues that such an Internet tax would be justified since it more fairly distributes resources in a mutually advantageous manner. The author uses the theories of John Rawls and David Gauthier in articulating his arguments in favor of the Internet sales tax.

Finally, Chapter 14 examines ethical questions related to one of the most common forms of communication used in e-business, e-mail. The author examines the numerous ways in which e-mail differs from other forms of business communication. As an asynchronous, distant, and text-based form of communication, the author demonstrates how e-mail communication can be misused, manipulated, and give rise to misunderstanding. Other problems, such a potential privacy infringements or the encroachment upon the personal time of employees, can also stem from improper use of e-mail as a business communication tool. As such, the chapter argues that it is important that companies are sensitive to the potential unethical behaviors that widespread e-mail usage can give rise to in the workplace. After carefully delineating all of the potential ethically problematic uses of e-mail communication, the author considers how corporations might foster ethically appropriate usage of e-mail. Here, the author suggests some ways in which the development of internal corporate culture and the articulation of clear policy guidelines could alleviate the potential for unethical behavior.

Acknowledgment

As with all books, this volume would have never come to fruition without the aid of many people. I would especially like to thank the people at IGI Global for all of their help during the publication process. In particular, I wish to acknowledge Jan Travers and Joel Gamon for providing aid and advice that was instrumental in the early stages of project, and Elizabeth Ardner, my primary editorial contact at IGI Global, who skillfully guided me throughout the various stages of submission and review.

My department and campus colleagues have also provided me with encouragement and aid throughout the process, and I greatly appreciate their ongoing support of my academic endeavors. Thanks go to my Dean at the Trumbull Campus of Kent State University, Dr. Wanda Thomas, as well, for providing me with a course reduction that afforded me additional time to work on this project.

I would also like to thank the chapter authors for their dedication and patience in working with me on this endeavor. All of the authors were extremely conscientious and responsive during the submission and review process and I thank them for being so easy to work with and for producing such high quality articles. Gratitude is due to the editorial advisory board for the book as well. The editorial advisory board members were responsible for blind reviewing all of the submissions to the book and the aid that their critical input provided cannot be underestimated. Some chapter authors served double duty by acting as reviewers as well as authors, and I thank them for their willingness to take on several tasks at once. Other reviewers working on the project included Dr. Shin Kim, Dr. Lisa Harris, and Dr. David Schmidt, and I wish to explicitly acknowledge the importance of their contribution to the project.

Finally, on a personal level, I would like to thank all of my friends and family for the tremendous support that they have provided me with during this project, particularly since I was often too busy working to let them know how much their support mattered at the time. None of it would be possible without them.

Daniel. E. Palmer
Kent State University, USA

Section 1
The New Landscape of Business:
Markets and Morals in the Age
of E-Business

Chapter 1

The Transformative Nature of E–Business:
Business Ethics and Stakeholder Relations on the Internet

Daniel E. Palmer
Kent State University, USA

ABSTRACT

The growth of various forms of e-business, from Internet sales and marketing to online financial processing, has been exponential in recent years. It is no exaggeration to say that nearly all forms of business involve elements of e-business today. Internet technologies provide businesses with the potential to more effectively research, market and distribute products and services, to more efficiently manage operations, and to better facilitate the processing of business transactions. However, e-business activities can raise ethical issues, as the new forms of technology and business practices utilized in e-business have the potential to pose significant moral risk as well. As such, both scholars and business persons have a responsibility to be aware of the ethical implications of e-business and to endeavor to promote ethically appropriate forms of e-business. The aim of this chapter is to aid in those enterprises by mapping out some of the major ethical issues connected to e-business. In doing so, this chapter seeks both to serve as a general introduction to this volume and to provide a conceptual framework for understanding and responding to many of the ethical issues found in e-business.

INTRODUCTION

E-business may broadly be defined as "the use of Internet-based computing and communications to execute both front-end and back-end business processes" (Hsu, Kraemer, & Dunkle, p. 9). In this sense, as Kraemer, Dedrick, and Melville (2006)

note, e-business includes any use of electronic forms of communication or "the Internet to conduct or support activities along firm and industry value chains" (p. 17). Such activities can include everything from marketing and sales to supply chain management and research and development. Thus, while the term e-commerce is usually used to more narrowly refer to the process of buying and selling of goods and services over the Internet (Holsapple & Singh,

DOI: 10.4018/978-1-61520-615-5.ch001

2000), e-business in the broader sense refers to any aspect of business that includes an electronic component. As such, while e-commerce perhaps represents the more well known public face of e-business, e-business currently involves much more than just this facet. Indeed, what makes e-business a particularly fertile realm of research is the extent to which aspects of e-business have been integrated into the operations of nearly all areas of business. S. Tamer Cavusgil nicely summarizes the holistic nature of e-business in remarking that Information technology and the Internet have transformed business, and this transformation isn't just about conducting business online. It's about integrating e-business capabilities into every aspect of value creation, such as procurement and customer relationship management. Right now, myopically e-commerce has transformed into e-business. This is no longer about exchange of services or information over the Web – it's about the total transformation of business services and product offerings. (2002, p. 26)

It is precisely because e-business is totally transforming the means by which businesses operate that it becomes so important to address the ethical issues involved in e-business. Ultimately, ethics is concerned with the principles that govern the interaction of persons, and seeks to discern the standards that will best facilitate human flourishing, promote the general welfare, and provide for respect for individual rights. Business, as Robert Solomon (1992) reminds us, can and should contribute to the aim of human flourishing as well a produce profits, otherwise it has no legitimate function in our society. Only a vision of business that connects it to the common good can ultimately justify business. As with all forms of business then, it is important to develop e-business in ways that are ethically justifiable. Given the central role that business plays in contemporary societies in facilitating both individual well-being and social goods, and given that e-business is becoming a crucial element of

most business practices, it is essential that those involved in business are sensitive to the ethical implications of e-business practices. The risks of doing otherwise are simply too great.

This chapter seeks both to introduce readers to the ethical issues involved in e-business and to provide a framework for investigating and responding to those issues. In doing so, it first illustrates the ways in which e-business is transforming business, and the ethical significance of these changes. The chapter then identifies and analyses a group of core issues around which many of the ethical questions about e-business revolve. It also uses the stakeholder model of business ethics as a means of conceptualizing the ethical challenges and responsibilities of e-business. In doing so, the aim is to provide a conceptual foundation for researchers, students, and business persons interested in ethical issues in e-business. The chapter also aims to provide a general introduction to the ethics of e-business that serves to situate the various essays on more specific aspects of ethics and e-business that follow in this volume.

BACKGROUND

The growth of e-business has been consistent and vigorous over the last few decades, to the point that e-business is an intrinsic element of most business operations. Numerous statistics attest to the phenomenal growth in e-business in recent years. For instance, data on e-business shows that:

- Since 1990 U. S. Census Bureau statistics show that in the United States, "retail e-commerce sales reached almost $127 billion in 2007 … an annual gain of 18.4 percent. Rapid growth in e-retail has been the norm. From 2002 to 2007, retail e-sales increased at an average annual growth rate of 23.1 percent, compared with 5.0 percent for total retail sales." (2009, p. 3).

- The same U. S. Census data estimated a total for 2008 of $132.4 billion in retail online sales in the United States, accounting for 3.4 percent of all retail sales (2009, p.4).

- The continued growth in online billing use is illustrated by the fact that in the U.S., "consumers used a financial institution or biller's Web site to pay 42% of their monthly bills in 2007" ("Consumers Paying More Bills Online," 2008, p. 15).

- A Globalization and E-Commerce (GEC) project survey from 2002 showed that 58% of all firms surveyed, from a global sample, made use of e-commerce for advertising and marketing purposes, 51% did so for exchanging operational data with business customers, and 48% used e-commerce for exchanging data with suppliers. (Kraemer, et al., 2006, p. 36).

These statistics illustrate only a few of the ways in which e-business has flourished, but they demonstrate well the general trend to increasing reliance upon e-business models. And, as Coltman, Devinney, Latukefu, and Midgley (2007) point out, while the dot.com bubble of 2000 may have put a damper on the enthusiasm for unlimited investment in Internet based companies, the value that businesses find in using web based technologies for business applications remains solid. E-business has been integrated so thoroughly into the operations of most businesses that it is no longer even possible for them to imagine existing apart from their e-business components. It is important to keep in mind that the phenomenal growth in e-business is not limited to wealthy developed nations either, as developing countries are rapidly adopting e-business models as well (Kraemer et al., 2006). On the global scale, the rise of e-business can be seen at the same time as both being driven by the

forces of globalization, as well as contributing to the further expansion of globalization.

Given the diversity of activities that e-business involves and the global extent of its reach, it would be very surprising if e-business did not generate ethical questions. In a very basic sense, any new technology, process, or organizational structure is bound to raise ethical issues (Ferre, 1995). And, since e-business involves new forms of all of those things, it is only natural to explore the ethical implications and limitations of the adoption of such technologies and practices. In this sense, the ethics of e-business is just part and parcel of the general progression of applied ethics as it continually seeks to analyze emergent forms of behavior. It is not surprising then that as e-business began to become more widespread, questions about its ethical implications began to appear on the radar screen of many persons interested in business ethics. At the theoretical level, as business ethics itself was a well established field of research by the time that e-business became a significant force, it was only natural that business ethicists would turn their attention to the rapidly growing arena of e-business. Early treatments of the ethics of e-business tended to do so in a piecemeal fashion, focusing upon particular ethical issues related to e-business as they emerged. Often, these efforts involved recognizing the manner in which the new technologies involved in e-busines were affecting traditional areas of business ethics. Richard DeGeorge (2000), for instance, early on recognized the importance that issues arising from increased use of informational technologies would have for business ethics.

However, while business ethicists were beginning to pick up here and there on the ethical issues involved in e-business, a correspondent wave of interest on the part of the public in the ethics of e-business was also occurring. In part, this was because e-business practices were raising important social and legal issues: the early legal battles over Napster and online file sharing, the many cases of online fraud that the public and authorities began to have to respond to, and the initial expression of concerns over the outsourcing of IT workers well illustrate this phenomenon. And, as is invariably the case, the legal and social issues invariably included moral elements as well. Second, as e-business became more prominent, its potential to become a nearly omnipresent force in our lives began to raise questions in the minds of many about how this would transform the application of certain traditional moral notions used to arbitrate personal relationships, such as those of privacy and trust (DeGeorge, 2000). Third, the increasing reliance upon electronic communication and data for nearly all forms of business raised, in many people's minds, concerns about what was at stake if e-business was put to misuse or resulted in moral hazard: the implications of such mishaps would seem to have vast ripple effects on many aspects of our lives. Those who have experienced identity theft, for instance, are well aware of its ability to impact nearly all facets of a person's life (Howard, 2007). These and other public concerns that began to be expressed in the media and other public forums were demonstrating the pressing need to think carefully and critically about the ethics of e-business for pragmatic reasons as well.

In a number of ways then, the ethics of e-business has become an unavoidable topic, one that all thoughtful persons will have to face as e-business continues to shape and influence the world in which we are all involved and increasingly interconnected. In order to address these issues correctly, however, a systematic approach is necessary, as hasty generalizations and knee-jerk responses are likely to obfuscate more than elucidate. The aim of this chapter is to aid in this endeavor by first clarifying the central changes that e-business has brought about, and then to illustrate the main ethical issues that a full treatment of the ethics of e-business must address. It is the contention of this chapter that the time is particularly ripe for providing such a schema, as e-business is now solidly enough established that

many of the ethical issues have become readily apparent and prominent enough that all parties are aware of the need to address such issues. When this chapter is taken together with the remaining chapters in this book that take on more detailed examinations of many of the specific ethical issues of e-business, this volume presents a comprehensive and systematic treatment of the ethics of e-business.

THE TRANSFORMATIVE NATURE OF E-BUSINESS

As Kracher and Corritore (2004) argue, one need not maintain that there is a special e-business or e-commerce ethics in examining the unique ethical issues in e-business. That is to say, it is not as if e-business is so distinct from other forms of business that it requires us to develop a whole new conceptual apparatus to deal with the ethical issues raised in e-business. However, e-business has transformed the nature of many business activities so that the manner in which many ethical issues arise is unique and the scope of their manifestation is different than in other business contexts (Kracher & Corritore, 2004). What is primarily needed in developing an ethical account of e-business then is a better understanding of how e-business has transformed business in ways that leads to the manifestation of ethical issues in new ways, and an account of what the key conceptual issues involved in understanding the ethical implications of these issues are. This section will be devoted to the former task, while the next will address the latter.

While there are many ways in which e-business is transforming the landscape of business, for the purpose of analyzing the ethical implications of e-business, there are several that are of particular significance. This section delineates three of the major transformative aspects of e-business that one must understand in order to arrive at a complete conceptualization of the ethics of e-business.

The Expansion of Access

One of the most recognized aspects of e-business models is that they offer nearly any person or organization the prospect of developing businesses online (Davis & Vladica, 2007). Unlike brick-and-mortar businesses that typically require a fair amount of physical infrastructure, involve significant maintenance costs, and, at least initially, have place bound market restrictions, e-business offers the potential for nearly any person to engage in commercial activities with relatively low start-up and maintenance costs, and yet provides the potential to immediately reach relatively large market segments. In this sense, some people have seen e-business as the great leveler, offering the potential to equalize the competitive playing field and democratize the world of business as never before, while revolutionizing the nature of business itself. Such claims are likely overstated, and despite an initial wave of mass entry into the world of e-commerce, in practice e-business has followed an adaptive pattern in which "existing firms incorporate the new technologies and business models offered by the Internet to extend or revamp their existing strategies, operations, and supply and distribution channels (Dedrick, Kraemer, King, & Lyytinen, 2006, p. 62). Nonetheless, it is still true that the expansion of access to markets offered by e-business has provided a much greater range of opportunities for persons and businesses to enter markets formally inaccessible, and in doing so, has transformed the way in which businesses must operate.

Four aspects of this transformation in access are of particular ethical significance. First, a number of successful e-businesses act as intermediaries in providing platforms for individuals to directly engage in commercial transactions on the Internet. Most famously, this mode of Internet commerce is illustrated by the incredible success of the online auction site eBay. Other examples would include Amazon.com's Marketplace, Craigslist, and peer-to-peer lending sites such as Prosper. As a result, many more persons are engaging in commerce than under traditional business models, and there is a correspondingly greater need for establishing trust and other means of fostering ethical behavior among these disparate groups of people. Second, this increased access leads to increased potential for competition among many different players on the Internet. While such competition can work to the favor of consumers in driving costs down, it can also raise ethical issues. For instance, manufactures who sell products directly online may now compete directly with their intermediaries, raising questions of loyalty (Stead & Gilbert, 2001). Third, this access also involves a much greater ability to access information about consumers and other persons, and the increased accessibility of information has raised numerous concerns about privacy and consumer protection. E-business techniques that involve things such as data mining, the buying and selling of consumer information, and tracking of consumer behavior online are all aspects of this transformation that have raised ethical issues. Finally, models of business that involve facilitating the direct interaction of buyers and sellers from all over the globe have been a central facet of e-business, which tends to vastly increase the number of cross-cultural business interactions. Such cross-cultural interactions can raise a number of ethical issues, as different cultural norms may come into conflict and intercultural communication can be fraught with the potential to offend or result in miscommunication.

The Lack of Common Mechanisms of Enforcement

The last point about global access raises another point concerning how e-business transforms business practices in a way that gives rise to ethical issues. For in rendering nearly universal access to markets on a global scale, e-business not only provides greater opportunities for businesses and individuals to participate in commercial activities, it also alters the manner in which business activities can be regulated. While global business is not a new phenomenon, e-business models have radically accelerated the extent and range of global business transactions.

Since e-business easily extends beyond established political jurisdictions and often involves parties operating in locals with both different legal requirements and distinct social or cultural norms, the reliance upon standard legal or cultural norms as the primary means of mitigating unethical business behavior is becoming more and more strained. Indeed, a number of the essays in this volume point to this very difficulty, as e-business transactions take place in a virtual world that does not conform to the social and legal boundaries of the physical world. Examples of the resulting ethical issues that this transformation can result in are well illustrated by the multitude of copyright disputes that have come about as a result of the virtual dissemination of various forms of media across the world on the web and in concerns over tracking and prosecuting perpetrators of Internet fraud across international jurisdictions.

In looking at this transformation of the business landscape being brought about through e-business, at least two issues come to the fore. For one, given the transnational nature of e-business, much more thought will be needed to be given to how to maintain an ethical climate for business on the part of politicians, business persons, and NGOs concerned with issues of business ethics. While governments will have to clarify and expand relevant aspects of international law, business and industry groups will also need to work to establish and disseminate clear guidelines for ethical business practices. Lisa Newton (2002) argues persuasively for the need to establish strong global ethical standards, such as found in the Caux Principles, in an era of technologically based global capitalism. Second, even with such efforts, it is doubtful that laws or business codes alone will be enough to guarantee ethical behavior on the Internet. In the world of e-business, ethics will be that much more important precisely be-

cause it will not be possible to rely upon purely legal, regulatory or policy mechanisms to enforce ethical behavior. In this regard, it will be necessary to establish a strong basic commitment to moral principles among those who participate in e-business. Along these lines, several papers in this volume take up this issue, arguing for the necessity of trust and honesty for the long-term effective functioning of e-business.

The Changing Nature of Organizational Structures

Finally, a third aspect of e-business models is that they have allowed business organizations to change many aspects of their traditional organizational structures. As noted by Introna and Petrakaki (2007), "never before in the history of business have organizations been subject to as much change" (p. 181). Indeed, this change has led to the notion of the 'virtual organization' as a model of e-business and which is characterized by its speed, flexibility, and fluidity (Introna & Petrkaki, 2007). Some features of this new model of business include the idea of operating outside of traditional organizational boundaries, using technology to disperse company operations widely, utilizing networks of temporary associations rather than relying on fixed structures for carrying out business objectives, and making much greater use of collaborative engagements to strategically leverage complimentary assets. In many ways, e-business is more flexible, more dispersed, and less stable than traditional business organizational models.

While these features allow e-business greater ability to respond quickly and efficiently to market opportunities, they also raise ethical concerns in a number of ways. For instance, employee outsourcing and the reliance on temporary workers is much more common under these e-business models. Indeed, at the extreme end, Amazon. com's Mechanical Turk now allows businesses to essentially micro-outsource even the smallest of tasks. Such practices raise questions about the ethical nature of the emerging model of employment and the protection of employee welfare and rights. More ethical issues arise in consumer relations as well, as consumers's interaction with businesses becomes both less direct and more difficult to mediate. Knowing how and where to respond to perceived problems is often more difficult for consumers in e-commerce than it is with consumers of traditional business organizations. In a similar way, the diffuse nature of e-business models makes it more difficult for organizations to enforce ethical standards across their organizations and operations.

STAKEHOLDER RELATIONS IN CYBERSPACE

As illustrated in the previous section, e-business is transforming business practices and organizations in a number of ways that have the ability to impact business practices in an ethically significant manner. In this sense, stakeholder relations in e-business are becoming more diffuse, more complex, and more flexible than under previous business models. For these very reasons, e-business is raising ethical issues in new ways. It is not that the moral norms applicable to e-business are distinct from those discussed in other areas of business ethics, but rather that the new forms of relations and means of conducting business that arise in e-business call for a new understanding of the application of these norms. Further, as argued above, there is both a pragmatic need for businesses to be aware of stakeholder interests as well as a moral imperative that e-business be developed in ways that maximize stakeholder interests. For these reasons, this chapter argues that the stakeholder approach to business ethics can effectively be extended to deal with the ethical issues of e-business. This section aims to delineate a framework for a stakeholder model of business ethics would provide a mechanism for analyzing

and responding to the sorts of ethical issues in e-business noted above.

The Elements of Stakeholder Theory

While stakeholder theory has a rich history and there are many different forms and aspects of stakeholder theory, for the purposes of this chapter, there are two essential features of stakeholder theory that need to be emphasized. First, the basic idea behind stakeholder theory is that the organizational or managerial functions of a business have the potential to impact numerous different parties. Such 'stakeholders' may include employees, consumers, suppliers, competitors, community members, and even the environment. Second, stakeholder theory rejects the idea that businesses can narrowly focus upon shareholder interests to the exclusion of the manner in which their activities impact these other stakeholders (Jones, Wicks, & Freeman, 2002). Such a focus, stakeholder theorists argue, is flawed from both a business and an ethical standpoint. On the pragmatic side, stakeholder theorists maintain that neglecting the interests of other stakeholders in business is strategically shortsighted and can thus actually inhibit the advancement of traditional corporate objectives. But stakeholder theorists also argue that in so far as stakeholder interests can be affected by corporate actions, managers also have a moral obligation to respect their rights and interests (Jones, Wicks, & Freeman, 2002). For both of these reasons, stakeholder theorists maintain that there is an intrinsic connection between business strategy and ethics (Freeman & Gilbert, 1988). Developing stakeholder theory in practice involves several different elements; including most importantly stakeholder identification, or determining who the relevant stakeholders are, and stakeholder analysis, which involves determining the nature of the moral obligations owed to stakeholders and balancing various stakeholder interests in managerial decisions (Palmer, Stoll, & Zakhem, 2008).

Moral Norms and Stakeholder Interests in E-Business

The chapters found later in this volume take up much more detailed discussions of the moral norms relevant to understanding particular areas of the ethics of e-business. In many ways, they offer examples of the kind of stakeholder analysis that this chapter argues ought to guide ethical considerations of e-business. The thesis being advanced in this chapter is that a proper understanding of stakeholder relations in e-commerce is essential, for both pragmatic and moral reasons, given the transformative nature of e-commerce. Providing such an analysis will involve identifying the relevant stakeholders affected by e-commerce activities, analyzing the interests of those stakeholders, and finding ways to balance those interests in maximizing the potential of e-commerce to contribute to the common good. Since, as shown above, e-business both involves a greater and more complex range of stakeholders, and since those stakeholder relations take place in a virtual environment not subject to the same constraints as more traditional business environments, exploring the ethics of e-business is essential.

The concern of this chapter is merely to situate these discussions by providing a framework to understand how e-business is transforming business and by suggesting that the stakeholder framework can afford a way of analyzing and responding to the new ethical issues involved in e-business. Nonetheless, without going into detailed areas of application, this chapter concludes with a consideration of some moral ideals that ought to guide stakeholder analysis in considerations of the ethics of e-business. The claim is that a commitment to these norms will allow those engaged in e-business to remain sensitive to the implications of their activities for their stakeholders and will also aid in promoting more ethical stakeholder relations. In particular, this section concludes by arguing that there are at least two basic moral ideals that ought to guide considerations of stakeholder relations in

e-business. Each of these moral norms, it is argued, is particularly important in the diffuse, complex, and frequently shifting world of e-business. These ideals, it is argued, will be essential to developing e-business in ways that properly respect the interests of diverse stakeholders.

The Commitment to Transparency

By vastly expanding the reach of business and dispersing the means by which business is conducted across global networks, e-business relies upon networks of interactions that are no longer grounded in personal relationships. E-business operates in a virtual environment which has nearly unlimited reach and which provides a platform for business transactions that requires none of the customary bonds, personal, social, and political, in which traditional forms of business were founded. For this reason, the great virtue of e-business is the capacity it offers to nearly all players to expand and develop business opportunities in seemingly unlimited directions. This virtue, however, comes with a potential risk. By separating business relations from traditional personal, legal, and social relations, e-business also can make it more difficult for those involved to discern the nature of the transactions involved and their implications. In doing so, it can make it more difficult for persons to fairly evaluate and respond to the information involved in e-business transactions. It also can allow unscrupulous persons or businesses to hide behind the anonymity of the Internet in carrying out ethically suspect actions. These concerns point to issues of both moral psychology and rational deliberation that need to be accounted for within the ethics of e-business in developing stakeholder relations on the Internet.

The concern with rational deliberation turns on the fact that the more complex information becomes and the more difficult it is for agents to discern the consequences of their options, the harder it is for them to rationally determine what is in their interests. Since business can only work to advance the interests of stakeholders in so far as they are able to make choices that truly reflect their own rational aims, activities which inhibit rational choice will, by their very nature, inhibit just market transactions. That is, in so far as persons are unaware of information about, or the potential consequences of, their actions, their ability to use business as a means of advancing their own interests is reduced, and the potential for business to contribute to the mutual flourishing of all parties is diminished. If forms of e-business weaken the abilities of parties to understand their choices, it thus presents a potential moral hazard within business. As such, the need for transparency in maintaining the ability of agents to rationally deliberate in e-business is essential.

The issue with moral psychology points to the fact that people are often more willing to engage in morally problematic behavior towards others when their communication is indirectly mediated, as with computer communications, rather than face to face (De Angeli & Brahnam, 2008). Since the Internet generally allows people to engage in mediated forms of relations, and often anonymous ones as well, it is not surprising that the potential for unethical behavior can be greater in e-business than in traditional person-to-person forms of business transactions. Lacking a personal or social relation to ground transactions, Internet users are often more likely to engage in behavior that is morally problematic than they would in face to face relations. The Internet, in this sense, can serve as a kind of moral buffer, obscuring the impact of our behavior upon real individuals. The danger then is that despite making the world a more inter-connected place, e-business can also weaken the very kinds of connections that served as a moral foundation for previous forms of business, by weakening the sense of moral responsibility that comes from close personal or social connections.

There is likely no going back to the days in which business relationships could find moral grounding in a simple handshake or community standing. However, if e-business is to continue to develop in ways that are conducive to a broad range of stakeholder interests, there does need to be some moral commitment that serves as a ground for business relations. Despite stereotypes to the contrary, business of necessity must be grounded on the moral commitments of those involved in business relationships. For instance, in order to confidently engage in business relationships, persons need to have a sense that those they deal with are honest and trustworthy. In previous ages, such commitments were primarily grounded in either personal relationships or specific cultural or legal norms. However, we have seen that e-business operates in ways that move us beyond these circles of influence. As such, the need for moral commitment to the fundamental fairness of the system by which business transactions are carried out becomes even stronger in the environment of e-business. In a sense, the means by which relationships are transacted becomes of greater moral significance than the relationships themselves. The fundamental moral commitment that then must become a priority for all e-business is therefore a dedication to transparency in stakeholder relations. Only when all parties have assurance that the basic rules of interaction are transparent can they have confidence in the system as a whole. For instance, the outrage that many consumers express when they find out that Internet companies are using their information in ways that they were not aware when they entered into business with those companies nicely illustrates the consequences upon trust when failures of transparency occur. The commitment to transparency will only become more imperative as e-business moves in further directions.

Implementing the commitment to transparency involves special sensitivity to the actual manner by which persons engage in e-business practices. For one, it is important not to confuse the utilization of particular means, such as online consent or disclosure forms, of providing legal assurance with the commitment to transparency itself. Indeed, in many cases such mechanisms can actually hinder rather than foster transparency. For instance, the lengthy, confusing, and often easily circumvented legal disclaimers commonly utilized by commercial web sites are typically both too difficult to understand and to easy to bypass with little attention to encourage real transparency in business transactions. A real commitment to transparency requires means of positively fostering an understanding among all parties about the nature of the transactions involved and of any potential use of information stemming from those transactions. E-business must strive, in this regard, to go beyond mere legal compliance in fostering an open environment for business on the Internet.

Second, the commitment to transparency is essentially a communicative function, as it involves a commitment to rendering information understandable. Effective communication, particularly intercultural communication will thus be an essential component of the ethics of e-business. Again, e-business allows people to engage in business transactions outside of any pre-established legal, cultural, or social boundaries, and thus easily brings people together that do not necessarily share a common background. Under such conditions, and when combined with the anonymous nature of Internet business platforms, the commitment to transparency will, more than ever, involve a commitment to ethical modes of communication. In this regard, managing stakeholder relations properly in e-business entails a commitment to effective modes of communication. The importance of intercultural communication for business in an era of globalization has already been stressed by a number of researchers (Limaye & Victor, 1991). In a similar manner, the importance of ethical forms of communication and inter-cultural communication needs to be stressed in the age of e-business if the commitment to transparency is become a fundamental component of e-business.

Business relations take place in a world far more complex than that of the local butcher, baker, and brewer famously discussed by Adam Smith, and e-business only further complicates the world of business. In such a world, the commitment to transparency is absolutely essential. Even Smith realized in his defense of markets that the parties involved in business could only advance their interests in so far as the parties involved were committed to honesty and fair play (Werhane, 2002). In order to engage in fair and honest transactions though, agents need reliable information about the nature of their engagements. In the world of e-business, the need for transparency in promoting fair and effective markets is more important than ever. As such, the commitment to transparency should be a guiding ideal of all of those working in e-business, as a manner of grounding fairness, efficiency, and justice in stakeholder relations.

Respect for Persons

Just as a commitment to transparency will be necessary to ground trust in e-business, a commitment to the ideal of respect for persons will be necessary to assure that e-business takes seriously the interests of the stakeholders involved. As noted previously at a number of points, the virtual nature of e-business has the potential to create a distance between the parties involved in business transactions. Such distancing can have the tendency to obscure the fact that it is real persons whose interests are at stake in e-business, just as in any other form of business. As Richard De George (2002) notes in discussing what he calls the myth of amoral computing and information technology, some people have been led to believe "that human beings are relieved of responsibility to the extent that computers are involved" (p. 268). People have a tendency, as it were, to forget the persons behind the technology. Doing so masks the fact that technological interactions are still interactions between and about persons. Virtual forms of business still ultimately rely upon and affect real persons, and the good and the bad in e-business still must be judged in terms of the bearing that the activities involved have upon the of real persons. The ideal for respecting persons is thus perhaps even more important in e-business than it is brick and mortar business, since in the former the personhood of those involved can be obscured in a way that is not typically possible in the latter. What respect for persons entails is of course a complicated matter, but the Kantian ideal of making sure that persons are not treated as mere means is a good starting point. This Kantian notion of respect for persons, among other things, requires that business practices be carried out in ways that are not coercive or deceitful, and that business be developed in ways that contribute to the development of human beings rational and moral nature (2002).

What specifically does the ideal of respect for persons in e-business entail in practice? A few points readily come to mind. For one, it entails a commitment to basic stakeholder rights. In the diffuse world of e-business, where stakeholders such as employees and consumers may be spread across the globe, and where business interactions may take a multitude of shifting forms, is essential that business persons not let their sense of responsibility for stakeholders erode. For instance, the reliance in many forms of e-business upon temporary or outsourced workers should not be used as an excuse to deteriorate workers' rights or as a means of coercing employees to engage in behavior that they otherwise would not consent to doing. Likewise, consumers concerns should not be treated as less significant simply because communication with them is mediated and at a distance. Second, respect for persons entails, as Bowie's comment above indicates, that e-business must not be deployed in ways that weaken human capacities for moral and rational development. In this regard, the use of deceptive or manipulative forms of e-business marketing and advertising techniques should certainly be discouraged, but so too should we be weary of the dissemination of e-business models that weaken respect for privacy or property rights, or that appeal to children in ways that inhibit their rational and moral

development. Ultimately, respecting for persons demands the e-business be developed in ways that fosters the moral and rational advancement of all parties involved.

The notion that business can and should be grounded in moral ideals may seem too idealistic to some, and no doubt ideals, by their very nature as ideals, will never be perfectly realized. However, the cost of fostering business environments not grounded in and through moral commitments is too great. The huge social and economic impact of business failures in cases such as Enron, World-Com, and more recently AIG and the financial industry more generally, remind us that business carried out without moral constraint can have significant negative impacts on large numbers of persons. Moral ideals serve to remind us of the moral heart of business and steer business back toward its moral purpose. In this respect, all of us interested in harnessing the potential of business for contributing to the common good have an interest in promoting moral ideals to business leaders, students, and researchers. In some sense, this entire volume is presented in light of the ideal that e-business can and should be pursued in ethically responsible ways.

FUTURE RESEARCH DIRECTIONS

The emphasis of this chapter has been on providing a conceptual framework for discerning the ethical issues involved in e-business. The focus has been on clarifying the general issues involved, rather than upon evaluating specific forms of e-business or responding extensively in a detailed way to moral controversies about particular aspects of e-business. In this sense, the intent of this chapter was merely to provide an overview of the ethical landscape of e-business, a basic map of the territory, if you will. A complete account of the ethics of e-business would involve a thorough treatment of the many specific ways in which ethical issues can arise in e-business; it would, so to speak, fill in the fine details of the rich and diverse terrain only alluded to here. By necessity such an endeavor will involve both a careful application of the sorts of conceptual matters invoked here as well as a sensitive treatment of the empirical aspects of the various technologies, processes, and models involved in e-business. Such research will also, again of necessity, be interdisciplinary. Business ethics, as a form of applied ethics, cannot be done in a conceptual vacuum, and research from such fields as business, legal studies, communications, sociology, psychology, and computer science is necessary to properly apply the relevant ethical concepts.

In many ways, the essays that follow in this volume represent exactly the kind of research that will best illuminate the ethical aspects of e-business. The proceeding chapters flesh out the map only traced here by investigating many of the particular ethical issues of e-business in a detailed manner, with much attention to the conceptual and empirical nuances involved in each case. The authors approach ethical issues in e-business from a number of different disciplinary perspectives, and they focus their attention upon a diverse array of issues and problems, but each does so with an eye toward careful analysis and an aim toward better understanding. Taken together, the essays go a good way toward filling in the details of the map. They certainly provide a good treatment of many of the most pressing ethical issues found in e-business. This is not to say that the essays included in this volume provide an exhaustive account of the ethical issues involved in e-business. Some of the ethical issues of e-business that have been discussed in the literature are not treated in this volume, and others are, no doubt, certain to arise as new forms of e-business develop. Nonetheless, the model of intellectually rigorous and interdisciplinary research found in these essays should serve as a guide to all future scholars of ethical issues in e-business. Future research in the field will be much more fecund if it builds upon the analyses found within these essays by

appealing to insights from an ever greater number of disciplinary perspectives and investing an even greater range of e-business practices.

CONCLUSION

This chapter has offered a general overview of the ethics of e-business, examining some of the fundamental ways in which e-business is transforming the landscape of business and the ethical implications of these changes. As the reach of e-business continues to extend, the ethical challenges it raises will continue to be an important area of concern for scholars, students, business persons, and consumers. By providing a conceptual framework for situating the ethical issues involved in e-business, this chapter aims to aid all of those involved in e-business in better understanding and responding to these ethical challenges. The ethical issues delineated here are central to any proper account of the ethics of e-business, and are intended to provide the foundation for a comprehensive approach to the ethics of e-business that is sensitive to all of the stakeholders involved. The chapters that follow further extend this endeavor by exploring in greater depth many of the specific aspects of the ethics of e-business.

REFERENCES

Bowie, N. E. (2002). A Kantian approach to business ethics . In Frederick, R. E. (Ed.), *A companion to business ethics* (pp. 3–16). Malden, MA: Blackwell Publishing.

Cavusgil, S. T. (2002, March/April). Extending the reach of e-business. *Marketing Management,* pp. 24-29.

Coltman, T., Devinney, T., Latukefu, A., & Midgley, D. (2001). E-Business: Revolution, evolution, or hype? *California Management Review*, *44*(1), 57–86.

Consumers Paying More Bills Online. (2008, September 1). Point for Credit Union Research and Advice. *Credit Union National Association, Inc.*, p. 15.

Davis, C. H., & Vladica, F. (2007). The value of Internet technologies and e-business solutions to micro-enterprises in Atlantic Canada . In Barnes, S. (Ed.), *E-commerce and v-business* (2nd ed., pp. 125–156). Amsterdam: Elsevier. doi:10.1016/B978-0-7506-6493-6.50009-5

De Angeli, A., & Brahnam, S. (2008). I hate you! Disinhibition with virtual partners. *Interacting with Computers*, *20*(3), 302–310. doi:10.1016/j.intcom.2008.02.004

Dedrick, J., Kraemer, L. K., King, L. J., & Lyytinen, K. (2006). The United States: adaptive integration versus the Silicon valley model . In Kraemer, K. L., Dedrick, J., Melville, N. P., & Zhu, K. (Eds.), *Global e-Commerce: Impacts of national environment and policy* (pp. 62–107). Cambridge, UK: Cambridge University Press. doi:10.1017/CBO9780511488603.003

DeGeorge, R. T. (2000). Business ethics and the challenge of the information age. *Business Ethics Quarterly*, *10*(1), 63–72. doi:10.2307/3857695

DeGeorge, R. T. (2002). Ethical issues in information technology . In Bowie, N. E. (Ed.), *The Blackwell guide to business ethics* (pp. 267–288). Malden, MA: Blackwell Publishing.

Ferre, F. (1995). *Philosophy of technology*. Athens, GA: The University of Georgia Press.

Freeman, R. E., & Gilbert, D. R. (1988). *Corporate strategy and the search for ethics*. Englewood Cliffs, NJ: Prentice Hall.

Holsapple, C. W., & Singh, M. (2000). Toward a unified view of electronic commerce, electronic business, and collaborative commerce: A knowledge management approach. *Knowledge and Process Management*, *7*(3), 151–164. doi:10.1002/1099-1441(200007/09)7:3<151::AID-KPM83>3.0.CO;2-U

Howard, K. (2007, March 14). *The damaging effects of identity theft*. Retrieved July 15, 2009, from http://www.ezinearticles.com/?The-Damaging-Effects-of-Identity-Theft&id=488838

Hsu, P, & Kraemer, L., K., & Dunkle, D. (2006). Determinants of e-business use in U.S. firms. *International Journal of Electronic Commerce*, *10*(4), 9–45. doi:10.2753/JEC1086-4415100401

Introna, L. D., & Petrakaki, D. (2007). Defining the virtual organization . In Barnes, S. (Ed.), *E-commerce and v-business* (2nd ed., pp. 181–191). Amsterdam: Elsevier. doi:10.1016/B978-0-7506-6493-6.50011-3

Jones, T. M., Wicks, A. C., & Freeman, R. E. (2002). Stakeholder theory: The state of the art . In Bowie, N. E. (Ed.), *The Blackwell guide to business ethics* (pp. 19–37). Malden, MA: Blackwell Publishing.

Kracher, B., & Corritore, C. L. (2004). Is there a special e-commerce ethics? *Business Ethics Quarterly*, *14*(1), 71–94.

Kraemer, L. K., Dedrick, J., & Melville, N. P. (2006). Globalization and national diversity: e-commerce diffusion and impacts across nations. In In K. L. Kraemer, J. Dedrick, N. P. Melville, and K. Zhu (Eds.), Global e-Commerce: Impacts of national environment and policy (pp. 13-61). Cambridge, UK: Cambridge University Press.

Limaye, M. R., & Victor, D. A. (1991). Cross-cultural business communication research: State of the art and hypotheses for the 1990s. *Journal of Business Communication*, *28*(3), 277–299. doi:10.1177/002194369102800306

Newton, L. (2002). A passport for the corporate code: From Borg Warner to the Caux Principles . In Frederick, R. E. (Ed.), *A companion to business ethics* (pp. 374–385). Malden, MA: Blackwell Publishing.

Palmer, D. E., Stoll, M. L., & Zakhem, A. (2008). Introduction . In Palmer, D. E., Stoll, M. L., & Zakhem, A. (Eds.), *Stakeholder theory: Essential readings in ethical leadership and management* (pp. 15–25). Amherst, NY: Prometheus Books.

Solomon, R. C. (1992). *Ethics and excellence: Cooperation and integrity in business*. Oxford, UK: Oxford University Press.

Stead, B. A., & Gilbert, J. (2001). Ethical issues in electronic commerce. *Journal of Business Ethics*, *34*, 75–85. doi:10.1023/A:1012266020988

U.S. Census Bureau. (2009, May). *E-Stats*. Retrieved July 9, 2009, from U.S. Census http://www.census.gov/estats

Werhane, P. H. (2002). Business ethics and the origins of contemporary capitalism: economics and ethics in the work of Adam Smith and Herbert Spencer . In Frederick, R. E. (Ed.), *A companion to business ethics* (pp. 325–341). Malden, MA: Blackwell Publishing.

Chapter 2
The New Paradigm of Business on the Internet and Its Ethical Implications

Susan Emens
Kent State University, USA

ABSTRACT

The advent of the internet and the subsequent growth in e-business has forced organizations to think be-yond the traditional business model and develop a new paradigm in order to compete effectively in such a dynamic environment. This new paradigm has permeated all aspects of how an organization conducts their operations and manages relationships from purchasing to sales. With these changes come ethical implications for both buyers and suppliers. This chapter will begin to explore those ethical implications as a consequence of this new paradigm.

INTRODUCTION

In order to fully understand the new paradigm of business on the internet, it is important to have some perspective on the magnitude of the changes that have taken place in relation to the traditional business model. Therefore, this chapter begins by exploring the traditional business model along with relevant theory and application. It progresses to explain how and why e-business required a shift

away the traditional model and presents several alternative e-commerce models. Next, the ethical implications of this evolution will be investigated. E-business in its broadest sense involves the exchange of product, services and information via computer networks, including the internet (Ja-Shen Chen & Ching, 2002). E-commerce, while often used inter-changeably with the term e-business, will be used in this chapter to mean a subset of the e-business framework, narrowly focusing on the transactional aspect of the exchange process between buyers and sellers. In order to maintain the chapter's focus on

DOI: 10.4018/978-1-61520-615-5.ch002

ethical implications, the chapter will include an illustrated example of one facet of e-business, online procurement.

The objectives of this chapter are to:

- Provide a clear understanding of the e-commerce business models in business to business (B2B) transactions and their benefits to organizations who consider using them.
- Explore the ethical implications that arise as a result of organizations conducting business using an e-commerce model.
- Suggest ways in which organizations can mitigate the impact of these ethical issues while still taking advantages of the benefits offered by the e- business model.

BACKGROUND

The business model is considered the blueprint for how any organization transacts business. It has been widely studied and the exact definition of what constitutes a business model is varied. It has been described from a very broad viewpoint as the architecture for the product, service, and information flows which includes a description of the various business actors and their roles in the value chain (Timmers, 1998; Osterwalder & Pigneur, 2002), as an explanation of how companies work (Magretta, 2002) and as a representation of a system comprised of different elements and relationships between them (Hoppe & Kollmer, 2001). In its totality, it is the entire system for delivering utility to customers and earning a profit from that activity (Slywotzky, 1996). Others have taken a much narrower focus by defining the business model in a specific context such as the internet (Mahadevan, 2000) or other technological innovations (Pateli & Giaglis, 2005). Meanwhile, Samavi, Yu, & Topaloglou (2009) offer a divergent view of the business model by framing it in terms of relationships among stakeholders to suggest

that corporate strategy and the business model be analyzed together using an integrated approach.

Despite the variety of definitions that exist for the business model, each includes common elements representing the functional aspects of an organization. These elements have been described as parts of value chains (Timmers, 1998; Osterwalder & Pigneur, 2002), strategic actions (Samavi et al, 2009), and supply networks (Cullen & Webster, 2007). In summary, the basic functional aspects of the business model consist of:

- Logistics
- Support activities
- Human Resources
- Technology development
- Procurement
- Sales and Marketing
- Operations
- Corporate infrastructure

Along each of these functional points in the business model, information is integrated and value is added. This then serves as the foundation for the implementation of the business process. Although there are numerous theories that underlie the construct of the business model, their exploration is beyond the scope of this chapter. Rather, this chapter will explore the model in terms of business practices, specifically e-commerce.

Paradigm Shifts

The business model for any organization will continue to work until an input triggers changes (Samavi et al, 2009). As such, business models should be viewed as dynamic, enabling organizations to change and adapt in order to stay competitive. This change may occur in any or all elements of the business model (Schweizer, 2005). For example, many companies, whose business model in the 1950's was largely integrated and self sufficient, were in the 1990's, outsourcing many functional areas like logistics or human resources to create greater efficiencies (Rogers, 2008). Technology can be viewed as one such input. However, in order for input changes to be impactful, they must be perceived as a signal and not merely noise in the environment. Therefore, it is important for managers to not only recognize the changes taking place, but to also accurately assess what type of technological change has occurred and the extent to which it will affect the organization.

Technological innovations that can be characterized as changes in architectural structure, such as the internet, are called disruptive technology. Disruptive technology is defined as changes that can alter an established product's performance, create new ways of providing value to customers, or provide for the emergence of new markets (Callaway & Hamilton, 2008; Bower & Christensen, 1995). The uncertainty surrounding the innovation contributes to its disruptive ability. Consequently, organizations vary in their ability to adapt and incorporate such changes. Christensen & Raynor (2003) studied e-commerce as an example of how a disruptive technology impacted the banking, finance and education industries. They found that by disrupting the product life cycle of traditional products, e-commerce provided opportunities for organizations to redefine product offerings and create new value for customers (Christensen & Raynor, 2003). As the internet continues to transform the fundamental strategies for conducting business, it is critical that organizations identify and take advantage of the key performance attributes of this disruptive technology (Lee, 2001).

The internet and the subsequent rise in the use of e-commerce by organizations have created a shift in the paradigm. The term *paradigm* can be defined as a way of thinking or as a theoretical framework of some kind (Paradigm, n.d.). Therefore the business model, defined in this chapter as a framework of the system for how business is transacted, would itself be one type of paradigm. Thomas Kuhn in his book, *The Structure of Scientific Revolutions* (1962), first used the term paradigm shift to describe a change in basic assumptions within the ruling theory of science (Dibona, Cooper, & Stone, 2005). Since that time, the term paradigm shift has been widely adapted in a variety of business contexts to mean replacing the former way of thinking or organizing with a radically different way of thinking.

The degree to which this shift has taken place varies from business to business. Some organizations, like Barnes and Noble or Toy R Us, have radically transformed their traditional business model to accommodate separate online business entities (e.g. barnesandnoble.com), while others choose to incorporate just one aspect of the e-business model (e.g., online procurement) (Cullen & Webster, 2007). Thus, the paradigm shift from the traditional business model to one of an e-business model is largely dependent upon the degree to which an organization is able and willing to adapt their organizational structure.

E-Commerce Business Models

E-commerce in its infancy was described simply as doing business electronically (Timmers, 1998). As its level of awareness and acceptance grew, so has the definition of e-commerce. Today, e-commerce implies a more extensive use of internet technology beyond the mere existence of corporate websites serving as repositories of information. Instead, the internet and its architectural infrastructure are viewed as a market that serves as a digital intermediary to facilitate product and information exchanges (Dai & Kauffman, 2002). This business to business (B2B) market functions as an electronic hub which brings together buyers and sellers to conduct business transactions, exchange information, and share knowledge (Cullen & Webster, 2007). Its use can range from organizations that conduct commercial transactions with business partners or buyers exclusively on the web to those who employ its functionality in conjunction with brick and mortar operations (Mahadevan, 2000).

Table 1. E-commerce business models

Business Model	Definition
E-Shop	Additional outlet seeking demand
E-Procurement	Additional inlet to seek suppliers
E-Auction	Electronic bidding: reverse and progressive
E-Mall	Acts as a portal to collection of e-shops
3rd Party Marketplace Providers	Intermediary providing transaction support to multiple businesses
Virtual Business Communities	Facilitates communication between members
Value-chain Service Provider	Provides support for specific function in the value chain such as electronic payment or logistics
Value-chain Integrators	Integrates and combines several levels of the value chain
Collaboration Platforms	Provides tools and information environment for collaboration between enterprises
Trust Service	IT support for privacy management issues (i.e. encryption)
Info Brokerage	Business information and consultancy services
Peer to Peer	Sharing files or computer resources directly between individual computers
Web 2.0	Second generation of services that allows people to collaborate and shore information (i.e. social networks, blogging/podcasts)

Out of any paradigm shift comes the formation of new ways to conduct business. As organizations seek to integrate e-commerce into their organizational structure, it becomes necessary for their existing business model to change as well. The result is the creation of a new landscape of e-business models that incorporate e-commerce along each of the functional points of the organization. The e-commerce business models described in Table 1 demonstrate how an organization's model is dependent upon the extent to which the organization chooses to use e-commerce to facilitate exchanges. For some organizations, such as E-Bay or Amazon, their entire model is based on e-commerce. Still others have chosen to supplant specific functional areas deemed customer critical (i.e. procurement) with e-commerce functionality.

A review of the literature reveals several different ways in which e-commerce business models can be analyzed and described. One method of distinguishing models is based on the value added services measured in terms of the degree of innovation (Timmers, 1998) and the integration of functions and partners necessary to create value (Osterwalder & Pigneur, 2002; Timmers, 1998). Table 1 lists the eleven generic business models and their description first identified first Timmers (1998) and updated to include the latest ways in which organizations have incorporated the peer to peer (Kracher & Corritore, 2004) and the Web 2.0 (Oreilly, 2007) e-commerce technology.

Other models are classified based on the type of transaction (e.g. buying or selling) that is facilitated by e-commerce (Cullen & Webster, 2007; Mahadaven, 2000) or on the type of services provided (Osterwalder, 2002; Dai & Kauffman, 2002). In each of these models, three broad structures emerge; portals, market makers and product and service providers. Portals are gateways to other sites, funneling customer attention into these web sites in a targeted manner or aggregating data from a large number of providers. Examples of these include eBay, iVillage, and WebMD. Market makers facilitate the transactions between buyers and sellers and provide value through security, trust and the reduction of product search and transaction costs. Product and service providers deal directly with their customers in a business transaction (Mahadevan, 2002).

Table 2. Types of transactions

Type of transaction	Description	Example
One to One	One buyer – One seller	Individual Trading
One to Many	One Buyer - many sellers Many Buyers – one seller	Reverse auctions Progressive auctions
Many to Many	Aggregation of many buyers and many sellers	Marketplace

From these three structures a variety of transaction models can form based upon the number of agents involved and the type of transaction taking place (buying or selling). Table 2 lists a summary of the types of transactions found in e-commerce business models. Much of the discussion in this chapter will be based on the ethical issues that emerge from these models.

An example of how the e-commerce business model has transformed the one to one transaction to a one to many can be found in the procurement process and the determination of order winners and order qualifiers. Order qualifiers are defined as those characteristics, such as price, quality, or service that potential buyers perceive as minimum standards of acceptability to be considered as a candidate for purchase (Aitken, Childerhouse, Christopher & Towill, 2005). In contrast, order winners are those characteristics of an organization's goods or services that differentiate them from the competition and will cause the end customer to buy that specific product (Aitken et al, 2005). These designations and the procurement decision-making process itself should stem from the underlying strategic plan of the firm.

Traditional procurement processes have followed the decision tree model of first narrowing the field of prospective suppliers through a qualifying stage. In the qualifying stage, buyers would conduct in-person meetings, then subsequently rate suppliers as either qualified or unqualified based on a pre-determined set of criteria. While order qualifiers could be virtually any characteristic the buyer deemed of value, there is a general set of characteristics in the area of supply chain management that are consistently identified. These characteristics are viewed as core competencies that all firms should possess at least at minimum acceptable levels to be considered as an order qualifier (Morash, 2001). They are:

- Customer Service
- Quality
- Information System Support
- Low Logistics Cost
- Distribution flexibility
- Productivity
- Delivery Speed

The consideration set for the buyer then becomes the list of organizations who meet the criteria as order qualifiers. It follows then, that those characteristics of an organization's goods or services that differentiate them from the competition in the consideration set would become the order winner criterion. In many cases, the order winner is then determined as a weighted ranking of those attributes or through personal negotiations (Sandholm, Levine, Concordia, Martyn, Hughes, & Jacobs, 2006).

In the traditional business model, the procurement process just described has been criticized for its inherent weaknesses such as lack of speed and transparency and low levels of competition among suppliers. As a result, buyers were left to question whether or not they were able to negotiate the best deal. The tendency for buyers to question whether they negotiated a good deal along with increased globalization and technological advancement created favorable conditions for the procurement process itself to go online (Sandholm et al, 2006). The paradigm shift which transformed the business model from traditional face to face meetings (one to one) to an electronic one-to-many (one buyer-many sellers) e-business model addressed many of the cited weaknesses. For instance, with reverse auctions, sellers typically receive instantaneous feedback from the reverse auction program as to their rank or bid relative to other sellers (Gattiker, Huang & Schwarz, 2007).

Once the firms enter the consideration set their focus shifts to price. Companies who regularly bid in reverse auctions develop strategies as to how to win the bid. Proponents of the reverse auctions cite that in order to become the low costs bidders, establishing closer relationships with the supplier's own supply chain are necessary (Chafkin, 2007). However, communication with the buyer typically occurs after the auction rather than during. Further, the one- to- many business model provides the ability to convey only quantitative data (i.e. price) under most circumstances (Gattiker et al, 2007).

Benefits of E-Commerce

The key driver underlying an internet-based model is the value it offers the organization. Mahdevan (2000) saw the formation of the e-commerce business model as a unique blend of the revenue stream, value stream and logistical stream. Examined in this context, the e-commerce business model provides benefits in each of these areas.

Regardless of whether the e-commerce business model is classified as a one -to-one or a one-to-many relationship, the internet serves a common function which is to create an environment that provides for the aggregation of buyers and sellers (Hunter, Kasouf, Celuch, & Curry, 2004). As a result, organizations can enhance their revenue streams through economies of scale realized through the increase number of potential customers (Lee, 2001). Costs are also further reduced as a result of increased competition. Likewise, costs normally associated with order taking and processing are lower due to automation of this process.

Benefits derived from the value stream can be viewed from both a marketing and customer service perspective. Organizations can showcase products faster, yet substantially reduce the costs of promotion associated with print media, catalogs and mailing. Buyers also perceive added value through reduced search costs as well as after sales services such as order tracking (Hunter et al, 2004).

The biggest area of focus for e-commerce business models has been in the area of supply chain management. The key benefit of e-commerce to the logistical stream is in the reduction of transaction costs from both the buyer and seller perspective. E-commerce streamlines the procurement process in a number of ways. First, it reduces the paperwork and communication inefficiencies that occur in the ordering process (Hunter et al, 2004). Second, it decreases the level of information asymmetry by using procurement techniques such as online bidding (Mahadevan, 2000). Finally, through the process of disintermediation (Mahadevan, 2000), organizations have been able to shorten the supply chain. For example, by decreasing the length of the logistics stream companies who employ a one to one business model (e.g. Dell Computers) are able to lower costs while maintaining and/or increasing customer responsiveness.

Ethical Issues, Controversies, Problems

In the internet's infancy, e-businesses concentrated on acquisition for growth at the expense of customer loyalty (Osterwalder & Pigneur, 2002). Building trust and the need to establish customer loyalty was underscored by the dot-com bust of the 90's. In business to consumer transactions, in particular, performance history, mediation services, third party verification and authorization are all examples of efforts by organizations to build positive relational dynamics to the e-commerce transaction (Osterwalder & Pigneur, 2002). Yet as businesses adjusted to the paradigm shift brought about by the growth of e-commerce, new issues began to emerge in the area of ethics.

While the technology of the internet acts to facilitate the interaction between buyers and sellers, it has also transformed the basic characteristics of that relationship. The business relationship has changed from relational to transactional. In the marketing context, transactional relationships are often viewed as discrete, arms length exchanges whereas relational exchanges are focused on establishing, developing, and maintaining long term relationships between the buyer and seller (Zahay, Peltier, Schultz, & Griffin, 2004). One of the consequences of this transformation from a relationship orientation to one that is more matter of fact is the loss of the personal touch (Cummings, 2008). A study by Stanford Institute for the Quantitative Study of Society (O'Toole, 2000) described the change as possibly even more severe than those found in traditional transactional dealings – morphing from never seeing the person again to not even knowing their name.

For organizations using the e-commerce business model, several ethical implications that have been identified and studied include access to information and privacy (Coles & Harris, 2006), security concerns (Johnson, 2001), intellectual property (Kracher & Corritore, 2004), and control of unsolicited information (Bynum, 2001). One issue in particular that has arisen due to the transactional nature of the buyer-supplier relationship that will be explored more fully here is the trust that develops between the buyer and seller as a result of long term relationships. Due to the lesser degree of face to face contact in e-commerce relationships, trust and loyalty between suppliers and buyers are viewed as important elements (Osterwalder & Pigneur, 2002). However, the change that has occurred in the nature of the relationships resulting from organizations' use of e-commerce transactions have given rise to concerns that trust and loyalty may have been adversely affected by the paradigm shift. This chapter focuses on the ethical implications occurring in these models as they apply to one functional point of the supply chain, the procurement process.

One application for online procurement is the use of the reverse auction technique for procurement. Reverse auctions are a process for competitive bidding between suppliers with the purpose of driving prices down as opposed to competition among buyers, which drives prices up as in forward auctions such as E-bay (Muscatello & Emens, 2007). Ethics dealing with the process of the reverse auction has been widely studied (Hur, et al. 2005; Jap, 2003; Handfield et. al. 2002; Muscatello & Emens, 2007; Marcoux, 2003) from the standpoint of the process itself (i.e. phantom bidding and sniping). However, ethical issues can be found in the nature of the buyer supplier relationship as well. For instance, in a typical single source supplier buyer relationship, interorganization integration or coupling can occur leading to the need for increased collaboration in project management, real time information exchanges and shared data formats (Dai & Kauffman, 2002). For this relationship to be effective, trust on both the part of the buyer and supplier is essential. When organizations introduce the reverse auction as the mechanism for procurement, the business model is transformed from one-to one interaction to one-to- many. The result is a transactional relationship that discourages the sharing of information that could ultimately lead to innovation or joint cost reductions. Not only is this in conflict with good supply chain management practices of long term collaborative supplier buyer relationships (Tassabehji, Taylor, Beach, & Wood, 2006), but it could also serve to undermine the supply chain as a whole (Stein, 2003).

E-commerce applications like reverse auctions also raise ethical concerns in the relationship between the buyer and other organizational stakeholders such as shareholders. The implied trust that exists based on the shared goal of maximizing shareholder wealth could be compromised if less capable suppliers are used solely because they are the lowest bidder (Tassabehji et al, 2006).

Procurement in the traditional business model often results in a more restricted marketplace thus allowing quality and other determinants to play a larger role in the decision process (Cullen, 2007).

Trust also comes into play from the supplier's perspective. The very design of the reverse auction mechanism requires the sharing of pricing information of other bidders in order to achieve the goal of auctioning the price of the product downward (Muscatello & Emens, 2007) This transparent marketplace is viewed by suppliers as transactional and impersonal, sending inconsistent messages to suppliers about collaboration and win-win partnership arrangements (Tassabehji et al, 2006). Consequently, buyers are perceived as not trusting suppliers' prices and lacking commitment to the relationship.

Price, however, is not always the sole criterion in determining which suppliers win the bid via the reverse auction. In a study conducted by Tassabehji et al (2006), information asymmetry in the form of unclear bid specifications eroded the trust of both the bid winners and losers. The implication of the reverse auction is that the lowest bid will win; however, switching costs and supplier proximity often play a role in determining the winner. In these instances, Tassabehji et al (2006) found a sense of betrayal to exist on the part of the lowest bidder. In addition, the order winners also felt a sense of betrayal, as terms of the contract were often adjusted after the bidding process ended. As a result of the changing dynamic from relational to transactional, 70% of suppliers reported they would not make any investment in customers using reverse auctions, while 83% reported they would not share cost savings with customers who use the reverse auctions (Tassabehji et al, 2006).

Order Winners vs. Order Qualifiers

While the main focus of Tassabehji et al's (2006) study was on the element of trust in the buyer–supplier relationship, it alluded to the role that order winner criteria plays as part of that relationship.

The introduction of reverse auctions as a procurement tool has given rise to ethical concerns as to how order winners and order qualifiers are determined. Order qualifiers, as noted previously could be virtually any characteristic the buyer deemed of value, there is a general set of characteristics in the area of supply chain management that are consistently identified.

As discussed earlier in this chapter, firms began relying less on in-person interviews and negotiation with suppliers and more on electronic procurement. Once the firms enter the consideration set their focus shifts to price. In companies who regularly bid in reverse auctions, strategies are developed as to how to win the bid. Proponents of the reverse auctions cite that in order to become the low costs bidders, establishing closer relationships with the supplier's own supply chain is necessary (Chafkin, 2007). However, communication with the buyer typically occurs after the auction rather than during. Further, the one- to- many business model provides the ability to convey only quantitative data (i.e. price) under most circumstances (Gattiker et al, 2007). Consequently, all bidders are treated alike and at an arm's length.

From the consideration set, an order winner is determined. Given that the reverse auction is a time-bound event, the order winner is, theoretically, the lowest bidder at the time of its conclusion. Based on this description, it would seem clear that the reverse auction mechanism has resulted in a shift from price as an order qualifying criterion to price as being the determinant for the order winner. However, the mechanism of the reverse auction, as well as buyers stating that non-price factors would come into play in determining the winner, has created confusion and dissatisfaction in the marketplace (Tassabehji et al, 2006).

The opacity of order-winning criteria underpins the ethical implications of the e-commerce business model. Tassabehji et al's (2006) findings reveal that suppliers perceived numerous ethical issues including the lack of pre-qualification of other suppliers, unclear details of bid specifications, and the presence of phantom bidders. The absence of social cues found in face to face dealings further compounds the erosion of trust by suppliers that the relationship with the buyer is founded on a win-win premise (Gattiker et al, 2007).

It is important for the reader to note that the exploration of the ethical issues resulting from the use of an e-commerce business model in this chapter and specifically in the functional area of procurement is not meant as a condemnation of the e-commerce business model itself. Indeed, the benefits derived from the utilization of e-commerce in any or all aspects an organization's business model are numerous and should not be understated. These benefits include greater efficiencies, reduced costs, increased transaction speed, and the ease of access to larger markets, to name just a few.

Solutions and Recommendations

The importance of ethics to business transaction is evidenced by the growing number of formal guidelines set forth by regulatory agencies and industry associations. For example, the U.S. Federal Procurement Standards require corporations that do business with Federal agencies to establish and meet the Federal guidelines on ethics (Muscatello & Emens, 2007). Professional societies such as the Institute for Supply Chain Management and the Mining and Metallurgical Society of America as well as corporations like Boeing and General Electric require their members to abide by a code of conduct when dealing with suppliers.

The e-commerce business model represents a paradigm shift that with it brings ethical implications to an organization's business practices. Examination of just one functional area of the supply chain, procurement, clearly indicates numerous ethical considerations that should not be ignored. While some have argued that e-commerce does not create a special set of ethical issues from those found in traditional business models (Kracher &

Corritore, 2004), recommendations have been put forth that can assist organizations in maintaining previous relationships developed using long term transactions as well as building new relationships via the e-commerce business model.

In the same vein, Peace, Weber, Hartzel, & Nightingale, (2002) have recommended the establishment of eBusiness Principles that address the basic ethical values most at risk in eBusiness: privacy, truth and cooperation. Recognizing trust to be at the core of any ethical relationship, Peace et al (2002) asserts that companies must be fully committed to facilitating its development when interacting with suppliers. Additionally, this research effort has uncovered the following recommendations which further address these issues for organizations utilizing the e-commerce business model:

- Build rapport using websites and customer specific videos (Cummings, 2008)
- Reduce information asymmetry found in reverse auctions with clearer purchase specifications and minimize changes to specifications after the auction closes (Tassabehji et al, 2006).
- Develop trust through mediation services, third party verification, clear and explicit privacy policy, and safe shopping guarantees (Osterwalder, 2002).
- Provide cues through website design and seals of approval indicating trustworthiness (Kracher & Corritore, 2004).
- Design online auction mechanism to safeguard supplier relationships and reduce suspicions of opportunism (Jap, 2007).

FUTURE RESEARCH DIRECTIONS

The breadth of ethical considerations pertaining to e-commerce is considerable. This effort contributes to the body of work exploring the paradigm shift in the business models by examining the ethical implications for one functional area, procurement. Additionally, this study was limited to examining trust as an element of ethical behavior in the buyer-supplier relationship.

The ethical implications explored in this research should serve as a roadmap for future empirical research that addresses other aspects of the business model such as logistics, human resources, operations, sales and marketing. Equally important ethical issues of trust can arise in each of these areas. Marketers, for instance, should consider the internet's impact of direct selling to the exchange relationships in their distribution channels. Analysis of the impact that ethics has on the value chain in the transition from relational to transactional is another area fruitful for exploration.

CONCLUSION

This chapter has explored the ethical implications resulting from the paradigm shift in business models resulting from the rise of e-commerce. It has met the original objectives set forth at the beginning of this chapter by providing a clear understanding of the transition from traditional business to the e-commerce business model found in B2B transactions. The benefits to organizations who consider using them have been explained. While the list of benefits is by no means exhaustive, it is clear that the advantages that e-commerce offers to business will contribute to its continued use and growth.

As the nature of relationships changed from relational to transactional, the resultant ethical implications for organizations were delineated. Most of the issues centered on trust between buyer and supplier. These were explored in the context of one functional area of the business model, procurement. Further issues were developed in the manner in which organizations determined order winners and order qualifiers and the subsequent impact those decisions have on trust between

buyers and suppliers. Finally, ways in which the organization can mitigate the impact of these ethical concerns while still taking advantage of the benefits offered by the e-commerce business model were suggested. While it is certainly not an exhaustive study, it serves as a starting point for ethical inquiry into other aspects of the e-commerce business model and should provide a foundation for future research efforts.

REFERENCES

Aitken, J., Childerhouse, P., Christopher, M., & Towill, D. (2005). Designing and managing multiple pipelines. *Journal of Business Logistics*, *26*(2), 73–95.

Bower, J. L., & Christensen, C. M. (1995). Disruptive technologies: Catching the wave. *Harvard Business Review*, *73*, 43–53.

Bynum, T. W. (2001). Ethics and the Information Revolution . In Spinello, R., & Taviani, H. (Eds.), *Readings in Cyberethics*. Sudbury, MA: Jones and Bartlett.

Callaway, S. K., & Hamilton, R. D. III. (2008). Managing Disruptive Technology — Internet Banking Ventures For Traditional Banks. *International Journal of Innovation & Technology Management*, *5*, 55–80. doi:10.1142/S0219877008001242

Chafkin, M. (2007). Reverse auctions a supplier's survival guide. *Inc.*, *29*(5), 27–30.

Christensen, C., & Raynor, M. (2003). *The innovator's solution: Creating and sustaining successful growth*. Boston, MA: Harvard Business School Press.

Coles, A., & Harris, L. (2006) Ethical Consumers and E-Commerce: The Emergence and Growth of Fair Trade in the UK. *Journal of Research for Consumers, 10*.

Cullen, A. J., & Webster, M. (2007). A model of B2B e-commerce, based on connectivity and purpose. *International Journal of Operations & Production Management, 27*(2), 205–225. doi:10.1108/01443570710720621

Cummings, B. (2008). It's a 24/7 world. *Wearables Business, 12*(1), 57–60.

Dai, Q., & Kauffman, R. J. (2002). Business models for internet-based B2B electronic markets. *International Journal of Electronic Commerce, 6*(4), 41.

Dibona, C., Cooper, D., Stone, M. (2005, October). *Open Sources 2.0: the continuing evolution.* Sebastopol, California: O'Reilly Press.

Gattiker, T. F., Huang, X., & Schwarz, J. L. (2007). Negotiation, email, and internet reverse auctions: How sourcing mechanisms deployed by buyers affect suppliers' trust. *Journal of Operations Management, 25*(1), 184–202. doi:10.1016/j.jom.2006.02.007

Handfield, R. B., Straight, S. L., & Sterling, W. A. (2002). Reverse auctions: How do supply managers really feel about them? *Inside Supply Management, 13*(11), 56-61.

Hunter, L. M., Kasouf, C. J., Celuch, K. G., & Curry, K. A. (2004). A classification of business-to-business buying decisions: risk importance and probability as a framework for e-business benefits. *Industrial Marketing Management, 33*(2), 145–154. doi:10.1016/S0019-8501(03)00058-0

Hur, D., Mabert, V. A., & Hartley, J. L. (2005). Getting the most out of e-auction investment. *Omega, 35*, 403–416. doi:10.1016/j.omega.2005.08.003

Ja-Shen Chen, & Ching, R. K. H. (2002). A proposed framework for transitioning to an E-business model. *Quarterly Journal of Electronic Commerce, 3*(4), 375.

Jap, S. D. (2003). An exploratory study of the introduction of on-line reverse auctions. *Journal of Marketing, 67*, 96–107. doi:10.1509/jmkg.67.3.96.18651

Jap, S. D. (2007). The impact of Online Reverse Auction design on Buyer-Supplier Relationships. *Journal of Marketing, 71*, 146–159. doi:10.1509/jmkg.71.1.146

Johnson, D. (2001). Is the Global Information Infrastructure a Democratic Technology? In Spinello, R., & Taviani, H. (Eds.), *Readings in Cyberethics.* Sudbury, MA: Jones and Bartlett.

Kracher, B., & Corritore, C. L. (2004). Is There a Sepcial E-Commerce Ethics? *Business Ethics Quarterly, 14*(1), 71–94.

Lee, C. S. (2001). An Analytical framework for evaluating e-commerce business models and strategies. *Internet Research: Electronic Networking Applications and Policy, 11*(4), 349–359. doi:10.1108/10662240110402803

Mahadevan, B. (2000). Business models for internet-based E-commerce: An anatomy. *California Management Review, 42*(4), 55–69.

Marcoux, A. M. (2003). Snipers, stalkers and nibblers: Online auction business ethics. *Journal of Business Ethics, 46*, 163–173. doi:10.1023/A:1025001823321

Morash, E. A. (2001). Supply chain strategies, capabilities, and performance. *Transportation Journal, 41*(1), 37.

Muscatello, J., & Emens, S. (2007). Do Reverse Auctions Violate Professional Standards and Codes of Conduct? In Parente, D. H. (Ed.), *Best Practices in Online Procurement Auctions* (1st ed.). Hershey, PA: IGI Global.

Oreilly, T. (2007). What is Web 2.0: Design Patterns and Business Models for the Next Generation of Software. *Communications & Strategies, 1*(1), 17.

Osterwalder, A., & Pigneur, Y. (2002, June 17-19). *An e-Business Model Ontology for Modeling e-Business.* Presented at the 15th Bled Electronic Commerce Conference e-Reality: Constructing the e-Economy. Bled, Slovenia.

Paradigm - On Definition. (n.d.). *Criticism Of Kuhn's Paradigms, Revolutions, Leaps Of Faith, Criticism Of Kuhn's Relativism*. Retreived August 4, 2009, from http://science.jrank.org/pages/7948/ Paradigm.html#ixzz0P1OrmeJJ

Pateli, A. G., & Giaglis, G. M. (2005). Technology innovation- induced business model change: A contingency approach. *Journal of Organizational Change Management, 18*(2), 167–183. doi:10.1108/09534810510589589

Peace, G., Weber, J., Hartzel, K., & Nightingale, J. (2002). Ethical Issues in eBusiness: A Proposal for Creating the eBusiness Principles. [from Business Source Complete database.]. *Business and Society Review, 107*(1), 41. Retrieved April 20, 2009. doi:10.1111/0045-3609.00126

Rogers, B. (2008). Contract sales organisations: Making the transition from tactical resource to strategic partnering. *Journal of Medical Marketing, 8*(1), 39–47. doi:10.1057/palgrave.jmm.5050119

Samavi, R., Yu, E., & Topaloglou, T. (2009). Strategic reasoning about business models: A conceptual modeling approach. *Information Systems & e-Business Management, 7*(2), 171-198.

Sandholm, T., Levine, D., Concordia, M., Martyn, P., Hughes, R., & Jacobs, J. (2006). Changing the game in strategic sourcing at procter & gamble: Expressive competition enabled by optimization. *Interfaces, 36*(1), 55–68. doi:10.1287/inte.1050.0185

Schweizer, L. (2005). Concept and evolution of business models. *Journal of General Management, 31*(2), 37–56.

Slywotzky, A. J. (1996). *Value Migration: How to think several moves ahead of the competition*. Boston, MA: Harvard Business School Press.

Tassabehji, R., Taylor, W. A., Beach, R., & Wood, A. (2006). Reverse e-auctions and supplier-buyer relationships: An exploratory study. *International Journal of Operations & Production Management, 26*(2), 166–184. doi:10.1108/01443570610641657

Zahay, D., Peltier, J., Schultz, D., & Griffin, A. (2004). The role of transactional versus relational data in IMC programs: bringing customer data together. *Journal of Advertising Research, 44*(1), 3–18. doi:10.1017/S0021849904040188

Zhu, K. (2004). Information transparency of business-to-business electronic markets: A game-theoretic analysis. *Management Science, 50*(5), 670–685. doi:10.1287/mnsc.1040.0226

Section 2
Anonymity, Trust, and Loyalty on the Internet

Chapter 3
The Anonymity of the Internet:
A Problem for E-Commerce and a "Modified" Hobbesian Solution

Eric M. Rovie
Agnes Scott College, USA

ABSTRACT

Commerce performed electronically using the Internet (e-commerce) faces a unique and difficult problem, the anonymity of the Internet. Because the parties are not in physical proximity to one another, there are limited avenues for trust to arise between them, and this leads to the fear of cheating and promise-breaking. To resolve this problem, I explore solutions that are based on Thomas Hobbes's solutions to the problem of the free rider and apply them to e-commerce.

TRUST AND THE INTERNET

A firm handshake and a face-to-face meeting over dinner and drinks used to be the model for business transactions. Buyers and sellers met face-to-face in stores, restaurants, board rooms, and even on front porches to arrange transactions, and it would have seemed bizarre to buy anything 'sight unseen,' much less to buy from a person you couldn't see or hear. But the Internet has changed the face of the world, and commerce is no exception: transactions occur between parties who have never met, will never meet, and do not ever need to speak to each other using anything more than a keyboard. These new,

electronic, possibilities for business and commerce are great, but they can also come at a significant cost: a sense of trust that is (for some) generated in a person-to-person (rather than a machine-to-machine) transaction. In this paper, I will argue that the perceived anonymity of the Internet raises problems for traditional models of commerce, but that the problem can be resolved in a Hobbesian fashion by creating an appropriately authoritative framework under which e-commerce can operate, and by having avenues for recourse should a transaction be unsatisfactory to either (or both) parties. I will attempt to shine a light on a more broadly philosophical problem (the problem of the 'free rider') by looking at the hazards of e-commerce, and I will argue that Hobbes' recognition of this problem,

DOI: 10.4018/978-1-61520-615-5.ch003

and his solutions to it, are useful for steering us clear of these hazards. My primary concern here is philosophical: I offer advice to participants in e-business using Hobbesian precepts without an in-depth analysis of how the business end of the advice might be cashed out. But I think it is clear that business can learn much from philosophy, and vice versa.

We would be wise to begin by noting some of the essential differences that make the problems facing e-commerce different from those that plague standard versions of commercial interaction. To begin with, I should clarify a crucial point: in this paper, I focus on the issues that plague e-commerce and not with the more broad category of e-business (which includes e-commerce but also includes electronic facets of the internal operation of a business) itself. I include all forms of electronic commerce under the broad heading of e-commerce, including sales from vendor to vendor, vendor to consumer, and consumer to consumer. I also include both fixed price sites and auction sites, and third-party hosted transactions. This means that my argument applies equally when ACME Widgets sells parts to General Industrial Incorporated, or when Annie sells Bill a hand-made scarf or a used copy of *London Calling* on Ebay or Amazon's Marketplace. I realize that the scope of the transactions is greatly different, but I do not think the principles that ground them should be.

The thrust of my argument here is that there is a philosophically interesting problem that faces e-commerce, and that this problem is based on the distinct environment of e-commerce: the distance and anonymity of the Internet. The problem is not completely unique to e-commerce, of course, because it would plague any form of mail-order sales and even, to a lesser degree, sales by phone and television. It is amplified in e-commerce, however, because there is rarely any contact with another human being, person-to-person, apart from email. When a customer buys a product from a vendor in a traditional setting, there are certainly going to be concerns about the transaction, but I argue that

concerns are greatly increased in an e-commerce setting for a number of reasons. These reasons include, but are not limited to, the following:

1. The lack of face-to-face connection with another person.
2. The lack of familiarity with the seller and her business practices.
3. The lack of repeat transactions with the same seller.
4. The reputed anonymity of the Internet.
5. The lack of a personal filter on conversations that take place over the Internet.

The lack of face-to-face, interpersonal interaction may impact the feelings of trust that each member of the transaction will have. According to recent psychological research (and echoing common folk psychological attributions) there is something important about having a 'trustworthy face', and visual perception of facial cues works to reinforce or create attributions of traits like 'trustworthy' or 'confident' or 'aggressive' in stranger (Oosterhof & Todorov, 2008, p. 126). The inability to examine a face, according to some views of this work, might impede the ability to trust, or at least cause the trust to develop more slowly than it would under face-to-face conditions (Wilson, Straus, & McEvily, 2006).

Secondly, there is the lack of familiarity with the seller. Being unfamiliar with a seller is not a situation limited to e-commerce: it would, presumably, apply to every first-time pair of transactors. But combining this with the 'lack of face' makes it more problematic: not only do parties not see their partner, but they also know little more than what each party posts on its website. In most cases, without doing much digging, it would be hard to ascertain which online sellers are reliable and which are scam artists.

This leads us to the third problem, where buyers may find themselves moving from seller to seller to get the best price (instead of simply committing to one location as central) and, ultimately, may not be able to develop mutually trusting relationships with any one seller. Again, this problem is not exclusive to e-commerce, but may lead to more 'one-off' transactions between parties, rather than a long-term economic relationship between two parties.

The fourth problem will loom large in my later discussion of trust, but I can point to a general issue here: people will be more likely to act in ways that are not as socially acceptable if they believe they are acting anonymously. Anonymous tip lines encourage people to turn in criminals knowing they won't be subject to retribution, and anonymous chatrooms and message boards allow people to say things they normally might refrain from saying in a face-to-face conversation. According to some social psychological research, hostile remarks may increase by as much as six times if a participant believes she is commenting anonymously (S. Kiesler, Siegel, & McGuire, 1984). Anonymity provides a shield behind which individuals feel free to say what they feel, and often those feelings are amplified by the sense of security they have in their anonymity. This leads one prominent scholar of the psychology of Internet behavior to note "when people believe their actions cannot be attributed to them personally, they tend to become less inhibited by social conventions and restraints" (Wallace, 1999, pp. 124-125).

This point connects to the final problem, that conversations between parties over the Internet might be subject to less restraint and less of a 'filter' than conversations that take place face-to-face or over the phone. It would seem to be much easier to lose one's temper, or say unnecessarily cruel things, to a blinking cursor than to a responsive voice or an emotive face. Despite the fact that rules of Internet etiquette ("netiquette") have been around dating back to at least 1995 (for an early draft of a code of Internet etiquette, see http://tools.ietf.org/html/rfc1855), it is not uncommon for 'flame wars' and malicious comments to be generated from across seemingly anonymous computer screens. The combination of these five factors (in various forms depending on the circumstances) can lead to unique situations that are not as common in traditional face-to-face or phone-to-phone transactions. And this can lead to a breakdown in a central facet in a business relationship: a failure in trust.

It should be noted here, also, that the anonymity of the Internet is not absolute: participants are subject only to as much anonymity as their technology (and Internet Service Providers) will allow (Wallace, 1999). A person with the appropriate level of tech-savvy (or the right court order) will be able to crack through almost any veil of Internet anonymity, and many 'anonymous' Internet participants give away much of their anonymity by using real names, personal details, or personal e-mail addresses in their on-line identities. There seems to be little guarantee that any Internet exchange would remain permanently anonymous, although, presumably, most do remain that way.

TRUST AND E-COMMERCE

The relationship between trust and commerce is deceptively simple: for a transaction to happen, some level of trust must exist between the parties in the exchange. If Ron tells Katie "pay me $30 today, and I'll bring the bike to your house tomorrow," Katie needs to have some level of trust that Ron will hold up his end of the deal, or Katie may have just given away $30 for nothing. In business, that trust may come in various forms, from warranties, work orders, and receipts, to interpersonal relationships (Katie is far more likely to trust Ron's cash-for-bike-tomorrow exchange if she knows and trusts Ron as a person) or even through legal institutions and business organizations (tort law, the Universal Commercial Code, etc) Regardless of the form, trust is crucial for such exchanges to be able to occur. As Carson puts it in a discussion of the ethics of advertising and sales, "Deceptive advertising is also harmful in that it lowers the general level of trust and truthfulness essential to a flourishing society and economy. The law alone cannot secure the level of honesty and trust in business necessary for people to be sufficiently willing to enter into mutually beneficial market transactions" (Carson, 2002, p. 41).

The 'deceptively' simple aspect of the trust relationship in commerce, however, stems from the desires of both parties to succeed at the expense of the other. The parties are, of course, competing with one another in the exchange. Ideally, both parties will walk away from the transaction in an improved position. If only one party walks away in an improved position, but the other party is not made any worse (known in economics as a Pareto improvement) and no further improvements can be made, the situation is economically (Pareto) optimal and efficient. But, in a practical sense, most participants in economic exchanges may not care about (or know about) Pareto improvements, and might be inclined to view the transaction as a "me versus them" battle, particularly if there isn't a trust-informed relationship already present between the two parties. If I desire a basket of fresh peaches, and I venture to a local fruit stand, while I may not have any strong desire to see the fruit vendor put out of business (in fact, I may want him to stay open because of the quality and convenience his stand provides me), I might not object if he grossly undercharged me.[1] The trust we need to give (and get) in commercial transaction is, at the very least, a little puzzling if not fully paradoxical, but it seems fully necessary to have something upon which we can rest our hopes for mutual fulfillment of the goals of the transaction. This need for trust is made even more problematic if we consider the possibility, raised by Eric Uslaner (2002), that trust is drastically reduced when parties are on unequal social ground, and that many participants in commerce have lost much of the trust they might have had with successful trade partners. In other words, the bigger and more successful you are in your business endeavors, the more likely it might end up being that people fail to trust you.

If we take, as a central premise in the argument for commerce, that trust (of some sort) is necessary for transactions to occur, and add to it the problem of the anonymity of the Internet for trust, we face an argument that looks something like this:

1. Trust is one essential component to successful commercial interactions between parties.
2. Trust is best obtained in situations where parties have some familiarity with one another, either through personal (face-to-face) interaction or through repeated business dealing.
3. Therefore, face-to-face interaction makes commerce between parties more likely to occur and/or succeed.

But since most e-commerce transactions do not occur through personal (face-to-face) interactions, this leads us to what I call the Problem of E-Commerce, or the Anonymity Problem:

4. E-Commerce, in general, does not utilize face-to-face interactions and is, generally, 'anonymous' in most relevant senses, and is, therefore, lacking in the materials that develop trust.
5. Therefore, e-commerce, is less likely to occur and/or succeed than the traditional face-to-face transaction.

Patricia Wallace has noted that all e-commerce transactions, but particularly 'grassroots' e-commerce transactions (in situations where individuals transact with one another rather than with or through a corporation) are open to abuse, fraud, and even violence (Wallace, 1999, pp. 244-245), and this certainly hampers to overall level of trust in the e-marketplace. Successful e-commerce would have to be mostly free of fraud, dishonesty, and bad faith, but anonymity and trust gaps also seem to encourage such shady dealings. So, how can we salvage e-commerce in the wake of this problem? Or, should we simply write off all successful e-commerce as aberrant? I think the answer to this question can be drawn from the work of Thomas Hobbes.

HOBBES'S ARGUMENT

Thomas Hobbes's impact on social, political, and moral philosophy cannot be understated. His answers to the problems of social and political co-ordination, his explanations of moral motivation, and his views on moral authority should be central parts of any discussion of these topics. Hobbes's arguments are most often used as the basis for political theories, but they need not be used exclusively so. His insights on human behavior and moral psychology, for instance, impact the theory of games, and his moral philosophy was greatly influential to several traditions of moral theory, notably contractarianism. Despite the fact that his arguments were developed in response to the dire strife of the English Civil War, his insights are still appropriate to contemporary issues. Social norms, in general, may rest heavily on Hobbesian arguments, providing a philosophical answer to the question "Why observe non-legal rules and norms?" Of course, Hobbes did not explicitly address the problem of e-commerce, because he lived (1588-1679) roughly five hundred years before such a thing even existed, but his thoughts on other topics can guide us in the direction of a useful Hobbesian solution to the Anonymity Problem, as he has done for other problems of the social sciences (Hollis, 2002). In what follows, I will provide an extremely brief sketch of several key elements of Hobbes's argument as developed in *Leviathan* (Hobbes, 1996).

For Hobbes, the crucial reason to enter into civil society, to consent to give up absolute freedom in favor of being governed, is because a life outside of civil society is a brutal and terrifying constant struggle. The 'state of nature', as Hobbes famously puts it, is a state where the life of all man is "solitary, poor, nasty, brutish, and short" (p. 89). One of Hobbes's central theses is that the state of nature is, for all practical purposes, a state of constant war and strife. Even when there is not active combat going on, all parties in the state of nature must be ready to go to battle with the rest of the world, and this is an exhausting and terrifying way to live. The real reason why self-interested agents would give up the freedoms of the state of nature for the structure and discipline of government is to get a reprieve from the constant struggles of life in the state of nature.

How might these struggles be abated, given the apparent human drive to be in constant conflict? For Hobbes, this problem is resolved by having all parties give up their rights, provided that all other parties do so as well, in the form of large-scale social contract. This contract will be protected by a powerful and people-authorized political leader (or assembly of leaders), an 'artificial man' who keeps the peace and protects the interests of the citizens, called the sovereign. For Hobbes, this means the collective will of the people is "united in one Person" and this serves as the "Generation of the great Leviathan, or rather (to speak more reverently) of that Mortal God, to which we owe under the Immortal God, our peace and defense" (p. 120). The Leviathan will be a feared and powerful political leader whose power and strength is so awe-inspiring and fear-inducing that his rules, whatever they may be, will surely be followed. The Leviathan will provide both law and order, and will do so with an iron fist, if necessary. Hence, the chaos and uncertainty of the state of nature is replaced by order and structure under the rule of the sovereign. Put even more simply, the sovereign replaces the free-for-all that exists prior to civil society with an order that is (even if brutal and cruel) better and more predictable than the disorder of the state of nature. We consent to be governed because even bad government is better than the brutality and uncertainty of the state of nature. This allows us to have agreements, mutually beneficial arrangements and, most importantly, contracts.

Chapter XIV of *Leviathan* sets up a classic philosophical defense of the contract and, by extension, commerce. Hobbes argues that for mutual performance of the two ends of a contract, there needs to be some sort of coercive force to ensure that each participant follows through on their end. It would be foolish, for instance, for one party to provide their service to another before receiving the payment because "he which performeth first, does but betray himself to his enemy" (p. 96). But, short of a simultaneous exchange, this would seem to leave commerce and other sorts of contracts to be difficult to achieve. Hobbes has the solution to this problem at the ready: the sovereign, and his awe-inspiring power. Since both parties to a contract are aiming to see their desires fulfilled without paying any costs, both would be willing to take what they want and leave their end of the contract unfulfilled. According to Hobbes, the third Law of Nature is "*that men perform their Covenants made*" and this will be achieved, when men are unwilling to do so on their own, by the implementation of "some coercive power, to compel men equally to the performance of their Covenants, by the terror of some punishment, greater than the benefit they expect by the breach of their Covenant" (pp. 100-101). In other words, there must be some force (the sovereign) in place to force parties to abide by contracts made. Without such a force, it would be foolish to expect the other party to your contract to abide by his end, and the whole structure and institution of contracts would collapse on itself in a self-interested muddle. Without contracts, we cannot be assured of any social cooperation at all, but without a sovereign to enforce the contracts made, we cannot consider contracts to be of any practical use at all: they would just be meaningless words, said for show, but broken at will when it would be in one's interest to do so. [2] And, if contracts are meaningless, commerce and trade are meaningless as well. He describes the state of nature as being a world where "there is no place for Industry, for the fruit thereof is uncertain"

(p. 89). Hobbes's entire political theory is based on the structure of a social contract between the sovereign and his citizens, and this becomes one version of a moral and political theory known as contractarianism.

For contracts and agreements to work, however, all parties must be expected to hold to them under all circumstances. If one party provided payment, but the other failed to deliver the paid-for service, the contract would be broken. Clever contractors might regularly fail to follow through on their contracts, which could put the whole institution of contracting at risk. This is an issue that Hobbes takes quite seriously, and is often referred to as the Free Rider Problem. Hobbes never directly refers to 'free riders' but speaks of a character called "The Foole", who has "said in his heart, there is no such thing as Justice" (p. 101). The Foole, Hobbes argues, would pretend to be playing by the rules of society, and would appear to be following the laws, but would, in secret, be violating the rules at every beneficial opportunity. The Foole wants to receive the benefits of living in a civil society (namely, being out of the brutal state of nature) without truly paying the costs (by following the rules, which may be inconvenient to his desires). Of course, he can only do this secretly, for if he were caught violating the rules, he would be subject to severe punishment, so the Foole appears to be a law-abiding citizen.

So what, if anything, is there to deter individuals acting as Hobbesian Fooles? Quite simply, the answer is the fear of being caught.[3] All Fooles will face the worry that their violations of the laws of civil society will be found out by their fellow citizens, or by those in society who enforce the laws, and this worry should be enough to keep them in line for at least two main reasons: the sovereign's fearful wrath, or the punishment of being cast out of civil society to perish back in the state of nature (p. 102). Whenever a person is tempted to act like a Foole, they should be reminded of the dangers of being found out, and the hazards of being excommunicated from society.

This, says Hobbes, should be enough to discourage their violation of the social order, and give us reason to trust in those who make contracts with us. Although it might appear that it would benefit someone, in the short term, to be a Foole, the long-term consequences might include the collapse of the very social networks and systems that support the Foole's ability to free ride, and this should be enough (from a Hobbesian view) to provide a disincentive for free riding.

These, then, are the core elements of the Hobbesian argument for social order, which tell us a story about the genesis of contracts and the requirements for commerce and trade to occur. In the next section, I move to connect the Hobbesian argument to the Anonymity Problem.

HOBBESIAN SOLUTIONS TO THE ANONYMITY PROBLEM

The Anonymity Problem, as we saw earlier, is essentially a problem of trust, and is exacerbated by the presumed anonymity of the Internet. In some ways, the free-floating world of the Internet is akin to Hobbes's state of nature, where there are not universal protections for all participants, particularly for e-commerce partners. Even from a legal and social perspective, the Internet has really become a sort of 'Wild West' frontier where boundaries, social norms, and legal institutions are tested and shattered on a regular basis, and this makes trust difficult to develop online. Hobbes clearly had the problem of trust in mind when he was developing a theory that would protect contracts and preserve social order in the face of predominantly self-interested agents. He provides us answers to the questions "Whom should I trust?" and "Why should I trust them?" by invoking the twin fears of the sovereign and the state of nature, and provides an argument for even the most selfish and egoistic agents to follow the rules imposed by them by social forces. Hobbes has provided what some social theorists (Hollis, 1982) have called a

'bottom-up' argument for social order, by arguing that social order is generated by individual agents, acting in particular (and predictable) ways, and Hobbes is claiming that because humans are predictably self-interested, we can control their behavior by providing the right incentives (or disincentives, as the case may be) for their actions. If we can take Hobbes's argument and apply it to the Problem, and particularly premise 4 from our initial argument ("E-Commerce, in general, does not utilize face-to-face interactions and is, generally, 'anonymous' in most relevant senses, and is, therefore, lacking in the materials that develop trust."), we might be able to avoid the unpleasant conclusion of the argument, namely, that e-commerce is more likely to fail than other forms of commerce.

The anonymity of e-commerce, and of the Internet in general, seems to support some Hobbesian claims about human nature. When one perceives oneself to be anonymous, or unable to be identified, it would seem one would be more inclined to act in socially unacceptable ways than if one were well-known to the people affected by the socially unacceptable action. Additionally, the absence of facial cues and the increased physical distance can lead us to feel "safer and more immune to a counterattack" (Wallace, 1999, p. 126) than we would feel if we were having a person-to-person interaction. So, for instance, it is much easier to launch an expletive-laden outburst at a passing motorist, or give them the middle finger, than it would be to do the same thing to a nearby pedestrian or the passenger in the bus seat next to you. And it would even easier to launch an expletive-laden outburst at someone we assume is hundreds of miles away from us, staring at a computer screen. Keisler and Sproull, studying small groups trying to reach consensus on difficult tasks, found a higher frequency of insults, personal attacks, hostile comments, and name-calling in the groups who were communicating by computer only. They found that computer mediated discussions seemed to generate a considerably larger amount of hostility than face-to-face discussions, so much so that on at least one occasion, the anonymous computer discussants had to be escorted from the research facility for fear of violence (S. Kiesler & Sproull, 1992). And, as Hobbes would be careful to inform us at this point, it will be much easier for us to play the role of the Foole if we are anonymous agents in society. So, while our anonymity is helpful in some respects, it might also encourage us to act in socially unacceptable ways (including cheating parties in e-commerce transactions) while online.

I think we can address some of these worries (at least from the perspective of e-commerce) about anonymity by considering some of the Hobbesian safeguards against socially problematic behavior. This is particularly important given the size and distance of the Internet community. If the Internet had, for example, a Leviathan, or a reasonable set of community norms that, if broken, would lead to expulsion from the community, we might be spared these worries (for a proposal of such "Netiquette" see Johnson, 2009), although it might be difficult to enforce an Internet ban. It is true that there are some of these safeguards already in place, in some cases, to protect e-commerce from Fooles and Free Riders. These safeguards can be roughly analogous to the Hobbesian safeguards of the fear of the sovereign and fear of being rejection by fellow citizens. The first type we could call roughly "institutional" remedies to violations of e-commerce norms, and would roughly be equivalent to the punishments that would arise were the sovereign aware of the transgression, although that need not specifically be remedies with the law itself, provided they were remedies offered by some large, powerful body or institution. The second type we could call "social" remedies to violations of e-commerce norms, and they do not feature a formal punishment from an authority figure, but instead are centered on various forms of stigmatization from non-authoritative fellow

participants in commerce. I will briefly examine a few examples of each type of remedy.

The 'institutional' type of remedy to e-commerce problems is quite common, and most individual buyers and sellers are probably likely to utilize something like it if they run into problems with a transaction. If a party to a transaction can appeal to an authoritative body to help resolve the problem, whether that body is a legal system, a business institution, a financial network, or some other body with the power and authority to offer a resolution to the situation, they are making an institutional appeal. This kind of appeal may include governmental intervention (local, state, Federal, international law) for violations of laws and codes dealing with commerce (The Universal Commercial Code, The United Nations Convention on Contracts for the International Sale of Goods), private intervention on behalf of consumers (the Better Business Bureau, local news stations and consumer advocates), financial institution remedies (credit card and bank card guarantees and charge backs), or business guarantees (Ebay's Trust and Safety Team, Amazon Marketplace's Guarantee program). A party who did not receive goods, services, or payment appropriate to their electronic transaction may have several institutional outlets to choose from for the appropriate correction. To take just one example, an item purchased from a seller on Amazon's Marketplace (new and used goods being sold by third parties, not by Amazon.com itself, with Amazon collecting fees from sellers) is subject to the "A-to-Z Guarantee" which protects purchasers if the goods are not delivered on time, are damaged or defective, or if the goods are returned but the appropriate refund has not been given. But, failing that, a consumer may also have recourse from her credit card or bank, and may be able to stop payment or receive a refund if an item is defective or undelivered. Clearly, these 'institutional' remedies are only loosely connected to the Hobbesian version, where the 'institution' in question is an all-powerful Leviathan, but they fit the profile of an authoritative set of powers to keep individual agents in line.

The 'social' remedies that are available often use informal social networks, with no specific power or authority over the parties involved, but can provide useful information to other potential buyers and sellers about the habits of their prospective partners. Many e-commerce websites (notably Ebay and Amazon Marketplace) ask for buyers to leave 'feedback' ratings on sellers, and allow them to leave short comments about their transactions, and often have mechanism built in to prevent excessive or unwarranted negative feedback. When a buyer examines a seller's item on one of these sites, they have some information about the seller, based on their feedback rating, the number of sales they have made on the site, and the comments that have been made by buyers about the seller. A potential buyer might avoid buying from a seller who is relatively new to the site, or who has a less than satisfactory feedback rating, or who has had a recent history of unsatisfied customers. An additional advantage is that these systems are built into the framework of the communities themselves, and the information is handily available for all users of that community. Outside of the e-commerce sites themselves, there are other possible sources for 'social' remedies. These include social networks, (Facebook, Twitter and MySpace), internet discussion groups, chatrooms and forums, and private companies (like the Better Business Bureau or Consumer Reports) that track and post relevant data about customer complaints. Some websites allow consumers to enter their own information about problematic transactions and create, essentially, their own webpage about the problems they have encountered, while others simply allow the date and information about the claim to be noted. Professional consumer advocates in the mass media sometimes bridge the gap between institutional and social remedies by allowing both the sharing of information and, sometimes, providing legal and media pressure to help customers who have been wronged.

This pair of solutions, "institutional" and "social," roughly paralleling the twin Hobbesian solutions to the problem of the Foole (the sovereign's wrath and the fear of social ostracism), provide us with a framework to protect trust, override fears of Fooles and free riders, and allow us to be as comfortable with e-commerce as we would normally be making a purchase directly. Combining the institutional solutions with social solutions gives consumers and sellers resources to protect their economic well-being if a transaction goes bad, and also provides them with research possibilities before making transaction decisions, allowing them to avoid potentially troublesome business partners. Clearly, the most Hobbesian solution for The Problem of Anonymity would be an all-encompassing global authority that would enforce these e-commerce contracts, but something of that sort seems to be quite unlikely to occur in the near future, so this version of a "modified Hobbesian solution" serves as an attempted solution to the Anonymity Problem. It is 'modified' in the sense that it takes the ideas behind Hobbes's own argument and recasts them in light of current social and political realities.

One final point might be noted here, to highlight the Hobbesian strain of thought as it relates to problems like this. One could take Hobbes's argument against Fooles farther than just one's e-commerce partners, and apply it broadly, to one's role as participant in the Internet. The Internet is a very large global community, with a fluidity of membership that makes it appear to change at each moment, but like many loose collectivities, there is constancy. Violations of social norms on the Internet that are not specifically related to e-commerce (spamming, trolling, e-bullying) are just as subject to the Hobbesian argument as violations of e-commerce norms are, although there are fewer 'institutional' solutions to non-commerce violations available. If we take Hobbes's social contract argument seriously, as I suggest we do, and apply it to our participation in the Internet, I think Hobbes provides us with an argument not only for following the basic rules of e-commerce (pay on time, don't falsely advertise, etc) but also an argument for following the basic norms of civility and decency as part of the larger community of Internet participants. We are, as participants in a geographically unbounded world of information and ideas, truly 'citizens of the world,' as Diogenes of Sinope was said to describe himself (Book VI, Laertius, 1985). The Internet enables us to be cosmopolitan citizens of a global world, and if we accept that we have 'contracted' to treat our fellow citizens (even those we only meet on the Internet) with respect, we can apply our Hobbesian social contract argument to our behavior on the Internet in general, and not merely to our purchases and sales. We could call this a contractarian view of Internet participation, and I think it would serve to curtail problematic behavior on the Internet, provided that reasonable standards of 'acceptable' Internet behavior were to exist. I think more needs to be said to make this brief argument complete, but I will not do so here.

FUTURE RESEARCH DIRECTIONS

Philosophers, social scientists, and business scholars should watch the development of e-commerce closely. Despite the fact that the Internet is now a crucial and well-integrated part of our lives, e-commerce is still in its relative infancy, and much more study needs to be done to determine whether it will follow the same paths as traditional commerce. Sociologists, psychologists, and economists working on issues in the methodology of trust should be particularly interested in the patterns of 'institutional' and 'social' remedies. In the brief final section that follows, I offer a few other suggestions and make some concluding notes.

CONCLUSION

If the argument I have developed works as I think it does, we have been able to avoid the challenge of premise 4 in the initial argument by offering a supplementary premise (4*) to salvage e-commerce. That supplementary premise is something like this:

(4*): In situations where trust is not readily available for partners in commerce, 'institutional' and/or 'social' remedies (which can include such things as laws, business guarantees, consumer advocacy groups, etc.) can be used to supply the missing elements of trust for both parties in the transaction.

This supplementary premise allows us to replace the conclusion of the argument (5) with:

(5*): E-Commerce is no less likely to occur and succeed than traditional business, provided the appropriate institutional and social remedies are applied to bolster the missing trust in the relationship.

Hence, the Anonymity Problem is rendered a non-problem, by involving a strain of Hobbesian thought to a very 21st Century problem. This solution is made more philosophically rich if we take the contractarian view of Internet participation, under which we have agreed, as cosmopolitan citizens of a global world order, to follow certain rules to keep our online 'society' running well. But even without this deeper view of Internet participation, we can still make strong claims about what should be done to solve the Anonymity Problem. We would be wise to help undercut future problems by, for example, encouraging better and more comprehensive legislation (an 'institutional' solution), locally and internationally, to protect buyers and sellers from fraud. Rooting out Internet scammers from countries with little or no legislation against their actions will engender more trust across the global Internet community. The development of better, open access online resources and databases ('social' solutions) that will allow consumers to share information about their e-commerce partners will create a more informed set of buyers and sellers. All of this will, hopefully, render the Problem solved, and make a better, and more efficient and effective, e-marketplace.

REFERENCES

Carson, T. L. (2002). Ethical Issues in Selling and Advertising . In Bowie, N. E. (Ed.), *The Blackwell Guide to Business Ethics* (pp. 186–205). Malden, MA: Blackwell Publishing.

Hobbes, T. (1996). Leviathan (Rev. student ed.). Cambridge, UK: Cambridge University Press.

Hollis, M. (1982). Dirty Hands. *British Journal of Political Science*, *12*, 385–398. doi:10.1017/S0007123400003033

Hollis, M. (2002). *The Philosophy of Social Science: An Introduction (Revised and Updated)*. Cambridge, UK: Cambridge University Press.

Johnson, D. G. (2009). *Computer Ethics* (4th ed.). Upper Saddle River, NJ: Prentice Hall.

Kiesler, S., Siegel, J., & McGuire, T. W. (1984). Social psychological aspects of computer-mediated communication. *The American Psychologist*, *39*(10), 1123–1134. doi:10.1037/0003-066X.39.10.1123

Kiesler, S., & Sproull, L. (1992). Group Decision Making and Communication Technology. *Organizational Behavior and Human Decision Processes*, *52*, 96–123. doi:10.1016/0749-5978(92)90047-B

Laertius, D. (1985). *The Lives and Opinions of Eminent Philosophers*. In C. D. Yonge (Eds.), Retrieved from http://classicpersuasion.org/pw/diogenes/

Oosterhof, N. N., & Todorov, A. (2008). The Functional Basis of Face Evaluation. *Proceedings of the National Academy of Sciences of the United States of America*, *105*, 11087–11092. doi:10.1073/pnas.0805664105

Wallace, P. M. (1999). *The psychology of the Internet*. Cambridge, UK: Cambridge University Press.

Wilson, J. M., Straus, & McEvily, B. (2006). All In due time: the development of trust in computer-mediated and face-to-face teams. *Organizational Behavior and Human Decision Processes*, *99*, 16–33. doi:10.1016/j.obhdp.2005.08.001

ENDNOTES

[1] In a very recent study of honesty in England and Wales, for example, more than two-thirds of participants admitted to keeping quiet after being undercharged in a shop. Full results of the recent "Honesty Lab" study performed by researchers at Brunel University were not available at the time of this writing, but preliminary comments can be found at http://www.guardian.co.uk/science/2009/sep/07/survey-lawyers-honesty-public-attitudes.

[2] Hobbes does offer at least one other argument (other than fear of the sovereign) as to why parties to a social contract should keep their promise to maintain order, although it is not often invoked by Hobbes's defenders. He argues (XIV, p. 93) that it would be 'absurd' to contradict what one maintained in the beginning by promising, and that some version of the law of non-self-contradiction would apply to social contractors. This would imbue Hobbes's perceived egoism with some semblance of 'morality' that is sometimes overlooked because of the primary importance of the fear of punishment

by the sovereign. I find this argument to be often overlooked and compelling, but it need not be accepted here for the purposes of my argument.

[3] I leave out, once again, the possibility that aversion to self-contradiction would also be a reason to avoid such behavior, although I am sympathetic to this reading of Hobbes, as noted in footnote 6.

Chapter 4
Anonymity and Trust:
The Ethical Challenges of E–Business Transactions

Andrew Terjesen
Rhodes College, USA

ABSTRACT

E-business offers an opportunity to create markets that transcend the "real-world" limitations on markets. However, E-business transactions carry an added level of distrust that is not present in "real-world" transactions due to the anonymity of the Internet. Attempts to deal with the problem of distrust using economic approaches have limited the markets that E-business is able to create and, consequently, the potential of e-business. Overcoming distrust requires an ethical approach to the problem of trust, which would mean that participants feel they ought to take a risk even when fraud is a possibility. Insisting that people adhere to traditional ethical principles does not address the root of the problem, which is the lack of personal contact that anonymity creates (which in turn undermines our motivation to be moral). A moral sentimentalist approach suggests a way to overcome the motivation problem without sacrificing all the added benefits of e-business.

INTRODUCTION

Two of the core principles of Adam Smith (1776) are that the division of labor is the key to the creation of wealth and that the division of labor is limited by the size of the market. In Smith's analysis, it was the addition of markets in the New Word that increased European wealth, not simply the fact that they had untapped resources. After all, what

good are all those natural resources if no one will buy the finished product? Although we may no longer accept all of Smith's economic principles, we should not forget this insight. New sources of wealth will be created (as opposed to just shifting wealth around) when we find untapped markets, especially those that let us exchange things that we couldn't exchange before. Something that can't be exchanged effectively has no value.

Today, E-business has begun to create markets that simply aren't possible in the world of bricks and

DOI: 10.4018/978-1-61520-615-5.ch004

mortar businesses and, therefore, allows us to tap into a source of possible transactions that could not be engaged in without the Internet. In a bricks and mortar store, a shopkeeper needs enough people coming into their store so that they can sell their goods and have at least enough profit to make it worth their time. It's difficult to run specialty shops like "Just Door Knobs" unless you live in an area with a very dense population. With the Internet, one can easily create JustDoorKnobs.com and focus on providing the best product possible. This much larger online market can generate enough sales to make this a viable business.

It is important to realize that some, if not most, of JustDoorKnobs.com sales will involve people choosing to buy online as opposed to going to their local hardware store. No new wealth has been created in the global market; it has merely been distributed differently. It's no different than Home Depot attracting the business that used to go to Local Hardware. It is true that this creates wealth for the owner of JustDoorKnobs.com, but what this chapter will focus on is the creation of value which wouldn't be possible without e-business. The most dramatic example of this value creation is the pornographic industry on the Internet. People who would never ask for such a product in person at a store will buy it on the Internet, and it would seem that even the most bizarre preferences can represent a sizable demand when you consider the Internet population as a whole. But there are also a lot of more mundane examples of how the Internet creates value.

Bricks and mortar stores can only afford to carry a certain selection of door knobs. Most people will just choose from the available models. Consumers who desire handcrafted customized door knobs will not have enough impact at any individual store to make it clear that there is sufficient demand to sustain a custom door knob business. JustDoorKnobs.com, on the other hand, can bring together door knob aficionados around the world and offer them unusual door knob styles (perhaps through a KnobFinder service similar to the way Amazon.com sells used books), or even advertise for someone who makes door knobs according to customer specifications. Merchandise that would otherwise languish on a shelf or in someone's workshop finds a happy buyer and everyone involved is enriched by the process. Value, that would have gone unrealized, is created using this Internet-based model.

The Internet allows people to take items that no one they know wants and offer it up to someone else halfway around the world who couldn't find it in a local store (turning "junk" into something with a cash value through eBay, for instance). Plus, the Internet offers ways to save on overhead so that prices can be lower. For example, Just-DoorKnobs.com can use a network of hardware stores to supply door knobs rather than store them in one physical location that has to be leased out. They can also save on advertising since the Internet makes it easier for the consumer to find them wherever they are and once the consumer finds them that consumer is pretty much browsing through the store's merchandise. This is an added advantage over bricks and mortar stores using targeted Internet advertising to lower their overhead. Lower prices indirectly create value because they free up money to be used for other goods, including goods we would not normally buy because we didn't have enough money to spare. It may seem unnecessary to recount all the benefits of e-business, but it is important to see that those benefits are also the cause of one of the most persistent problems in E-business transactions: a lack of trust in the system.

THE PROBLEM OF TRUST ON THE INTERNET

One of the ways that the Internet creates value is that it brings together strangers who are separated by vast distances so that they can trade with each other. But this also means that they will probably only deal with each other on a single occasion.

Typically, if either party is dissatisfied, it is too much trouble to physically confront the other party or to seek legal restitution. The Internet also lowers the start-up costs for any business since it's not necessary to buy a retail space, hire lots of employees, or advertise as heavily. However, that also means it is easy to disappear if customers are dissatisfied. Businesses can even operate under a completely new name the very next day (or operate under several names at once) without any change in business practices. In an environment with all of these features, it is not surprising that the problem of trust keeps E-business from realizing its full potential. Many people avoid buying and selling things online because they do not want to deal with the trust problem (Gefen, Karahanna, & Straub, 2003; Doolin, Dillon, Thompson, & Corner, 2008).

There is no agreed-upon definition of what trust is. Das & Teng (2004) provide a very thorough cross-disciplinary overview of the various definitions of trust that have been offered. Fortunately, for the purposes of this chapter, it is not necessary to settle on the precise definition of "trust" as all the proposed definitions seem to share certain features. What comes through in Das & Teng's review is that trust involves a situation where there is a risk that the party being trusted will not perform as expected and that there will be a cost to the trusting party should those expectations not be met. Even definitions that do not explicitly use the word "risk" have implicit references to risk through words like "confidence." When dealing with the question of whether it is commercially irresponsible to trust someone, Blois (2003) makes a distinction between weak trust and strong trust. A strong form of trust exists when one expects things to go as planned, even if there are "perceived opportunities and strong incentives" (Blois 2003, p. 186) to break that trust. In contrast a weak form of trust is when there is an expectation that things will not go wrong, but there is also no reason to believe that the trust will be violated. The strong form of trust is difficult to achieve on the Internet

because of all the opportunities and incentives to take advantage of people and so I will focus on this sense of trust in the remainder of this paper.

Since the main issue seems to be whether the other party has the capabilities and intentions necessary to fulfill expectations, the economic approach to the problem of trust has been to treat it as a version of Akerlof's (1970) lemons problem. The lemons problem is inspired by the market for used cars and the fact that it is often hard to find a good used car (one that is not a "lemon"). In the case of used cars, the seller knows the quality of the car, but the buyer does not. As a result, the buyer will pay the same for lemon as they would for a good used car. We see a similar situation with online transactions. The buyer does not have enough information about the reliability of the seller to determine whether the seller can deliver the product as promised. Nor does the buyer have enough information about the seller to determine if the seller's intention is merely to defraud the buyer. Similarly, the seller does not have enough information about the reliability of the buyer to make the same judgments. The effect with the market for used cars is that the price of a used car is based on the value of getting a lemon. This means people trying to sell good used cars will lose value and so many of them pull out of the market, the presence of lemons drive out the good cars. A similar situation could happen on the Internet as reliable sellers and buyers are not being properly valued due to the information asymmetry and so they go back to doing business in the bricks and mortar world. On the economic approach, the buyer and seller's mutual distrust is regarded as a problem of information. The common approaches to the problem of trust on the Internet reflect the popular solutions proposed to the lemons problem.

Name Recognition and Other Forms of Signaling

The solution to the lemons problem proposed by Spence (1973) involves what has come to be known as "signaling." Using the job market as an example, Spence demonstrated how one might reliably signal the necessary information to the other party in order to eliminate the information asymmetry. For example, a college degree can signal a level of dedication and competence that employers are looking for. On the Internet, there are also ways for e-businesses to signal their reliability and honorable intentions to the consumer in order to allay consumer fears about risk.

Studies of consumer trust on the Internet show that a number of things can serve as a signal to the consumer that the vendor can be trusted. Aside from some of the more obvious signals (such as providing contact information and displaying certifications such as the TRUSTe seal or Verified by Visa), website design proves to be one of the most effective signals of trustworthiness (Sultan, Urban, Shankar, & Bart, 2002). Given the perceived difficulty of website design, it's not surprising that it should be a factor that people use in determining whether something is a fly-by-night operation or a dependable business. Website design is one aspect of what has been called "familiarity," which is an important element in developing trust in a business (Noll, 2001). Familiarity is how well the consumer understands the workings of the company they are dealing with. Most signals that the E-business vendor can provide on their own (as opposed to getting from a third-party) will involve some aspect of familiarity.

Consumers are not going to spend time on the Internet investigating every possible E-business vendor, so they will gravitate towards websites that resemble their past shopping experiences in the real world. It is not a coincidence that many of the most successful e-tailers today are the so-called "clicks and mortar" stores. Barnesand-Noble.com, Target.com, and BestBuy.com all thrive because consumers are already familiar with them. Plus, clicks and mortar businesses are able to take advantage of signals that bricks and mortar stores use. A company that maintains brick and mortar stores is not going to disappear overnight. Moreover, a company that can afford to pay overhead on a physical store is probably doing good business (either online, offline or both) so it is assumed that it must provide good products and services. These signals can be faked (and have been by unscrupulous individuals long before the Internet), but it is not as easy as faking a website.

The success of clicks and mortar stores also highlights how familiarity can default to name recognition. Retailers that people have heard of are the retailers that attract business on the Internet. The pure play e-tailers (sometimes known as "clicks only") like eBay and Amazon.com continue to thrive because they have name recognition. Presumably, being an early entrant to e-commerce enabled them to attract customers, and through customer satisfaction, they were able to grow a loyal customer base. As Internet pioneers, the media and news coverage of their growing E-business also aided in making them household names. In the end though, even a successful E-business will often start by engaging in signaling through a physical presence. They may form partnerships with bricks and mortar stores (for example, as Amazon.com did with Target and Borders) and spend money on naming rights, which may signal that an E-business is doing well.

In the late 1990s, the conventional wisdom predicted that bricks and mortar stores would be driven out of business by pure play e-tailers. This hasn't happened. Not only did some of the bricks and mortar stores adapt to the E-business model, but also many pure play e-tailers failed to capitalize on their initial advantage in the medium. The triumph of familiar names shows how hard it is to overcome the problem of trust on the Internet. There are no signals that can be sent in a "clicks only" environment that are immune to fabrication. Certificates of quality can be forged, websites can be designed to look professional, and buying the naming rights to a Bowl Game may not be enough to convince consumers they should trust an Internet business. In the end, the strongest signals are physical signals. The familiarity, branding, name recognition, and other signals of trustworthiness that a new pure play e-tailer can produce will likely not be enough to successfully compete with similar signals produced by clicks and mortar stores (or those pure play e-tailers that have developed a dominant brand). Thus, the Internet's promise of a low cost of entry into a market is unfulfilled if a business is forced to rely on those physical signals to develop consumer trust.

Additionally, name recognition and other signals are only helpful to the buyer. There is another side to the trust equation: sellers still get regularly taken advantage of by disingenuous buyers. The e-commerce buyer is not expected to obtain and display the same kinds of signals as an e-business.

Community Policing and Trust

eBay appears to be one of the few solid success stories of pure play e-business. In getting their E-business off the ground, the company's founders had to find ways to deal with both sides of the trust problem. If too many people who made the high bid at the auction chose not to render payment or if items didn't match the seller's description (or didn't arrive at all), then people would abandon eBay in favor of bricks and mortar stores or flea markets. The agreement between a buyer and a seller on eBay couldn't be a legal contract, given the problems of legislation and enforcement of a global Internet, so another means of creating trust had to be found.

To deal with this problem eBay relied on another form of signaling: reputation. Using eBay's easily discernible ratings and comment system, someone who provided prompt service and merchandise exactly as promised would quickly receive a lot of positive feedback on the website. This feedback is easily accessed by potential buyers. Unreliable parties just as quickly accrue negative feedback. Since feedback can be left by either buyer or seller, it provides a way to develop a reputation for trustworthiness on both sides of the transaction. Other online sites,

like Amazon Marketplace, have also adopted the feedback mechanism as a way to deal with the problem of trust (although Amazon does not allow two-way evaluations by both buyer and seller). However, Amazon.com does not allow the buyer to rate their experiences with purchases that are fulfilled by Amazon. E-businesses that are third-party facilitators of Internet transactions make use of the feedback mechanism on their website, whereas those sites that are direct vendors can only be evaluated through separate websites like ePinions.com. It is not clear whether third-party facilitators need more guarantees of trust than direct vendors or direct vendors are resisting a mechanism that would actually increase the likelihood of new business.

Feedback and reputation signals are not fool-proof. Studies of the eBay feedback mechanism (Dellarocas, 2001, 2003; Masclet & Pénard, 2008) and other methods for creating reputation signals (Bolton, Loebbecke & Ockenfels, 2008; Bolton & Ockenfels, 2008) show the limits of reputation feedback in overcoming the problem of trust. To begin with, feedback can be left by anyone. Someone could use a different alias to boost their feedback. Sellers might go so far as to appear under a different username (or several usernames) to promote their product with unsolicited testimonial or to bad-mouth their competition. If someone is really weighed down by negative feedback, they could abandon that online identity and adopt a new one. The fraction of negative feedback on eBay seems small, which raises the question as to whether other forces might be pressuring people not to leave negative comments for fear of reprisals (Dellarocas, 2003). E-businesses that use feedback mechanisms, like eBay and Amazon Marketplace, have done much to try and mitigate these problems, but the anonymous nature of the Internet makes it impossible to completely cut off these manipulation and abuses.

Feedback mechanisms remain in a perpetual arms race between those who are trying to game the system and those who are trying to expose attempts to exploit it. For example, it is possible to trace usernames and ISPs in order to determine the origin of feedback. If a lot of positive feedback is coming from the same place, it becomes suspect. What this really highlights though is how the feedback mechanism is not like the process of seeking recommendations in the physical world. We ask friends, family, co-workers and other people we already trust for information about people we might do business with. On the Internet, it's comparable to someone in New York randomly calling someone in Tokyo to see who the best plumbers are in SoHo. One study of Amazon.com (Chen, Dhanasobhon, & Smith, 2008) highlights this problem as they show that high ratings are more likely to influence buyers when they are judged to be "helpful" by other visitors to the website. Feedback mechanisms work best if they operate within a community (such as regular Amazon.com consumers), but that trust will come at the cost of e-business's potential for expanding the market.

According to Bolton, Loebbecke & Ockenfels (2008), "reputation information trumps pricing in buyer deliberations" (p. 20). The signals they get about the reliability of an E-business leads online users to prefer to do business with that company even in a case where the price is much lower elsewhere. Regardless of whether this response is an emotional bias or a calculated tradeoff based upon the perception of risk, we once again see the field of E-business competitors cut. Even though pure play e-tailers utilize signals to enhance reputation, the problem of trust still limits market potential. Since reputation will trump pricing, it is not the case that the Internet will indirectly create value by offering things at the lowest prices possible. In fact, there seems little reason to offer it at a price much lower than the most reputable e-tailers. Moreover, since reputation wins out, consumers tend to stick with existing e-businesses, the Internet is not necessarily an easier medium in which to start up a successful business.

Marketing Circles and Other Forms of Screening

When dealing with information asymmetry, signaling is only one possible solution. One has to be wary of false signals, but the nature of the Internet is such that it is relatively easy to develop false signals no matter how sophisticated the signals become. Since signals are being created by those who have information that we want (such as whether we can trust them to deliver a quality product), we must always be worried that it is in their interest to send us a false signal. Stiglitz (1975) proposed a solution to information asymmetry which focused on getting someone to reveal the information you need. According to Stiglitz's theory of screening, the person who is under-informed can present options in such a way that the other party will reveal the desired information. For example, a health insurance company can offer plans with different deductibles and premiums. People who know that they are not in the best of health (based on their lifestyle or current symptoms) will choose plans that have lower deductibles, while those who are pretty confident about their general health will opt for higher deductible plans. While not a perfect system, screening in this manner can mitigate the risk that the insurer might experience. But how does one "screen" for trust? The traditional examples of screening reveal information based on the assumption that the individual can be trusted to engage in regular economic transactions.

Screening on the Internet might be set up in such a way that the parties involved in the transaction have a choice of contractual arrangements and the choices made reflect the information that each party is trying to screen for. To an extent, this is what eBay and other sites do. People join these sites on the understanding that they must adhere to certain community standards regarding how they should do business. However, the cost of promising to adhere to these standards is low as the penalty can be gotten around by setting up a new online identity. Moreover, any penalty is dependent upon someone taking formal action against the other person (as opposed to just never doing business with them again or leaving the eBay community entirely). The low fraction of negative feedback on eBay may indicate that people are more likely to avoid the "hassle" of enforcement.

In order to properly mimic screening contracts, there would need to be a participant choice that forces a party to reveal whether or not they plan to go through with the transaction. When there are concerns about trust with respect to bricks and mortar transactions, people are often required to put down deposits. However, the problem with Internet transactions is that both parties are wary, so even giving a deposit to someone would require some trust already exist. If it is a transaction that is facilitated by a third party, this might be easier. For example, Amazon has recently changed the policy for Marketplace transactions so that vendors must confirm shipment before receiving money from the payment the buyer made to Amazon. However, many E-business transactions involve direct vendors so that there isn't a third party to mediate the transaction. Something like PayPal comes close, but they cannot stop a vendor from selling on the Internet the way Amazon could ban someone from Amazon Marketplace. Consequently, screening alone is not used on the Internet. The closest example might be marketing (or "insider's") circles, where participants pay money (usually a monthly fee) to join a group that includes a membership list for business purposes. For a number of practical reasons this is not a very effective form of screening. Many of the benefits of a marketing circle can be achieved for free by forming an online community and dealing only with members of that community.

Regardless of whether one pays to join a marketing circle or uses a social network to screen potential business transactions, the effect is the same: a greater degree of trust is achieved. But once again, that trust comes at a cost to e-business'

potential. If one is afraid to venture beyond a particular community to do business, then one has artificially restricted the market and, consequently, restricted its potential for growth beyond the markets of bricks and mortar (or even clicks and mortar) stores. Instead of only doing business with your physical neighbors, you only do business with your online neighbors. Either way, it's a smaller scale than the Internet is capable of.

Marketing circles also reflect a danger in focusing too much on the wrong kinds of trust in business. Much has been written on the importance of trust in business, but it should also be recognized that "blind trust" in the sense of "I trust you because you are a member of my family/ clan/marketing circle" can lead to collusion and monopoly power. Husted (1998) points out that an overreliance on trust can lead to problems that were "the motive for developing more impersonal modes of human organization which did not depend upon trust" (pp. 245-246). There are various institutions that maintain people's confidence in the prevailing global system of trade so that most people do not feel it necessary to form a closed market to protect from exploitation. So far, there has not been an Internet mechanism or institution that has the same effect.

ANONYMITY AND THE PROBLEM OF TRUST

When businesses deal with ethical issues, they tend to treat them as practical problems that can be dealt with through incentives. As one author put it, "moral intentions can be substituted by economic incentives that should be designed in a way that even opportunistic actors will try to behave in a socially desirable manner" (Grabner-Kraeuter, 2002, p. 49). Since Internet anonymity provides a lot of incentives to be untrustworthy, the focus has been on creating disincentives to exploit someone (through Internet Crime Task Forces and similar agencies) and to create the impression that Internet transactions between two parties can be secured from a third-party's interference.

Economical Solutions Sacrifice Economic Benefits

Even the simplest economic transactions in the bricks and mortar world require a good deal of trust. For example, if one goes to a store and purchases an item that is kept behind the counter there will be a point at which one party has what they want and the other doesn't. The clerk could shut the register and claim they have not been paid yet. Or the buyer could grab the item and run before handing over any money for the exchange. Why is it then that a simple grocery store transaction does not end up being treated like a hostage crisis? To begin with, we tend to avoid businesses or people that seem untrustworthy. If we enter a store and it "doesn't look right" or the clerk seems "shady," we will go elsewhere. Trying to extrapolate this approach to E-business is a problem. Many of the cues we look for in a physical transaction are non-verbal cues that will not be conveyed in an online transaction and the remaining cues we could detect on line are prone to the signaling problems discussed in the previous section.

Another reason why buying a cup of coffee in the morning does not end in a standoff is that most of us have difficulty acting immorally towards another person, especially when we are dealing with them face to face(and why this is so will be discussed later). Nevertheless, some people lack this moral compunction. Their reason for being honest in buying a loaf of bread is not that they think honesty is good, but rather because of the cost of dishonesty. To put it in economic terms, we have found ways to internalize the externalities of dishonest behavior so that actors will usually see being honest as the rational action to take. Instead of trying to figure out how to show that someone is trustworthy, trust is created with the expectation

that most people will not cheat because it is too costly to do so.

Some of the means by which we internalize the costs of dishonesty are difficult to replicate in e-business. For example, you could only walk out without paying so many times before there is no grocery store left in your neighborhood to buy food from. The Internet eliminates the spatial restrictions that keep us in a relatively small community where our misdeeds would quickly become known to all. In addition, the storeowner has access to a local legal system that will punish you for theft. Issues involving jurisdiction and the ease with which one can hide one's identity on the Internet significantly reduce the chances of getting caught when acting dishonestly (for example, the difficulty of policing auction fraud is described in Chua & Wareham (2004)).

It is hard to physically walk off with someone's merchandise and get away with it. Instead, those who want to get away without paying for the goods resort to violence or fraud. Violence is difficult to accomplish through the Internet, so the *modus operandi* of online criminals is usually fraud. Many storeowners in bricks and mortar stores take steps to prevent people from offering false payment: they use pens to check for counterfeit bills; they require identification for payment by check or credit card; and nowadays electronic communication allows clerks to verify a check or credit card transaction immediately. E-business transactions use some of the same tactics. You must often provide a billing address and detailed information about the account you are using to pay. Credit cards now have security codes that are printed on the back of the card as a way to try and ensure that you have physical possession of the card (and didn't just read its number off a discarded receipt). For buyers who are worried about surrendering all that information to a seller there are online services like PayPal that will serve as an intermediary protecting both parties' information.

Using all of these measures has gotten us to the point where e-commerce is a regular part of some people's lives. Still, it's a lot easier to commit fraud on the Internet and to be taken advantage of. The measures that most companies and consumers take to ensure good faith and security bring e-commerce to a certain level, but the risks are still great enough that many people prefer to shop in a bricks and mortar store whenever they can. The shortcomings of the e-commerce economic approach become apparent when one considers the simple fact that being trustworthy doesn't always pay and no amount of disincentives are going to change that. As is pointed out in Urban (2003), while trust does pay in some instances on the Internet, it is not always rewarding. If someone is focused on short-term financial goals or is dealing with a short-term customer base (as is the case with most eBay auctions) then there is no great benefit to being trustworthy. Unfortunately, those are the kinds of e-transactions that, with enough consumer confidence, would help E-business realize its full potential.

From a purely economic standpoint, everyone should try to maximize their expected utility. The various measures designed to punish fraud are meant to lower the expected utility of fraud so that the best option is to transact business honestly. However, the *homo economicus* has every reason to want to find ways to avoid the costs being imposed upon them in order to achieve a greater expected utility. Thus, an arms race is created between the people trying to discourage fraud and those who want to commit it. For every method developed to punish fraud, people will try to find ways around it. More importantly, they should from the standpoint of economic rationality as it would maximize their utility if they did so.

A classic example of how e-business's potential has been stymied is digital distribution. Among other advantages, providing songs and other media over the Internet is a great way to reduce costs—there is no need to buy expensive packaging or pay a distributor. However, making it easier to distribute a copy of something makes it easier for everyone to distribute a copy of the same thing without an additional sale. One person can buy an e-book, but now they can share it with five of their friends. Instead of making more money by lowering the cost of their product, the distributor makes less money by lowering the demand for it in the legal market. Recognizing this problem, attempts were made to prevent copying and distribution after the sale. This has not ended the problem because any copy protection can be defeated with the right tools and sufficient effort. Another problem is that too much protection, or limiting the number of times something can be downloaded, can hinder the product's desirability. For example, DRM (digital rights management) does not work well with anything but certain versions of Windows Media Player running on a specific operating system, which means a segment of the market will be excluded. Limiting the opportunity for fraud has restricted the extent to which digital distribution improves on existing forms of distribution.

Anonymity and the Inadequacies of Traditional Ethics

The problem of trust in E-business is largely due to the degree of anonymity that the Internet provides. It is an impediment to enforcing any regulations regarding the Internet, and it interferes with our ability to gather information about the people we do business with. We can discourage people from exploiting the anonymity that the Internet provides, but only at the cost of potential e-business. The underlying reason is that the areas in which E-business holds the most promise are situations where enforcement is not possible and some level of trust is necessary to facilitate transactions. A whole-hearted embrace of digital distribution of media content across multiple platforms is unlikely to happen as long as the producers of the content are worried that they will be easily pirated. Though digital distribution has developed legal outlets, it is too early to tell if the media industry will still be thriving in a few decades with easily distributed content (and, as it stands, legal media downloading is largely happening through iTunes, which requires certain software and equipment to function well). In the example of media, the full potential is not being realized because of a lack of trust and the impossibility of serious enforcement. Other examples of unrealized potential could be found in online auctions of big-ticket items like houses, jewelry, and cars (which are still relatively rare) or online education (which is still regarded very suspiciously).

From an economic standpoint, in these new kinds of markets and transactions there is often not enough information or enforcement power to give us good reason to believe we can trust the person on the other end of the transaction, unless we believe that person is acting out of an internal motivation to be trustworthy. The crucial point is that novel Internet transactions are not do-or-die situations; there are other ways to achieve similar results, or they involve things that we could live without. If E-business is going to realize its potential it will be because people took an ethical approach when a mere economic approach was inadequate.

The most obvious difficulty with an ethical approach to the Internet is that anonymity is an impediment to ethical practice, as well. Being able to hide one's actions makes it easier for people to do things that would shame them if their action were public knowledge (De George, 2003; Wallace, 2008). It is much easier on the Internet for people to post hurtful comments and verbally bully and abuse others. The family and friends of "anonstarwarsfan" don't know their loved one is threatening to kill anyone who doesn't think the original trilogy is the greatest bit of moviemaking ever.

A less often addressed problem with ethical practice and anonymity is that anonymity dep-

ersonalizes those we do business with: they're faceless buyers or seller who don't resemble the flesh and blood people we interact with on a regular basis. As a result of this depersonalization, we have a harder time being motivated to treat them ethically. It's not just that the anonymity of the Internet lets us escape consequences; it's also that anonymity drains a situation of its moral character. Technically, you can't wrong a computer.

The irony of the problem of trust is that a lack of trust can only really be cured by the addition of trust. This is not as paradoxical as it sounds. In order to create a two-way reciprocal relationship of trust (or more generally an environment in which people take for granted that they can enter into such relationships with most people they meet, which is what we need for a healthy economy) it is necessary for someone to take the first step and engage in a unilateral act of trusting. After that, it is important that such unilateral trust is not consistently abused. The first step though is for someone to take a risk. Once again, this is why the economic approach would counsel against trust in novel Internet situations. Anonymity constantly undermines that unilateral trust as it would not be prudent to extend trust when it is so easy to be deceived. The only reason someone might be willing to take the first step in such an environment is because they feel they ought to do so.

Without going into a lengthy review of the state of ethical theory today, it should be enough to note that three major approaches to ethics have been employed in addressing issues in business ethics. The deontological approach (usually represented by the ideas of the philosopher Immanuel Kant) focuses on our moral duties as reasons why we ought to act a certain way. The consequentialist (or utilitarian) approach focuses on the consequences of our actions in order to calculate what the right thing would be to do in a situation. Finally, the virtue ethics approach draws our attention to the character traits that we need to acquire in order to be a good person (and therefore be someone who would do the right thing). None of these approaches offers us a conclusive reason why we ought to trust someone on the Internet.

Deontological ethics has always had a problem when dealing with issues that are part of the personal sphere. Do parents have a *duty* to love their child? It seems more appropriate to say that a parent has a duty not to harm their child and a duty to make sure that their needs are met. It seems awkward to say there is a duty to trust someone. Myskja (2008) has argued that Kant's theory does imply a duty to trust, but the he does not seem to be able to justify the level of trust required by novel Internet transactions. His argument is based upon the idea that mistrust shows a lack of respect for humanity. It is not obvious that respect for humanity would entail always extending a hand in situations where there is great risk as would be the case in the E-business transactions that promise to transcend the limits of physical markets. Even if we did assume that duties can include things like trust, we need to recognize that our duties are shaped by those with whom we deal. Everyone has a duty to his fellow humans not to harm them, but no one has a duty to make sure everyone is fed, clothed and sheltered (although most of us would agree we have a duty to our families and friends, or at very least our dependents to do so). Similarly, my duty not to harm my fellow humans does not extend to someone who is trying to harm me. Killing someone in self-defense does not violate a duty not to kill innocent people. In a situation of anonymity, those facts are obscured. Am I dealing with someone who is honest or not? In such situations, it is hard to motivate a duty to trust others.

If one does not believe the person on the other end of the computer is worthy of trust, then they will not believe they have a duty to trust them. To be fair to one of the most famous deontologists, Kant would argue that it is irrational to think "I am different than everyone else; I must be allowed to lie and cheat in order to keep others from doing the same to me." In other words, two wrongs don't make a right. According to Kant, such a mindset shows a lack of respect for the moral law which is the source of all our duties. The problem with this response is that it presumes one recognizes the inherent equality of all persons. But, once again, anonymity can be a stumbling block to that moral realization. I know who I am, but I know nothing for certain about anyone else I meet on the Internet, so, in an important way, I am special.

Consequentialism does not fare much better. The consequentialist argument for trust would be along the following lines: if people followed the practice of trusting people they met on the Internet (within reason, of course), then the overall benefits of such interactions would be such that they outweighed the harms done to individuals who were taken advantage of. The problem with such an argument is that it is actually very difficult to say whether the overall practice would work out so nicely. The argument would have to assume that fraud would not be common. But as we've already discussed, Internet anonymity makes it much easier to take advantage of someone, so that may not be a fair assumption.

The argument for trust on the Internet is further undermined by the fact that we could imagine a perfectly functional and happy world without e-business. This is in contrast to a world where no one ever told the truth or property was determined by whether you can physically defend yourself against attempts at theft. Worlds like that are clearly less good than worlds in which people tell the truth and respect property rights. Since trust on the Internet is not nearly as morally pressing as those other issues, it seems hard to argue that we must engage in a practice of trusting e-businesses and online vendors. To strengthen the argument, the consequentialist would have to prove that novel Internet transactions yield a world which people would recognize as inherently superior to one without e-business. That's a tall order.

Virtue ethics is the most difficult approach to discuss because it lacks the precision of the other two approaches (both of which appeal to rules that we have found inadequate to the task at hand). Instead of determining rules for action, the virtue ethicist is concerned with developing character traits. Virtue ethicists typically model themselves after Aristotle in claiming that we know what is good through the acquisition of a habitual character trait that is relevant to determining the appropriate action in each particular circumstance (Aristotle, trans. 1999). These character traits are defined with reference to a conception of the good shared by the members of the community that are trying to inculcate those virtues. For the sake of this discussion, I will assume a roughly Aristotelian notion of virtue is at work, although I think other forms of virtue ethics will find themselves dealing with similar issues as the Aristotelian model. The most obvious criticism of virtue ethics is that it is unable to deal with those who have not acquired the virtues – the people who are exploiting the anonymity of the Internet. As long as there are enough of those individuals lurking around, it is not prudent to take major personal and financial risks in Internet transactions. The virtue ethicist is unable to convince consumers to take risk beyond what can be regulated. That means most of the potential of E-business remains untapped. This is a practical problem and actually one faced by every ethical theory: why be moral if you think everyone else will be tempted to take advantage of you?

A more central problem with virtue ethics is that in order to have relevance to the moral issue of trust there needs to be something that can be called the virtue of "trustworthiness" (and maybe even the virtue of "trusting"). Aristotle never named a virtue equivalent to trustworthiness or trust. The closest he might have come is "friendliness," but that is probably a stretch when applied to e-business. If you scanned through other lists of virtues you might find something like trust or you might not. That's the problem, the virtues are not universally agreed upon and so we must rely on cultural context to determine whether a trust-related virtue is something that we want to inculcate. Even if we could settle on such a virtue,

it's not clear that it would recommend trust in a situation like Internet transactions. Just like the virtue of courage can sometimes recommend a strategic retreat, the trust virtue might recommend remaining guarded on the Internet. We can't rely on virtue ethics to promote trusting someone on the Internet as a moral obligation.

A NEW ETHICAL PARADIGM: MORAL SENTIMENTALISM AND E-BUSINESS

At this point, one might conclude that there is no way for E-business to solve the problem of trust. Despite our best efforts, the Internet will always be a risky place and the promise of E-business will remain largely unfulfilled. Traditional ethical theories appear inadequate because Internet anonymity makes it difficult to say someone ought to take a chance and risk being taken advantage of. As mentioned previously, a deontological approach might try to get around this problem by arguing we should not hesitate because someone else might take advantage of us; instead we should have faith that most people are like us, able to recognize the equality of all beings and to see the wrongness in taking advantage of someone else for our own benefit. Simply put, we should trust that most people recognize the right thing to do is not to commit fraud, and if that is so, it is not a great risk to engage in Internet transactions. To use Kantian language, every rational person understands the basic precepts of morality, and that understanding produces a respect for the moral law that defines their duties. We may not have a duty to trust everyone, but we do have a clear duty not to take advantage of anyone else. Recognizing that, we should feel comfortable with E-business transactions.

However, this is where Internet anonymity really does the most damage. Even if we all recognize the moral prohibition against fraud, what will make us act out of respect for that law?

The Internet enables us to hide our identity and form new businesses overnight. Given the ease with which we can disappear, how does respect for the moral law trump the clear advantage to ourselves of cheating when everyone else plays by the rules. If everyone thought that way, the system would quickly break down, but unlike the physical world, the breakdown of the virtual world would not be unbearable. Instead, it would seem that an endless cycle would be created, where as soon as we begin engaging in E-business transactions that require trust (because compliance can't be enforced), people will be exploited. Once this became known, people would back off from that transaction retreating to the forms of E-business that can be relied upon. We've seen that happen with e-tailing as the proliferation of e-tailers has been followed by a consolidation of a few reliable and established companies. The market expands somewhat with these established e-tailers, but not as much as it would have in a situation where people frequently did business outside the relatively small circle of "trusted" e-businesses.

The Importance of Personal Connection

The real problem is that we don't deal directly with people on the Internet. When we encounter flesh and blood people, we can't help but respond to some of their emotional states. The pain caused by taking advantage of someone is written on the faces of those who have been exploited. On the Internet, people are bits of information and we don't see their pain. Even a strongly worded email or text will lack the force of seeing the hurt in someone's eyes. Some (Noddings, 1984) would argue that our moral responsibilities are founded on our relationships. They would say that it is wrong to take advantage of someone once you recognize them as a fellow human being. The problem then with Internet anonymity is that it interferes with our ability to form real relationships. I do not want to make such a strong argument as that. Social

networking suggests you can form relationships on the Internet. But I do think some people may still exploit their Internet "friends" because they don't seem "real." It's not the relationships that really matter from a moral point of view; it's the emotions those relationships invoke.

Personal connection on the Internet is important for developing ethical behavior, but only insofar as connections make it easier for us to respond to the emotions of others. Otherwise we can do what people have done for millennia and discount their feelings because of our own desires ("sure, that woman is in pain, but I really need to make a million dollars"). Even worse, we might try to dehumanize them in order to prevent ourselves from having an emotional response to their plight.

The idea that emotions are integral to our moral judgments and behavior has been demonstrated by recent experiments in moral psychology (Greene, Nystrom, Engell, Darley & Cohen, 2004; Greene & Haidt, 2002; and Greene, Sommerville, Nystrom, Darley & Cohen, 2001). In these studies, it was shown that subjects who were asked how they would deal with a moral dilemma activated more of the emotional centers of their brains. It seems then that, when we are confronted with a question regarding how we should act, we turn to our emotions for an answer. It also appeared that subjects' emotional reactions shaped judgments about those situations. One sample dilemma (Greene et. al., 2001) involves a trolley that is barreling towards five people. A quick flip of the switch would divert the trolley so that it would only kill one person. Most subjects thought it would be okay to flip the switch and save the five at the cost of one person. However, when the same dilemma is presented but this time the trolley can only be stopped by pushing a very, very heavy person off of a footbridge and onto the tracks of the oncoming trolley (where the weight is assumed to be able to stop the trolley), a much smaller number of people judge this to be morally permissible. These studies have been interpreted to suggest that emotions can interfere with our moral judgments, but another possibility looms in these examples: in cases where our emotions are active, we are more likely to grasp the moral weight of the situation. It is wrong in both cases to save the five at the cost of the one, but it is only clearly wrong in the case of the footbridge push.

What is Moral Sentimentalism?

The solution to the problem of trust on the Internet lies in taking a different approach to ethics. Instead of adopting an approach based on principles that are determined by duties or consequences, we should look at ethics as a matter of sentiment. Although this has not been a major approach to ethics in philosophical circles, it does have its proponents (most notably the 18th Century philosophy David Hume and his good friend Adam Smith). Since this approach relies upon our emotions to determine what is right and wrong, it is usually called "moral sentimentalism."

David Hume (1979) held that we judged something to be morally right or wrong depending on whether it produced a feeling of approbation or disapprobation in us. When explaining the source of this feeling, Hume sometimes referred to the pleasure of the utility of an action which would understandably lead one to think that his theory is just a form of consequentialism. However, he also appealed to the pleasure of the immediate agreeableness of an action. According to Hume we approved of some things because they simply seemed to fit (and not for their consequences). Smith (1982) used language that was even less likely to be confused with consequentialism as he referred to the approval caused by the propriety of someone's actions. So, a general definition of moral sentimentalism would be that something is right because it evokes a positive response in us by its very nature.

A major criticism of moral sentimentalism is that it is a form of relativism (Copp, 2005, p. 13). After all, we have experienced situations where we have had a vastly different emotional

reaction to something than other people. We've even experienced vast mood swings ourselves. To make morality a matter of sentiment would seem to relegate it to a matter of simple preference. And there are some moral sentimentalists who do view morality this way, but Hume and Smith were not among them. In their works, they made frequent reference to moral principles that they thought applied cross-culturally and even across time. Just because plantation owners did not feel slavery was wrong did not mean slavery was not wrong in the antebellum South. Both Hume and Smith recognized we must take ourselves out of the equation when we try to make a moral judgment. In order to see a situation clearly, we needed to put our own emotions aside and take an impartial point of view. For Hume, this meant viewing the situation from the perspective of everyone involved except one's self to create what he called "the general point of view." Smith employed the notion of an "impartial spectator"—the imagined perspective of someone who has no interest in the situation—through which we would view the circumstances of others. In both cases, the idea was that one could remove the biases of self-interest from our emotional states and whatever was left behind was a pure reaction to the situation.

Once we see things clearly and impartially, we should recognize that exploiting people on the Internet is wrong; and more importantly, our judgment will still be an emotional reaction (just not based on our own emotions), so it should move us to avoid exploiting people, even if it were in our own interest to do so. This is not to suggest that no one would be able to do so, but only that they might be less likely to do so. On that basis we would have some occasion for trust. More importantly though, moral sentimentalism offers us a means by which we can talk about a moral obligation to trust and be trustworthy. Our interactions with others can create positive feelings that suggest a morality beyond the prohibitions of most moral codes. A moral sentimentalist would recognize that killing is wrong because of the pain caused by ending someone's life. But a moral sentimentalist would also recognize that a trusting relationship is immediately agreeable. Anyone who seriously contemplates what it is like when two people trust each other has to admit that there is something that feels good about that kind of relationship, so in a sense, we ought to reach out to people and create those bonds.

The Moral Sentimentalist Approach to the Problem of Trust

The fact that moral sentimentalism can make sense of a positive attitude towards trusting on the Internet as opposed to merely a prohibition on abusing trust is just one example of how a moral sentimentalist approach might be superior to the other ethical approaches with regard to the problem of trust. Adopting that approach would change the focus of the discussion surrounding these issues. Attempts to enforce expectations would take a backseat to attempts to encourage moral behavior.

Some of the methods already in place for creating trust online reflect a moral sentimentalist way of thinking. Online communities in which people monitor each other and help each other are a way to recreate emotional interactions on the Internet that shape how we behave. Those communities may still be pretty anonymous. If e-business's full potential is to be realized, then those communities need to develop into communities that have a very warm and human feeling. In addition, they need to be open enough to outside traffic that they do not become the kind of closed circles that limit the market; they need to feel like a place you can visit (like going to spend a day doing business in another town).

It might appear that one of the best ways to do this is to get rid of the anonymity of usernames and avatars and so forth, but that's not actually necessary. In fact, as Wallace (2008) argues, there are a number of positive effects of anonymity, such as allowing people to express themselves in a man-

ner that their local community might discourage, and these should not be ignored. Just like we go through the bricks and mortar world doing business with people we don't know by name, it should be possible to do the same on the Internet. What is important though is that the people we encounter there feel like real people when we interact with them. The most obvious way to achieve that is to move towards transactions where the participants are looking at each other (presumably through a webcam). When confronted with the image of a person, it would be much harder for someone to dismiss the impact of their actions upon the person at the other end of the transaction. Currently some e-businesses have tried to reintroduce the human element through chat services for customers. These services can be very helpful in dealing with trust issues involving information about the company or individual someone is doing business with, but they will not be as effective in personalizing the transaction to avoid trust issues involving the motivation of both parties to behave morally when there are no costs imposed on immoral action. Any Internet savvy person knows that Internet chats don't always involve strangers who are who they say they are, and so chats will not be able to undo all of the distrust that anonymity breeds. Using virtual avatars like in *Second Life* would also be inadequate for overcoming the effects of anonymity. Cameras recreating real time interactions are a different story.

Any other way to reintroduce the nonverbal cues that we look to when we judge someone's trustworthiness would be of interest from the moral sentimentalist approach, but cameras seem the most promising at the moment. One might have concerns that moving in this direction would roll back some of the advances of e-business. For example, this technology would require the company to maintain one or more persons who serve as the clerks of these virtual businesses. So, e-businesses would not reduce their overhead by as much. Plus, consumers or vendors might judge the other party based on their appearance.

And consumers might have been originally attracted to an online vendor simply because they did not want to be seen by anyone while making their purchase (due to social pressure or simply a desire to shop in their pajamas). Although these issues (as well as others related to appearances and the costs of maintaining them) might reduce some of the potential of e-business, they do not shut off possibilities if they are able to gain trust in situations where the bricks and mortar world is unable to complete such a transaction.

Social networking sites like Facebook, Myspace, Twitter and so on, as well as more personalized forms of online communication such as blogs, might also be a resource in developing more ethical behavior. They would help to create the kinds of communities mentioned before that would add an emotional dimension to the Internet. They also serve as a way to personalize people that we encounter on the Internet. These services restore some of the personality to the anonymous "faces" we encounter on the Internet and that would remind us that real people are at the other end of the transaction. As one author described it, feedback mechanisms are a form of "institutionalized gossip" (Bolton & Ockenfels, 2008). One of the reasons they may be relatively successful is that they replicate that aspect of human interaction and make the transactions feel warmer.

These are just some quick suggestions, but anything that makes our encounters on the Internet more like our encounters in the physical world would encourage moral judgment and behavior from a moral sentimentalist standpoint. The more shopping on Amazon.com feels like going down to the corner store, the less tempted we will be to renege on our online agreements. Since anonymity is at the heart of the problem of trust on the Internet, we need to find ways to remove it. Of course, what all of this does is increase the likelihood that E-business transactions regularly involve emotional reactions to each other. In order to yield a moral result, the moral sentimentalist (at least one like Hume or Smith) would argue that

we need to train ourselves to apply an impartial corrective to our experiences so that we don't get dragged around by our emotions.

CONCLUSION

The promise of E-business is that it can operate in ways that markets constrained by bricks and mortar are unable to. This also means that some of the mechanisms we rely upon to make the market competitive and efficient and fair are unable to operate. As a consequence, people are likely to distrust other actors in the potential marketplace. In such circumstances, one cannot approach the problem in a purely economic way and regard it as an informational or incentive problem that needs to be solved by massaging certain aspects of the market transactions. This is not to imply that an economic approach has no place in these discussions. First of all, our sentiments are shaped by the environment we live in and we need to recognize that economic approaches to bricks and mortar business are what make us relatively comfortable with them. If we didn't have confidence in the physical-world commerce system,, there would be nothing for our sentiments to build upon in the virtual world. Secondly, the point of this chapter has been that there is a trade-off we need to consider between economic approaches and E-business potential. Because of the very nature of the Internet, there will always be some potential E-business transactions that outstrip our ability to regulate them economically (and by external motivations) without limiting the potential of those transactions to create new markets or improve existing market mechanisms. It is an entirely separate question as to whether we are better off erring on the side of external regulation or on the side of realizing new potentials in e-business.

I assume, though, that we could think of some potential Internet transactions that would be worth trying to develop and could significantly increase the wealth of society. In those cases, it is important to recognize that ethical behavior (not purely economic behavior) will be what gets that market off the ground and maybe keeps it running indefinitely. In those circumstances, trust needs to come about in spite of Internet anonymity. The traditional ethical theories do not offer the strongest arguments for why someone should be willing to be the first to extend trust in those situations. Moral sentimentalism is a promising alternative ethical theory because it doesn't simply command or recommend that we trust for moral reasons; it explains how to evoke the judgment that we ought to trust in a given situation. By developing the Internet in such a way that e-businesses engage in transactions that evoke an emotional component similar to what we experience when doing business in the real world, the moral sentimentalist approach claims it will be possible to bring about the internal motivation that will enable us to work to realize some of the riskier aspects of e-business' potential.

REFERENCES

Akerlof, G. (1970). The Market for "Lemons": Quality Uncertainty and the Market Mechanism. *The Quarterly Journal of Economics, 84*(3), 488–500. doi:10.2307/1879431

Aristotle. (Trans. 1999). *Nicomachean Ethics, Second Edition.* (T. Irwin, Trans.) Indianapolis, IN: Hackett Publishing.

Blois, K. (2003). Is it Commercially Irresponsible to Trust? *Journal of Business Ethics, 45*(3), 183–193. doi:10.1023/A:1024115727737

Bolton, G. E., Loebbecke, C., & Ockenfels, A. (2008, April). *How Social Reputation Networks Interact with Competition in Anonymous Online Trading: An Experimental Study.* (CESifo Working Paper Series No. 2270). Retrieved July 19, 2009, from http://ssrn.com/abstract=1114755

Bolton, G. E., & Ockenfels, A. (2008, February). *The Limits of Trust in Economic Transactions - Investigations of Perfect Reputation Systems* (CESifo Working Paper Series No. 2216). Retrieved July 24, 2009, from http://ssrn.com/abstract=1092394

Chen, P., Dhanasobhon, S., & Smith, M. D. (2008, May) *All Reviews are Not Created Equal: The Disaggregate Impact of Reviews and Reviewers at Amazon.Com.* Retrieved July 23, 2009, from http://ssrn.com/abstract=918083

Chua, C. E. H., & Wareham, J. (2004). Fighting Internet Auction Fraud: An Assessment and Proposal. *Computer, 37*(10), 31–37. doi:10.1109/MC.2004.165

Copp, D. (2005). Introduction: Metaethics and Normative Ethics . In Copp, D. (Ed.), *The Oxford Handbook of Ethical Theory.* New York: Oxford University Press. doi:10.1093/0195147790.001.0001

Das, T. K., & Teng, B. (2004). The Risk-based View of Trust: A Conceptual Framework. *Journal of Business and Psychology, 19*(1), 85–116. doi:10.1023/B:JOBU.0000040274.23551.1b

De George, R. T. (2003). *The Ethics of Information Technology and Business.* Malden, MA: Blackwell Publishing. doi:10.1002/9780470774144

Dellarocas, C. N. (2001, October). *Building Trust On-Line: The Design of Reliable Reputation Reporting: Mechanisms for Online Trading Communities* (MIT Sloan Working Paper No. 4180-01). Retrieved July 24, 2009, from http://ssrn.com/abstract=289967

Dellarocas, C. N. (2003, March). *The Digitization of Word-of-Mouth: Promise and Challenges of Online Feedback Mechanisms.* (MIT Sloan Working Paper No. 4296-03). Retrieved July 19, 2009, from http://ssrn.com/abstract=393042

Doolin, B., Dillon, S., Thompson, F., & Corner, J. L. (2008). Perceived Risk, the Internet Shopping Experience, and Online Purchasing Behavior . In Becker, A. (Ed.), *Electronic Commerce: Concepts, Methodologies, Tools and Applications* (*Vol. 1*, pp. 324–345). Hershey, PA: Information Science Reference.

Gefen, D., Karahanna, E., & Straub, D. W. (2003). Trust and TAM in Online Shopping: An Integrated Model. *Management Information Systems Quarterly*, *27*(1), 51–90.

Grabner-Kraeuter, S. (2002). The Role of Consumers' Trust in Online Shopping. *Journal of Business Ethics*, *39*(1/2), 43–50. doi:10.1023/A:1016323815802

Greene, J., & Haidt, J. (2002). How (and where) Does Moral Judgment Work? *Trends in Cognitive Sciences*, *6*(12), 517–523. doi:10.1016/S1364-6613(02)02011-9

Greene, J. D., Nystrom, L. E., Engell, A. D., Darley, J. M., & Cohen, J. D. (2004). The Neural Bases of Cognitive Conflict and Control in Moral Judgment. *Neuron*, *44*, 389–400. doi:10.1016/j.neuron.2004.09.027

Greene, J. D., Sommerville, R. B., Nystrom, L. E., Darley, J. M., & Cohen, J. D. (2001). An fMRI investigation of Emotional Engagement in Moral Judgment. *Science*, *293*, 2105–2108. doi:10.1126/science.1062872

Hume, D. (1978). *A Treatise of Human Nature*. New York: Oxford University Press.

Husted, B. (1998). The Ethical Limitations of Trust in Business Relations. *Business Ethics Quarterly*, *8*(2), 233–248. doi:10.2307/3857327

Masclet, D., & Pénard, T. (2008, January). *Is the eBay Feedback System Really Efficient? An Experimental Study*. Retrieved July 24, 2009, from http://ssrn.com/abstract=1086377

Myskja, B. K. (2008). The categorical imperative and the ethics of trust. *Ethics and Information Technology*, *10*(4), 213–220. doi:10.1007/s10676-008-9173-7

Noddings, N. (1984). *Caring: A Feminine Approach to Ethics and Moral Education*. Berkeley, CA: University of California Press.

Noll, J. (2001). *The Importance of Confidence for Success in E-Commerce*. Retrieved July 19, 2009. from http://ssrn.com/abstract=288940

Smith, A. (1981). *An Inquiry Into the Wealth of Nations*. Indianapolis, IN: Liberty Fund.

Smith, A. (1982). *The Theory of Moral Sentiments*. Indianapolis, IN: Liberty Fund.

Spence, M. (1973). Job Market Signaling. *The Quarterly Journal of Economics*, *87*(3), 355–374. doi:10.2307/1882010

Stiglitz, J. E. (1975). Information and Economic Analysis . In Parkin, J. M., & Nobay, A. R. (Eds.), *Current Economic Problems* (pp. 27–52). Cambridge, UK: Cambridge University Press.

Sultan, F., Urban, G. L., Shankar, V., & Bart, Y. Y. (2002, December). *Determinants and Role of Trust in E-Business: A Large Scale Empirical Study* (MIT Sloan Working Paper No. 4282-02). Retrieved July 19, 2009, from http://ssrn.com/abstract=380404

Urban, G. L. (2003, March). *The Trust Imperative* (MIT Sloan Working Paper No. 4302-03). Retrieved July 19, 2009, from http://ssrn.com/abstract=400421

Wallace, K. A. (2008). Online Anonymity. In K. E. Himma. & H. T. Tavani (Eds.), The Handbook of Information and Computer Ethics (pp. 165-189). Hoboken, NJ: John Wiley & Sons, Inc.

Chapter 5
Trust, Loyalty, and E-Commerce

Leonard I. Rotman
University of Windsor, Canada

ABSTRACT

E-commerce has experienced a meteoric rise from technological curiosity to substantive institution in little more than a decade of meaningful existence. The annual value of its global transactions is measured in the trillions of dollars. However, the unique nature of e-commerce has created a host of challenges for those seeking to ensure its continued vitality. The most significant of these challenges is the maintenance of user trust. To this point, e-commerce has tended to look to traditional methods of regulation to govern its participants and their transactions. However, the unique character of e-commerce and the concerns it generates warrant consideration of non-traditional approaches to regulation as well. This chapter suggests that fiduciary law, with its focus on maintaining the integrity of certain important relationships in contemporary society, could be a useful tool in e-commerce regulation by facilitating the trust and loyalty that is foundational to its success.

INTRODUCTION

Since its origins in the early 1990s, e-commerce has become a mainstream component of contemporary business reality around the globe. E-commerce transactions have grown to the point that their annual value is now measured in the trillions of dollars. According to the United States Census Bureau E-Stats (2008), in 2006, $1.158 trillion in manufacturing

DOI: 10.4018/978-1-61520-615-5.ch005

shipments were attributed to e-commerce in the United States alone. Yet, despite its current success, the issue of public trust continues to plague e-commerce, particularly in the relatively small, but highly visible retail sales sector where it could potentially experience its greatest success.

Trust is an essential component of various forms of human interaction, from personal relationships to business transactions. Trust is essential to e-commerce because of the manner in which e-commerce is conducted and the potential for opportunistic behaviour that it creates. The relative anonymity of e-commerce provides a basis for opportunism that does not exist in more traditional forms of business interaction. This, in turn, creates a general sense of wariness among potential e-commerce participants.

To engender trust, e-commerce has tended to look to traditional methods of regulation, such as legislation, international agreements, and voluntary self-regulation (including web site privacy policies and third-party web seals or trustmarks) to govern its transactions. While these methods have had success, they do not entirely address unique circumstances or pay sufficient attention to the highly interactive nature of e-commerce transactions, particularly in the retail sector. These existing methods of regulation do not focus sufficiently on the fact that what makes e-commerce successful is the willingness of users to participate in it in light of the risks to their security and economic well-being. Perhaps that should not be surprising, insofar as positive law generally attempts to promote certainty by articulating broad principles applicable to all forms of interaction. However, to be truly effective, these broad principles need to be augmented, where appropriate, by the use of discretion and their situationally-appropriate application to unique circumstances, such as e-commerce, and the peculiar issues that arise under it. The failure of existing regulation to address this central component of e-commerce significantly affects user confidence in the system and is ultimately detrimental to its effectiveness.

The effectiveness and efficiency of law are premised upon competing notions. The first is that the law must provide a readily ascertainable basis for its acceptable standards of behaviour; the second is that the law must retain sufficient flexibility to respond to new and unique circumstances. Too much certainty leaves insufficient space for the discretion needed for the situationally-appropriate application of law. Law must, therefore, balance its desire for certainty with an appropriate measure of flexibility and discretion. This is evidenced by

the co-existence within numerous legal systems of both positive laws of general application and equitable principles designed to mollify the former and fill in their gaps. Law maintains its appropriateness in a wide variety of circumstances by virtue of the co-existence of these complementary, yet distinct, legal methodologies.

Equitable principles, like the fiduciary concept, extrapolate beyond proscriptive law by providing the context to judicial decision-making that is often lacking in common or civil law regimes. This facilitates the law's ability to respond to disparate fact situations. As the most doctrinally pure expression of Equity, fiduciary law supplements traditional bases of civil obligation, such as contract and tort, but only where the interaction in question is one of sufficient social or economic importance or necessity resulting in an implicit dependency and peculiar vulnerability of the beneficiary to the fiduciary (Rotman, 2005). Unlike traditional bases of civil obligation, which are designed to impose liability upon wrongdoers and to award relief to aggrieved persons, fiduciary law seeks to facilitate the construction and preservation of social and economic interdependency. The protection of trust and how the reposing of and caring for that trust affects human interaction is central to this conceptualization of fiduciary law.

The unique character of e-commerce and the concerns it generates warrant consideration of non-traditional approaches to regulation. This chapter suggests that fiduciary law, with its focus on maintaining the integrity of important social and economic relationships in contemporary society, could prove useful in e-commerce regulation by looking to its often-neglected human interaction component.

Certainly, the rise of e-commerce from its humble foundations in the early 1990s indicates that it has been immensely successful. However, when one examines some of the characteristics of e-commerce since its transformation from curiosity to substantive institution, certain initiatives to foster and maintain trust are most prominent. These include the development of web seals programs that provide electronic "seals of approval" for electronic retailer ("e-tailer") privacy policies, business practices, etc. These web seal programs are prime indications of the importance of trust and confidence, or, at least, the semblance of these, to e-commerce.

The services supplied by web seal providers are just some of the initiatives that seek to foster and enhance user trust and confidence in contemporary e-commerce. Others may be found in web site privacy statements made by e-tailers and in claims made by Internet Service Providers ("ISPs") who speak of the reliability, trustworthiness, and ease of use of their systems. Still others are observed in the use of increasingly sophisticated encryption technology to protect users' personal information. This recognition of the need to cultivate and promote user confidence demonstrates an awareness of the centrality of user confidence to the continued success of e-commerce.

Given the fundamental need for user trust and confidence in e-commerce and the extent to which attempts to protect it are already visible in the system, fiduciary law, which has trust and confidence at its very core, could be an effective regulatory tool. The purpose of fiduciary law is to facilitate important social and economic interactions of high trust and confidence where one party's reposing of trust in another results in significant power held by the trusted party and a resultant vulnerability of the entrustor to the use, misuse, or abuse of that power (Frankel, 1983, Rotman, 2005).

This chapter seeks to establish a conceptual basis for the application of fiduciary principles to e-commerce. It does not intend to demonstrate how fiduciary law is to be specifically applied to any particular form of e-commercial interaction. Indeed, it is impossible to accurately describe fiduciary law's application to any relationship, regardless of whether it takes place in "real" space or "cyberspace," without a solid understanding of fiduciary principles and the precise facts pertaining

to the specific interaction in question. Hypotheses of fiduciary applicability may follow only once the initial threshold of establishing the basic principles of fiduciary law and its fundamental purpose, as well as the potential difficulties likely associated with its application to e-commerce, are established. This chapter looks only to these threshold issues, leaving the remainder for another day.

The first part of the chapter sets the stage for a discussion of how fiduciary law may address concerns about user confidence generated by e-commerce. To accomplish this, it is necessary to understand what e-commerce is and some of the difficulties in its regulation. This examination also looks to the vital role that user confidence plays in maintaining the viability of e-commerce.

The second part of the chapter focuses on the purpose of fiduciary law and how it ensures that the mutual dependency that exists in many segments of contemporary society is not only protected, but enhanced. The understanding of fiduciary law generated in this section serves as the basis for its application to the specific concerns that e-commerce raises regarding user confidence.

WHAT IS E-COMMERCE?

Although e-commerce is generally associated with retail transactions taking place over the Internet, the reality is that this comprises only a small percentage of the value of e-commerce transactions. In 2006, retail e-commerce accounted for only $107 billion in sales, or 2.7 per cent of total retail sales in the United States (Unites States Census Bureau, 2008). E-commerce also extends to transactions conducted via other means such as automatic teller machines (ATMs), facsimiles, and electronic data interchange (EDI). It may reflect traditional commercial practices, including manufacturing and retail sectors, non-traditional uses engaged in electronically (*e.g.* on-line casinos, pornography sites), or it might refer to other forms of interaction whose non-electronic counterparts are not generally understood as commerce (*e.g.* information and data retrieval and transmission).

While some definitions of e-commerce focus on content, others focus on the means by which e-commerce is achieved. E-commerce may also be defined by contrasting it with non-electronic commercial interactions. Here, speed, convenience, the removal of temporal and geographic considerations, and the relative anonymity of e-commerce take centre stage. E-commerce has opened a new realm of commercial activity, particularly via the Internet, that allows individuals or businesses to view products or services, obtain information about them, and arrange for their acquisition from anywhere an Internet connection is available. All of these characteristics distinguish e-commerce from more "traditional" forms of business activity. Not surprisingly, these unique factors create equally unique concerns among users.

The very nature of e-commerce has resulted in the need to create new terms to describe the actors and activities taking place within it. Thus, we have witnessed the dawn of the "e-conomy" and "e-tailer" which, in turn, have been distinguished from their older counterparts by the invention of yet more terms designed to indicate the latter's non-virtual nature – for example, the adjectives "traditional," "conventional," or "bricks-and-mortar" to describe both non-electronic retailers and the economy in which they function. We have also seen the creation of distinctions within e-commerce, such as the B2B ("business to business") and B2C ("business to consumer") economies.

Along with the development of new terms to describe new methods of business, additional terms have been created to describe the confluence of e-commercial and traditional commercial actors. Distinctions between bricks-and-mortar enterprises and e-tailers have sometimes become blurred with the institution of "bricks-and-clicks" establishments – commercial enterprises that maintain a presence in both the traditional and electronic economies. Some of these "bricks-and-clicks" establishments are traditional retailers which have also become e-tailers by adding ".com" to their names and instituting their own electronic retail operations, while others are "bricks-and-mortar" businesses that have engaged in formal or informal partnerships or joint ventures with enterprises that originated as electronic commerce concerns.

Regardless of what form e-commerce takes, a common theme remains the difficulty of establishing and maintaining user trust and confidence.

ENGENDERING TRUST IN E-COMMERCE

Indications and expressions of trust abound in e-commerce. The names attached to some web seal programs – TRUSTe (http://www.truste. org), WebTrust (http://www.webtrust.org), and VeriSign (http://www.verisign.org) – are prime indications of the importance of trust and confidence or, at least, their outward appearance, to e-commerce. Certainly, these purveyors of trust place great emphasis on how their services boost user confidence in e-commerce.

TRUSTe's current slogan is "click with confidence." The company's very first sentence on its home page illustrates its awareness of the importance of trust to e-commerce: "Consumer trust – it's key to online business." VeriSign's home page prominently displays the slogan "Increase Transactions by Increasing Trust." BBBOnline uses the phrase "Start With Trust." This emphasis on trust in e-commerce is not new. Indeed, the accounting firm Ernst & Young demonstrated its appreciation for promoting user trust and confidence in an advertisement it placed in *Wired* magazine in July of 2000, (as cited in Ribstein, 2001), which stated:

Protect your customers' privacy and transactions or you may be the one who gets burned ... [if] your internet customers feel exposed they'll quickly take their business elsewhere. The CyberProcess Certification solutions that we offer, including WebTrust, help you build and maintain their trust. So you can grow your customer base and establish a competitive advantage. (n. 203)

Although trust is an intangible, it provides definite and tangible effects. The value of trust lies in its ability to sufficiently alleviate fears associated with an action that would render that action undesirable. Ribstein (2001) has described "trust" as "a kind of social glue that allows people to interact at low transaction costs" (p. 553). Regardless of whether trust actually lowers transaction costs, its primary value to e-commerce lies in its ability to encourage interaction that would otherwise not occur.

The virtuality of e-commerce renders abstract notions of trust far more important to its success than to bricks-and-mortar retailers, which can establish customer trust and confidence in more concrete ways (such as by the size, location, and appearance of their retail operations). The virtual existence of the on-line store does not allow it to engage in such practices; it may, however, more easily mislead consumers, as by the creation of sophisticated websites that belie an e-tailer's existence as a small "mom and pop" operation. Certainly, website designs can be manipulated far more easily and economically than the physical location of a bricks-and-mortar business. Additionally, an e-tailer can more easily furnish misinformation to prospective consumers, such as by altering the appearance of products or providing false information, than their bricks-and-mortar counterparts.

Frankel (2001) contextualizes the important role of trust in e-commerce:

The Internet is a wonderful innovation, allowing people around the world to communicate, trade, and obtain services. Convenient and rich in choices and opportunities, the Internet is tremendously attractive to buyers. Naturally, businesses are flocking to the Internet. The warning has been sounded that those who do not stake a claim in this incredible new communication medium will be left behind to perish. Yet, with all the enthusiasm, many buyers hesitate to take a serious plunge. Businesses are told repeatedly that they must obtain their customers' trust, yet find it more difficult to gain this trust in cyberspace than in real space. Trust has become a serious stumbling block to developing e-commerce ... (p. 457)

In spite of e-commerce's tremendous success in a relatively brief period of time, its success may well be brief if it is unable to quell user mistrust and non-confidence generated by fears about privacy, security, fraud, and the difficulties of enforcement, whether these are warranted or not.

User confidence in e-commerce is significantly affected by concerns over privacy and protection of personal information and interests. Concerns over privacy and the protection of personal information are, in turn, exacerbated by jurisdictional issues and related problems of regulation and enforceability. The "faceless" economy that has emerged from e-commerce has engendered many of the same concerns that the introduction of previously-unfamiliar practices did decades earlier – the use of automated tellers in banks or, much earlier, the introduction of self-service as a means of purchasing groceries, clothing, and gasoline – but on a much broader and enhanced scale.

These concerns among e-commerce users are not unwarranted. Indeed, not only privacy concerns, but personal safety and security issues have been raised by the use of person-to-person sites, like Facebook, that contain significant personal information, and online sites, like e-Bay, Craigslist or Kijiji, that sell goods or services through auctions or on-line classified ads.

Facebook was recently the subject of a complaint to Canada's Privacy Commissioner as a result of the identification of potential problems with Facebook's privacy policies and practices. The Commissioner subsequently released a scathing report which concluded that Facebook violated Canadian privacy legislation by indefinitely retaining the personal data of millions of global users – even after they had deactivated their accounts – and by sharing that information with almost 1 million third party software developers around the world. Facebook subsequently agreed to change the manner in which it collects, stores, and shares personal information, including requiring third parties to obtain express permission from users before accessing their data. Facebook also agreed to establish a clear policy for users wanting to close their accounts by indicating that they would be able to choose between "deactivation," which would allow Facebook to retain their data on file, and "deletion," which would erase all user data from Facebook's database (http://www.thestar.com/news/canada/article/687227).

Another recent issue of concern in e-commerce is the use of Craigslist ads to perpetrate scams. Examples include phony "items for sale" ads and fake real estate ads. One example of the former occurred in Tacoma, where a mother and son responded to a used car ad and were robbed of $3,600 and attacked with a hammer upon meeting with the ad poster (http://www.upi.com/Top_News/2009/04/22/Craigslist-sale-ends-with-hammer-attack/UPI-98501240421063/). The phony real estate ad scam is one of the broadest currently on Craigslist. Here, con artists use properties from legitimate real estate postings and then place ads on Craigslist offering to rent or sell those same properties at cut-rate prices. The phony ads are used to entice would-be renters or purchasers to provide rental deposits or down payments to purchase the properties in question (http://www.chron.com/disp/story.mpl/business/sarnoff/6427385.html).

More serious concerns over Craigslist emanate from the posting of ads to facilitate criminal activity or persons who respond to

legitimate ads for criminal purposes. Ads for massage services have been used by "customers" to sexually assault (http://www.ktvb.com/news/regional/stories/ktvbn-aug1909-craigslist_rape_suspect.f6cf39ab.html) and murder the advertisers (http://www.nydailynews.com/news/us_world/2009/04/16/2009-04-16_craigslists_masseuse_julissa_brisman_found_dead_in_posh_boston_hotel_second_kill.html). Babysitting ads posted on Craigslist have resulted in murder (http://wcco.com/local/katherine.ann.olson.2.461959.html) and the use of children in pornographic videos (http://www.huffingtonpost.com/2008/12/05/aaron-jay-lemon-babysitte_n_148799.html). The proliferation of crime-related scams on Craigslist has led to significant media attention (http://www.theglobeandmail.com/life/craigs-crime-scene/article682509/) and the creation of a website devoted entirely to them (craigscrimelist.org).

As a result of these and other occurrences, many individuals are less willing to engage in commercial transactions online, even if the party on the other side of the transaction is the same entity that operates out of bricks-and-mortar retail space. The virtuality of e-commerce is, quite simply, unnerving to some. Users wonder how they may ensure they will receive what they have contracted for when there may be no person to speak to and the product they have ordered cannot be touched or, in some cases, cannot be viewed. When this natural wariness is accompanied by threats to privacy and difficulties of enforcement, user aversion to e-commerce is increased and user adoption is correspondingly decreased.

Indicating the need to address user confidence concerns does not ignore initiatives that have already been undertaken by public and private organizations to bolster user confidence in e-commerce, including the introduction of web seals and privacy legislation. However, it is suggested that these existing methods do not concentrate sufficiently on the interactive component of e-commerce, thereby leaving untreated a significant element of that medium. For example, the use of contract law to provide relief may yield a remedy after the fact, but that remedy may not be sufficient for users. Indeed, the nature and extent of that remedy may not entirely or adequately address the specifics of the harm caused or create sufficient deterrence to perpetrators to discourage subsequent *mala fide* behaviour. It will likely also not address the reluctance of users who have been harmed by an e-commerce transaction or encounter to engage in further e-commerce transactions or encourage reluctant would-be users to engage in e-commerce for the first time.

Contract law's good faith standard of behaviour is of lesser magnitude than fiduciary law's requirement of the utmost good faith on the part of fiduciaries. While this is important, the most vital distinction between contract law and fiduciary law is that contract law is premised upon self-regarding behaviour, whereas fiduciary law prescribes other-regarding conduct on the part of fiduciaries vis-à-vis the fiduciary element(s) of an association. Implementation of the stricter fiduciary standard, which focuses on the actions of fiduciaries towards users, is more likely to increase the chance that reluctant users will engage in e-commerce than implementing a contractual standard that allows e-commerce operators to act entirely in their own self-interest and without due regard for the interests of users.

Fiduciary law extrapolates beyond common law considerations by prescribing duties of honesty, loyalty, integrity, selflessness, and the utmost good faith upon fiduciaries and emphasizing the uniqueness of individual interactions. Fiduciary law thus counterbalances individualistic ideas founded in contract with broader social and economic goals that are consistent with the construction and preservation of social and economic interdependency.

E-commerce conducted via the Internet is a facilitative mechanism for the electronic connection of users – not just their computers – for commercial purposes. Some means of recogniz-ing this interactivity is a necessary component of addressing user confidence concerns about e-commerce. This is where fiduciary law may prove especially useful, since it was specifically designed to govern interpersonal relations.

"Law," says Frankel (2001), 'offers trusted persons a "brand name" guarantee of their trustworthiness, which may be too costly for trusted persons to create or buy in the markets" (p. 474). In this sense, fiduciary law may provide a realistic and cost-efficient means of ensuring that consumers may place their trust and confidence in e-commerce, knowing that if something goes wrong, they will have legal recourse. One is far more likely to take risks where some form of insurance exists to offset such risk-taking than where there is no such risk reduction. This helps to explain why money-back guarantees are successful in encouraging consumers to purchase a product; the risk associated with its purchase is reduced by the ability to obtain a refund of the purchase price.

When considering the regulation of e-commerce, perhaps one might be forgiven for neglecting the human component of a medium that was enabled by technology and remains saturated with jargon that is reflective of its technological foundation. However, it is unwise to forget that while technology enabled e-commerce, user adoption is responsible for its present, and future, success. As Kahn and Cerf (1999) have suggested:

The success of the Internet in society as a whole will depend less on technology than on the larger economic and social concerns that are at the heart of every major advance. The Internet is no exception, except that its potential and reach are perhaps as broad as any that have come before. (p. 13)

For this reason alone, consideration ought to be given to the use of fiduciary law to promote user confidence in e-commerce.

THE PROBLEM OF REGULATING E-COMMERCE

The unique nature of e-commerce has created significant challenges for e-commerce regulation. The conceptual and practical difficulties associated with regulating e-commerce are reflected in *Digital Equipment Corp. v. Altavista Technology Inc.* (1997), where it was said (p. 463) "The Internet has no territorial boundaries. To paraphrase Gertrude Stein, as far as the Internet is concerned, not only is there 'no there there,' the 'there' is *everywhere* where there is Internet access." In addition, the Internet's very architecture, which was not constructed for commercial purposes, compounds the difficulties of regulating e-commerce.

Lessig (1999) emphasizes this problem by indicating that, until 1991, the National Science Foundation forbade the Internet's use for commerce. He further stresses that, at least initially, information sent over the Internet was completely unsecured, thereby making it difficult to protect (as well as to regulate). Indeed, the historical design of the Internet demonstrates why these problems of confidentiality and data protection arose.

As explained in *Reno v. American Civil Liberties Union*, (1997), the Internet is an "international network of interconnected computers" originating from a United States Department of Defence initiative from the late 1960s that sought to create a computer network that could survive a nuclear attack (pp. 849-850). The network created for this purpose used a decentralized system of information transmission known as "packet-switching." By dividing information into smaller packets of data following separate routes to the recipient's computer, packet switching makes the information sent less vulnerable to attack than a direct "point A-to-point B" protocol.

This decentralized process is further replicated in the various levels of providers of Internet service who supply the connectivity needed for the transmission of information. At the first level is the Internet Service Provider (ISP), which delivers the Internet to its users. Regional Network Providers (RNPs) provide the ISP with Internet connections across a wide area network. The RNP connects to the Internet "backbone" through Network Access Points (NAPs). Finally, the Internet backbone is maintained by service providers who operate the networks that route the information packets sent by users. The lack of a centralized point of control has created profound difficulties for the regulation of the system.

As a result of the Internet's decentralized system of operation, traditional methods of regulation may not be entirely effective vis-à-vis Internet activity. As Johnson and Post (1997) have argued, "the net weakens many of the institutions that we have come to rely on for a solution to the basic problems of collective action" (p. 1). The ineffectiveness of current forms of regulation is one of the unsettling effects that the Internet has had on user confidence.

One of the reasons for highly visible e-commerce regulation is that the perception of security risks may prevent users from engaging in e-commerce transactions without the need for actual proof of fraud, identity theft, or other unscrupulous activity. This explains why web seals are effective in reducing user wariness.

Like the money-back guarantee, web seals engender trust and facilitate e-commerce transactions by providing an independent assurance of the safety and security of a participating website. This is necessary because of the virtuality of e-commerce. While credit card information can be skimmed as easily in a restaurant or gas station as over the Internet, users generally seem less concerned about providing their credit cards to a waiter in a restaurant or to a gas station attendant for bill processing than providing the same information over the Internet to an "anonymous" e-tail site.

User fears that their personal information may be obtained by unauthorized parties as a result of engaging in e-commerce transactions over the Internet are not without merit. Lessig (1999) illustrates how consumers may become vulnerable once they log on to an e-tail website by drawing a comparison with similar activity in a bricks-and-mortar environment:

If you walked into a store, and the guard at the store recorded your name; if cameras tracked your every step, noting what items you looked at and what items you ignored; if an employee followed you around, calculating the time you spent in any given aisle; if before you could purchase an item you selected, the cashier demanded that you reveal who you were – if any or all of these things happened in real space, you would notice. You would notice and could then make a choice about whether you wanted to shop in such a store ... Whatever the reason, whatever the consequent choice, you would know enough in real space to know to make a choice.)

In cyberspace, you would not. You would not notice such monitoring because such tracking in cyberspace is not similarly visible. ... [W]hen you enter a store in cyberspace, the store can record who you are; click monitors (watching what you choose with your mouse) will track where you browse, how long you view a particular page; an "employee" (if only a bot[1]) can follow you around, and when you make a purchase, it can record who you are and from where you came. All this happens in cyberspace – invisibly. Data is collected, but without your knowledge. Thus you cannot (at least not as easily) choose whether you will participate in or consent to this surveillance. In cyberspace, surveillance is not self-authenticating. Nothing reveals whether you are being watched, so there is no real basis upon which to consent. (pp. 504-505

This situation has been described by Kang (1998) as akin to being "invisibly stamped with a bar code as soon as you venture outside your home" (p. 1198).

There are a variety of means by which e-tailers may monitor user activity on their websites. Applications such as "cookies" and "web bugs," as well as practices such as "clickstream monitoring," facilitate the tracking, or "cyberstalking," of users' activities on the Internet. Although each of these operates in distinct ways, they all use the Internet's architecture to monitor users' activities. While cookies, web bugs, and clickstream monitoring all have legitimate purposes, such as allowing for the recording of user preferences and the quicker loading of information, they may just as easily be used for nefarious purposes.

The concern is that the user is often unaware of the purposes this information is being used for, despite the existence of website privacy policies. Most users do not read such policies or are unaware of their existence. Additionally, while users can theoretically "choose" whether to allow much of this information gathering, such as by turning off cookies on the Internet security settings of their web browsers, the reality is that most sites will not function properly, or at all, if the user disallows cookie collection. The "choice," then, is to either use the site in question and allow the transmission of information, or not use the site at all.

Other, Internet-specific schemes, such as Trojan horses or phishing/spoofing, have been developed that yield no benefit to users, but which also take advantage of the Internet's unique architecture to collect information from unsuspecting users or result in users being charged for services they do not even know they are receiving.

Trojan horses are programs that overtly carry out one function while covertly engaging in another. A notorious version of this practice involved the sending of spam messages advertising free access to pornographic images upon downloading special "viewing software" (Fisher, 1997). On downloading this software, images appeared on the user's monitor, but, simultaneously and unknown to the user, the program muted the user's modem, terminated the existing connection to the user's ISP, and dialed a long-distance telephone number in Moldova that reconnected the user to the Internet. The designers of the scam benefited by receiving a portion of the charges from the long-distance call that the users were not even aware they were making. The "Moldovascam" resulted in approximately 800,000 minutes of long-distance phone time made to the small nation in the former Soviet Union in only six weeks, which was far in excess of normal usage.

A newer version of the Trojan horse is "phishing," or "spoofing," where fraudsters impersonate legitimate businesses, such as banks or other financial institutions, in an attempt to get individuals to convey their personal information for fraudulent purposes. Phishing expeditions target information such as bank account and personal identification ("PIN") numbers on the false premise of protecting users from the misuse of this information by others. A common example exists where a fraudster impersonates a legitimate bank by copying portions of its website in order to appear legitimate and sending out multiple random e-mails to potential clients of that bank with messages like "Important Fraud Information" or "Important Information About Your Account." The fraudster then misrepresents that one's personal information has been compromised, which has prompted the action to "protect" the personal information of its customers. Targets of this scam are then advised to "confirm" the numbers of their bank accounts, credit cards, and PINs by sending this information in a reply e-mail. This information is then used by the fraudster to either withdraw money from those accounts, to make unauthorized purchases on the credit cards, or to create bogus credit cards using the information surreptitiously obtained.

These schemes exist in addition to other Internet phenomena such as "spam," misdirections (clicking on one site and being directed to another), and the proliferation of false information.

What is largely ignored in all this discussion, however, is that the cyberspace within which e-commerce functions is not truly a place of its own (Goldsmith, 1998; Lessig, 1995). It is, rather,

merely a facilitative medium that electronically connects individuals.

Since the Internet ultimately connects people, attempts at monitoring or regulating activity taking place on that medium must account for the idiosyncracies of human behaviour. This includes the peculiar needs or desires that individuals have that may neither be "rational" nor "efficient" because they are premised upon perception, emotion, and other human frailties.

The characterization of human behaviour as neither rational nor efficient contrasts with the assumptions built into Law and Economics scholarship, which assumes that human behaviour is inherently rational and premised upon achieving efficient outcomes (Posner, 2002). However, as Gewirtz (1996) explains it, law and economics scholars:

... [a]ll too often ... assume much too much rationality in human affairs. They make extravagantly unreal assumptions about the degree to which individual human action is based upon rational calculation; build abstract rational models that leave out muddy but utterly real-world variables; view costs and benefits in quantitative rather than qualitative terms; and focus on instrumental rationality rather than on murkier questions of goals. (p. 1044)

A recent study by Berk and Hughson (2006) on the actions of contestants on the game show "The Price is Right" reveals that, even in closely controlled circumstances where optimal decisions are generally more evident to contestants than what exists in most situations, contestants do not always make optimal decisions and are inconsistent in their decisionmaking. The study considered the behaviour of contestants in bidding on items from "Contestants' Row" in order to move on to the next round of the game and the actions of contestants in the "Wheel Spin" bonus round to determine who advanced to the final "Showcase Showdown" round of the show.

On Contestants' Row, four contestants all bid on the price of prizes without going over the actual price (or be disqualified). Where the first three bidders all bid over the actual price of the prize, the final contestant can bid $1 and win the prize by default as the only bidder to bid under the price of the prize. Similarly, where none of the other bids are over the price of the prize, the fourth bidder can bid $1 over the next highest bid and win the prize. While bidding in this manner would have increased the fourth bidders' chances of winning by almost 50% (from 30.6% to 43.2$%), almost half of these fourth bidders failed to take advantage of this opportunity.[2]

In the Wheel Spin round, contestants have up to two spins of a large wheel containing 20 numbers, ranging from 5 cents to $1. They must attempt to reach a value of $1 from their spin(s) without going over that amount (and subsequently be disqualified). Where contestants accumulate a value of $1 exactly, they win a bonus. The contestant obtaining the highest value from their spin(s) without going over $1 advances to the Showcase Showdown. Contestants may elect to decline spinning the wheel a second time if they believe the value of their first spin (*e.g.* 95 cents) is likely to be higher than the values obtained by subsequent contestants. The study found that in circumstances where contestants had relatively high value initial spins, they would overwhelmingly (approximately 96%) not spin again and risk disqualification. Thus, they almost always made optimal choices in the Wheel Spin round. The end result of the study is that contestants did not bid optimally where it was easy to bid optimally (on Contestants' Row), but did spin optimally in the Wheel Spin game even though it contained many more variables than the Contestants' Row game.

The study concludes that there is no direct or logical relationship between the two games, even though the players of the Wheel Spin game must first be successful on the Contestants' Row game. While players who, for the most part, did not perform optimally in the initial game over-

whelmingly performed optimally in the second game (approximately 97.5%), the players who performed optimally on Contestants' Row did not perform optimally more often in the Wheel Spin than those who performed sub-optimally on Contestants' Row.[3]

This study debunks the myth of the rational actor by indicating that human behaviour cannot always be estimated using the assumption of rational or objectively optimal decisionmaking. It demonstrates that human behaviour is not always rational or optimal, even in situations where acting optimally is simple and the benefits to doing so is clear. Analogizing the findings from this study to the issue at hand reaffirms the proposition that it is preferable to focus on methods of regulation that look to the unique and situation-specific nature of human behaviour rather than relying upon generic and inapt generalizations. As a result, looking to fiduciary law as a basis for monitoring e-commerce activity appears to be an intrinsically worthwhile endeavour, since it is premised upon the peculiarities of human interaction.

Looking to fiduciary law does not mean that current laws must be discarded, resulting in an entirely new "cyberlaw" that is not at all reflective of the rationales or policy choices of existing legal regimes. Rather, it entails that new legal and regulative approaches are required, some of which may make use of existing methodologies (*e.g.* contract law, communications regulation) that are adapted to the peculiarities of both the medium and the activities that they are monitoring, while others may adopt novel legal applications, such as fiduciary law.

THE FIDUCIARY CONCEPT

Modern fiduciary law may be traced back more than 300 years in English common law and, further back, to principles of Roman civil law (Vinter, 1955, Johnston, 1988). Fiduciary law has been applied to a wide range of relationships throughout its existence to protect the reposing of trust by some in the honesty, integrity, and fidelity of others. Fiduciary relations are as prevalent between parties on an equal footing, such as partners or joint venturers, as among parties in an inherently unequal relationship, such as the relations between parent and child or guardian and ward (Weinrib, 1975; Rotman, 2005). Its flexibility makes it particularly useful in its application to unique fact situations. For these reasons, fiduciary law would appear to be an appropriate tool for monitoring e-commerce.

It must be recognized, however, that there has been an historical reluctance to apply fiduciary principles to commercial interactions in general. There appear to be three primary reasons for this reluctance. The first is that applying fiduciary principles to commercial interactions wrests control away from the participants (Finn, 1992). Secondly, fiduciary principles are said to be antagonistic to the speed and certainty that are deemed to be necessary components of commercial interactions (Millett, 1998). Finally, commercial transactions allow – and, indeed, expect – parties to act on the basis of self-interest, which is inconsistent with fiduciary law's requirement of selflessness (*Hodgkinson v. Simms*, 1994, p. 180). Nevertheless, it has also been recognized that "where the ingredients giving rise to a fiduciary duty are otherwise present, its existence will not be denied simply because of the commercial context." (*Cadbury Schweppes Ltd. v. FBI Foods Ltd.*, 1999, p. 164) Indeed, many forms of commercial interaction, such as partnerships, have long been understood to contain important fiduciary elements. Thus, there is nothing inherent in the fiduciary concept that ought to disallow its application to e-commerce. Yet, as a result of the historical reluctance to apply fiduciary law to commercial interactions generally, the idea of applying fiduciary principles to e-commerce suggested in this paper is novel.

In his judgment in *Securities & Exchange Commission* v. *Chenery Corp.* (1943), the prominent American jurist Felix Frankfurter made an insight-ful statement about the fiduciary concept that still ought to form the basis of any investigation into its application. As he explained:

... [T]o say that a man is a fiduciary only begins analysis; it gives direction to further inquiry. To whom is he a fiduciary? What obligation does he owe as a fiduciary? In what respect has he failed to discharge these obligations? And what are the consequences of his deviation from duty? (pp, 85-86)

Thus, while describing a person as a fiduciary is simple, the subsequent, and necessary, process of infusing that description with substance is far more onerous. Consequently, in the absence of a sound basis for the use of the term "fiduciary" and a solid understanding of its implications, the description of a person or a duty as fiduciary is meaningless.

Fiduciary law has undergone a great degree of scrutiny over the years. Many have attempted to develop a comprehensive definition of the fiduciary concept in order to bring a greater degree of certainty to its application. Yet, in spite of these attempts, no single definition has been able to carry the day (Finn, 1989, Rotman, 2005). As a result, a great degree of uncertainty surrounds fiduciary law and its application. Indeed, it has been described as one of the least understood of all legal constructs (*LAC Minerals Ltd. v. International Corona Resources Ltd.*, 1989). As a result, there can be strong disagreement over how the fiduciary concept is understood.

While many fiduciary law commentaries have focused on the parties to fiduciary interactions and the implications of applying fiduciary law to them, this paper adopts an operational vision of the fiduciary concept that regards fiduciary law as being premised upon the need to protect certain socially or economically important or valuable interactions of high trust and confidence and the resultant implications of applying fiduciary law to relationships rather than the parties to them

(Rotman, 2005). The notion that relationships rather than individuals are the prime concern of fiduciary law is of fairly recent vintage (Frankel, 1983; Finn, 1989; Rotman, 1996). However, the rationale supporting the need to protect such vital forms of interaction is well-recognized in fiduciary jurisprudence (*Meinhard v. Salmon*, 1928, *Hodgkinson v. Simms*, 1994).

Fiduciary law protects certain forms of interaction by imposing onerous duties upon those with power over the interests of dependent others. These duties require that the party possessing the power (called fiduciaries) act honestly, selflessly, with integrity and the utmost good faith in the interests of those dependent others (called beneficiaries). While acting in a fiduciary capacity, fiduciaries may not place their personal interests, or those of third parties, ahead of (or even on par with) their beneficiaries' interests. The beneficiaries, meanwhile, become vulnerable to the fiduciaries' discretionary use, misuse, or non-use of that power within the confines of their fiduciary relationship. While any given fiduciary interaction will necessarily result in an inequality in power between the fiduciary and beneficiary within its confines, there does not need to be an inherent inequality in the power relations of the parties for a fiduciary relationship to arise between them (Rotman, 1996).

The fiduciary concept is posited herein as a tool that facilitates the construction and preservation of social and economic relations of high trust and confidence that facilitate specialization and lead to informational and fiscal wealth. Central to this conceptualization of fiduciary law is the protection of trust and how the reposing of and caring for that trust affects human interaction (Rotman, 2005).

Fiduciary law maintains the viability of interdependent societies premised upon parties reposing trust and confidence in others and caring for the trust and confidence reposed by others. It accomplishes this by prescribing onerous duties on fiduciaries and imposing strong disincentives to breach those duties. By protecting socially and economically important relationships, fiduciary law secures the benefits associated with the effective continuation of valuable forms of human interaction.

Not all interactions are properly subjected to fiduciary norms. Relations that are appropriately subject to fiduciary scrutiny may be identified by their substance (*i.e.* their more-than-fleeting nature), important social or economic character, and the high trust and confidence existing within them. They are also conspicuous by the power held by one party over the interests of another that results in the latter's implicit dependency upon and peculiar vulnerability to the former within the fiduciary element(s) of their interaction. However, not all aspects of a fiduciary interaction are, themselves, subject to fiduciary regulation. Thus, while lawyers hold fiduciary duties to their clients, those duties pertain only to the lawyers' legal representation of those clients' interests.

Where trust is abused, the interdependent relations premised upon that trust are placed in jeopardy. Protecting the trust that underscores this interdependency is no small task. The common law is largely ill-equipped for this purpose, since common law principles are proscriptive, with much more modest and direct goals that focus on individual rights and their enforcement (Rotman, 2005). Fiduciary law's prescriptive approach, meanwhile, emphasizes the unique aspects of individual relations and facilitates an expansive understanding of the nature of obligations existing between parties – or what is sometimes referred to as their "spirit and intent" – that transcends their strict, common law limits.

To paraphrase Keeton (1965), the distinction between the common law and Equity is not just one of history, but one of attitude. The result of this distinction is described by Burrows (2002):

... what may not be a wrong when committed by a non-fiduciary may be a wrong when committed by a fiduciary. Hence undue influence or non-disclosure, while not in themselves wrongs, may

be wrongs where committed by a fiduciary because they may then constitute a breach of the duty to look after another's interests. This explains why compensation was awarded for a fiduciary's – a solicitor's – negligent misrepresentation in Nocton v. Lord Ashburton 50 years before the development of the tort of negligent misstatement, outside a fiduciary relationship, in Hedley Byrne & Co. v. Heller and Partners Ltd. (p. 4)

Fiduciary law's primary emphasis may be seen, therefore, to differ significantly from that of contract law, which is primarily concerned with the self-interested motivations of contracting parties. In contract law, the agreement itself is the centre of judicial attention. Fduciary law, as indicated above, is concerned with the relationship of the parties to each other and the circumstances under which one is alleged to have a duty to act in the other's interests. Where a contract is in place, contract law looks to the actions of all parties to the agreement. Fiduciary law is concerned only with the actions of the fiduciary. While the content of contractual obligations is principally established by the parties to the agreement, fiduciary obligations are mandatorily imposed by courts where required (Brudney, 1997; Rotman, 2005). More fundamentally, contract law's good faith standard of behaviour is exceeded by fiduciary law's requirement of the utmost good faith on the part of fiduciaries (Fitzgibbon, 1999).

Contract law also regards the motivations of the actors involved in contractual relationships differently than fiduciary law regards the motivations of parties to fiduciary interactions. While subjective motivations may be relevant in determining whether a contract exists or a duty of good faith has been breached, similar motivations are completely irrelevant to determining the existence of a fiduciary obligation or its breach (*Keech v. Sandford*, 1726; *Furs Ltd. v. Tomkies*, 1936; *Boardman v. Phipps*, 1967).

Commercial standards of reasonableness and other market-influenced pressures play an impor-

tant role in determining acceptable standards for contracting parties. Fiduciary law, meanwhile, is not dependent upon such pressures. This explains why Chief Justice Cardozo distinguished the level of conduct for fiduciaries from the morals of the marketplace or "the level … trodden by the crowd" (*Meinhard v. Salmon*, 1928, p. 546) and why La Forest J. stressed that "the marketplace cannot always set the rules" (*Hodgkinson v. Simms* 1994, p. 187).

Fiduciary law instills a greater degree of predictability in important social and economic interactions than does contract law by imposing norms that maintain the integrity of these associations (*Graham v. Mimms*, 1982). Under contract law, individuals must continually guard against the potentially harmful effects of non-trusting relations via the drafting of complex contractual terms. Foreseeing potential breaches of contract and attempting to protect oneself from them by anticipation is notoriously inefficient and creates significant opportunity costs and anxiety.

In addition to the above, contracting parties remain responsible for monitoring the self-interested activities of their fellow contracting parties that might destroy existing contractual arrangements. Fiduciary law eliminates the need for continuous monitoring of fiduciaries' activities by beneficiaries, since requiring such monitoring would obviate the reason for applying fiduciary law in the first place (Rotman, 2005). Since beneficiaries may properly rely upon the fidelity of their fiduciaries and need not monitor the latter's activities, beneficiaries' confidence in their fiduciaries is enhanced by the imposition of fiduciary, as opposed to contractual, norms.

As a means through which law imposes its highest standard of ethics on an infinite variety of persons involved in an equally indefinite number of circumstances, fiduciary principles cannot be defined with the explicitness generally demanded by legal actors. Indeed, fiduciary law's situation-specific nature makes even meaningful generalization difficult. Nonetheless, fiduciary law's protean quality and its focus on the specific facts of individual interactions render it particularly suitable for situations involving the peculiarities of human interaction. This is why, at least upon initial glance, it could well be a useful vehicle for monitoring e-commerce relations and protecting user interests.

Before moving to the application of fiduciary law to e-commerce, it should be noted that, in addition to disagreement over how the fiduciary concept is understood, which was noted earlier, there is also disagreement about the precise application of fiduciary law that could affect its application to e-commerce. This may be seen particularly in corporate law in relation to the existence of directors' duties to their corporations.

Academic commentators have long agreed that directors' duties are owed to "the corporation." (Berle and Means, 1933, Davies and Prentice, 1997). Yet, in articulating the further implications of this point of agreement, commentators have disagreed on whether the fiduciary duties owed to the corporation are, in fact, owed primarily to shareholders (Hansmann and Kraakman, 2001) or to a host of stakeholders, including shareholders, creditors, employees, and communities (Blair and Stout, 1999, Greenfield, 2005). Hansmann and Kraakman's view that fiduciary duties are owed only to shareholders would appear to conflict with an articulation of fiduciary law that would protect e-commerce users, which would fit more easily within the broader, stakeholder-centred regime proposed by Blair and Stout and Greenfield, among others.

However, while Hansmann and Kraakman contend that fiduciary duties are owed only to shareholders, case law has demonstrated quite the opposite. Blair and Stout (1999, p. 303) have shown that "a series of mid- and late-twentieth-century cases ... have allowed directors to sacrifice shareholders' profits to stakeholders' interests when necessary for the best interests of 'the corporation.'" While a broader articulation of Blair and Stout's argument and the rationale underlying

it is beyond the scope of this paper, it does reveal a fatal flaw in Hansmann and Kraakman's argument. Indeed, upon closer inspection, the bulk of domestic corporate law jurisprudence is more consistent with the progressive understanding of corporate law proposed by Blair and Stout and Greenfield than with the shareholder primacy approach articulated by Hansmann and Kraakman. Consequently, there is neither difficulty nor conflict in having fiduciary duties be owed to e-commerce users, where appropriate, rather than to corporate shareholders.

THE FIDUCIARY CONCEPT AND E-COMMERCE

When contemplating the application of fiduciary law to e-commerce, it should be kept in mind that fiduciary law does not attempt to establish boundaries or ascribe limits upon the types of relationships that may be found to be fiduciary. Rather, it looks to the quality of the interaction existing within the confines of a relationship, not whether it has been recognized as fiduciary in the past or who the actors involved in it are (Rotman, 2005; *Tate* v. *Williamson*, 1866; *Guerin v. The Queen*, 1984). Therefore, while e-commerce relations may not fit within so-called "classic" forms of fiduciary interaction, such as parent and child or solicitor-client, the categories of potential relations that may properly be described as fiduciary are no more limited than the categories of negligence at common law (*Laskin v. Bache & Co.*, 1971; *Frame v. Smith*, 1988). Thus, any relationship may, potentially, be found to have fiduciary components (*Lloyd's Bank Ltd. v. Bundy*, 1975; *Western Canadian Shopping Centres Inc. v. Dutton*, 2001). Yet, as described earlier, not every aspect of a fiduciary interaction is fiduciary.

In the example of lawyer-client relationships illustrated previously, lawyers do hold fiduciary duties to their clients. However, these duties pertain only to the lawyers' legal representation of their clients' interests. Consequently, a lawyer who is representing a person accused of a crime holds fiduciary duties with respect to that client's legal defense, but would hold no commensurate fiduciary duty relating to any investment advice provided to that client, even if acting on the advice turns out badly. The lawyer's fiduciary duty in this scenario does not extend to any activity unrelated to the lawyer's representation of the client's legal interests, such as the provision of investment advice, since the latter exists beyond the boundaries of the fiduciary elements of the lawyer-client relationship.

As with fiduciary relationships generally, not all of the various aspects of e-commerce relations may be properly characterized as fiduciary. When a user visits a website, searches through its offerings, and ultimately purchases goods, each element of that user's visit may generate different legal obligations. The characterization of the legal obligations that may be found to exist will be dependent upon any representations or warranties made by the operator, the presence or absence of exculpatory clauses or other liability-limiting devices, or the existence of user queries and the responses thereto.

Fiduciary obligations are not needed where other principles of civil obligation, such as contract and tort, are both available and applicable. Moreover, since fiduciary law imposes harsher sanctions on fiduciaries in default of their duties and provides a wider range of relief for aggrieved individuals than the laws of contract or tort, fiduciary duties are properly imposed only where they are truly warranted by the facts at hand. Although it is not possible to establish an interaction as fiduciary in the absence of specific facts, in general it may be said that ordinary e-commerce representations and warranties may confer contractual duties, the provision of defective or harmful goods may provide either contractual or tortuous obligations, while the collection, use, or dissemination of user information may impose fiduciary obligations.

The ability to track and monitor users and their activities and preferences make users highly vulnerable to the use, misuse, or abuse of such information. This applies to the gathering, storing, use, or dissemination of such information, such as through the use of cookies, web bugs, or clickstream monitoring described earlier. As illustrated above by Lessig and Kang, Internet users are peculiarly vulnerable to having their personal information collected, whether by ISPs, search engines, web site operators, hackers or others. Users may be unaware of the collection or monitoring of their personal information or the extent to which the information is being used. Further, there is not much they can do to limit the collection or use of their personal information. The nature of the information gathered and the corresponding vulnerability of the users from whom it was gathered is perhaps the clearest example of where fiduciary obligations may be owed in e-commerce. It is also, as indicated earlier, the largest threat to the future success and well-being of e-commerce.

Where fiduciary obligations may be owed as a result of the collection of user information, the collectors of that information may owe specific duties to inform users that their information is being gathered, to limit the information gathered, disclose what is being gathered and for what purposes, to protect the information once it is collected, and to not disseminate the information without user consent. In some circumstances, however, it may be claimed that there are legitimate reasons why such information must be disclosed to others. In one recent case, a complainant sought and obtained an order in the New York Supreme Court requiring the search engine Google to disclose information (name, physical address, e-mail address, IP address, phone number and "all other information that would assist in ascertaining the identity of that person or persons": see http://www.cyberslapp.org/documents/OrderGrantCohenPet.pdf) about an anonymous blogger who had posted a blog entitled "Skanks of NYC" – http://

skanksnyc.blogspot.com – that contained allegedly inflammatory information and unauthorized pictures of the complainant. The unmasked blogger subsequently planned to sue Google for breaching "its fiduciary duty to protect her expectation of anonymity" (http://www.nydailynews.com/gossip/2009/08/23/2009-08-23_outted_blogger_rosemary_port_blames_model_liskula_cohen_for_skank_stink.html. Indeed, arguments could equally be made that some individuals, such as corporate whistleblowers, have rights to keep their identities private. In addition to raising difficult issues of fiduciary obligation, this suit and counter suit generate significant privacy concerns that require greater analysis than can be entertained herein.

While it is not possible to establish the fiduciary nature of any interaction in the absence of specific facts, there is nothing inherent in e-commerce relations that would preclude them from being described as fiduciary. Fiduciary law's goal of maintaining the integrity of socially and economically important or necessary relations would appear to be particularly relevant to the relationships created under the auspices of e-commerce and the potential for harm created therein, particularly with respect to the collection and use of user information. Certainly the need to promote confidence in potential and existing users to facilitate and maintain their participation in e-commerce would make fiduciary principles attractive to achieve such ends.

Fiduciary law could, conceivably, provide a method of regulating e-commerce that is neither rooted in nor controlled by private parties or government, but is directed by the goal of fostering trust and confidence propounded by both. Moreover, it would not need to displace existing common law or statutory regimes to operate, but would work alongside them, providing its assistance where it would be both necessary and appropriate to the task.

Pursuing fiduciary regulation of e-commerce would not create the difficulties associated with the creation, ratification, and domestic implementation of international agreements or protocols. Fiduciary law's application across jurisdictional boundaries offers another practical legal basis to prefer it to other alternatives for regulating e-commerce. Moreover, the doctrinally-similar, if not identical, application of the fiduciary concept in jurisdictions emanating from the English legal tradition and many civil law jurisdictions traceable to Roman law provides it with a wider, complementary jurisdictional authority to entertain e-commerce disputes than domestic legislation.

Fiduciary law's application to e-commerce is potentially adoptable also in those jurisdictions that view law and Equity as possessing different, though complementary, functions, even if they do not share a common history with English or Roman legal traditions. This is so because the fiduciary concept's application to existing methods of regulating e-commerce serves to ameliorate the harshness of the common law and to fill in its gaps in the manner that Equity does alongside the common law.

An effective legal regime must balance its need for certainty with an appropriate measure of flexibility and discretion. The working arrangement between law and Equity provides both the establishment of fundamental notions of justice in absolute terms – which has the effect of maintaining the objectivity and legitimacy of law – while retaining the necessary flexibility within its application to specific circumstances to avoid inappropriate and excessive rigidity – which would detract from the legitimacy of the law. For e-commerce, positive laws of contract and privacy legislation are necessary to generate reasonable or legitimate modes of conduct based upon ascertainable standards. Yet, they do not adequately address all of the concerns raised by e-commerce. This is where fiduciary law fits as a means to address these additional concerns that remained untouched by the common law. The law governing e-commerce may thereby maintain its appropriateness in a wide variety of circumstances by virtue of the co-existence of these distinct legal methodologies.

CONCLUSION

This chapter is designed to provoke thought and debate by suggesting an alternative to existing methods of e-commerce regulation that may more adequately addresses its important interpersonal element. The chapter is not intended as a polemic, but simply explores the potential and suitability of fiduciary regulation of e-commerce.

The chapter suggests that fiduciary law could provide a greater sense of equity and fairness among e-commerce users that will enhance user confidence in the system. Adding to its attractiveness is that the fiduciary concept, as a creature of Equity, would not displace existing common law or statutory regimes, but would work in conjunction with them. Thus, it would complement and augment existing methods of regulation and remedy them where they are deficient rather than superseding them. Further, its common foundation would avoid problems associated with the conflict of laws from various jurisdictions.

Fiduciary law is a positive and powerful tool for monitoring socially and economically important interactions. Its protean quality facilitates fiduciary law's application to a wide variety of interactions. However, to retain fiduciary law's status as an important legal tool, fiduciary principles must be used judiciously and only where they are warranted by the circumstances rather than in any situation they may be squeezed into given the right amount of persuasion. As Sir Robert Megarry (1991, p. 11) once warned, "[t]he traditional beauty of a land flowing with milk and honey is marred by the realisation that it would be very sticky. What of a land awash with fiduciary relationships?"

Law is no more static than the societies which it governs; thus, the expansion of established spheres of law into new areas, or the creation of

entirely new spheres, may be required as circumstances dictate.

A number of years ago, Judge Frank Easterbrook and scholar Lawrence Lessig debated whether cyberspace required an entirely new law to govern it or if a modification of existing laws would be sufficient (Easterbrook, 1996; Lessig, 1999). This chapter grapples with similar issues regarding e-commerce. While it does not suggest that an entirely new legal regime is required to govern e-commerce, it does indicate the need to reconsider which laws are applied to it in light of e-commerce's unique qualities.

While contract law still has its place in the realm of e-commerce, it is not entirely effective vis-à-vis the interactive nature of e-commercial relations. Contract law may, therefore, benefit from the approach taken by fiduciary law, which can augment the limits of contract principles in the manner that Equity has traditionally smoothed out the common law's rough edges and filled its gaps.

There may well be other possibilities that are suited to the task of regulating e-commerce. Fiduciary law has been discussed herein because it possesses the requisite flexibility to adapt to the still-evolving nature of e-commerce, focuses primarily on the importance of the relationship rather than the individual actors in it, and is a fact and context-sensitive method of monitoring parties' interactions.

Since a large proportion of e-commerce interaction is contractual, fiduciary law may easily fit alongside existing contract standards and principles, as indicated by Mason J. in *Hospital Products* (1984, p. 454): "That contractual and fiduciary relationships may co-exist between the same parties has never been doubted. Indeed, the existence of a basic contractual relationship has in many situations provided a foundation for the erection of a fiduciary relationship." Applied in this manner, fiduciary law will not supersede contract law as a basis for governing e-commerce. Instead, it will augment contract law to make the law governing e-commerce more responsive to e-commerce's unique qualities and characteristics. This can only enhance existing users' confidence in e-commerce and offer encouragement to those who are presently too wary to participate in it.

Asking whether there is, or ought to be, a special law governing e-commerce or whether ordinary principles of contract law and other methods of regulation ought to be modified to become more appropriate to e-commerce is not the same thing. These inquiries do, however, share in common the idea that the straightforward application of existing legal rules without accounting for the unique characteristics of e-commerce is inappropriate. It is on this common ground that this chapter rests its ideological feet.

REFERENCES

Berk, J., & Hughson, E. N. (2006). *Can Bounded Rational Agents Make Optimal Decisions? A Natural Experiment.* (Robert Day School Working Paper No. 2008-7). Retrieved November 18, 2008, from http://ssrn.com/abstract=1281150.

Berle, A. A. Jr., & Means, G. C. (1933). *The Modern Corporation and Private Property.* New York: Macmillan.

Blair, M. M., & Stout, L. A. (1999). A Team Production Theory of Corporate Law. *Virginia Law Review, 85,* 247–328. doi:10.2307/1073662

Brudney, V. (1997). Contract and Fiduciary Duty in Corporate Law. *Boston College Law Review. Boston College. Law School, 38,* 595–665.

Burrows, A. (2002). We Do This at Common Law But That in Equity. *Oxford Journal of Legal Studies, 22,* 1–16. doi:10.1093/ojls/22.1.1

Davies, P. L., & Prentice, D. D. (1997). *Gower's Principles of Modern Company Law* (6th ed.). London: Sweet & Maxwell.

Fisher, M. J. (1997, September). Moldovascam. com: A Complicated Case of Electronic and Telephone Fraud Suggests Just How Vulnerable Internet Users May Be. *Atlantic Monthly, 280*(3), 19-22. Retrieved April 26, 2009, from http://www. theatlantic.com/issues/97sep/moldova.htm

Easterbrook, F. H. (1995). Cyberspace and the Law of the Horse. *The University of Chicago Legal Forum,* 207–216.

Finn, P. D. (1989). The Fiduciary Principle . In Youdan, T. G. (Ed.), *Equity, Fiduciaries and Trusts* (pp. 1–56). Toronto, ON: De Boo.

Finn, P. D. (1992). Fiduciary Law and the Modern Commercial World . In McKendrick, E. (Ed.), *Commercial Aspects of Trust and Fiduciary Obligations* (pp. 7–42). Oxford, UK: Clarendon.

Fitzgibbon, S. (1999). Fiduciary Relationships are not Contracts. *Marquette Law Review, 82,* 303–353.

Frankel, T. (1983). Fiduciary Law. *California Law Review, 71,* 795–836. doi:10.2307/3480303

Frankel, T. (2001). Trusting and Non-Trusting on the Internet. *Boston University Law Review. Boston University. School of Law, 81,* 457–478.

Gewirtz, P. (1996). On "I Know It When I See It." . *The Yale Law Journal, 105,* 1023–1047. doi:10.2307/797245

Goldsmith, J. (1998). Regulation of the Internet: Three Persistent Fallacies. *Chicago-Kent Law Review, 73,* 1119–1131.

Greenfield, K. (2005). New Principles for Corporate Law. *Hastings Business Law Journal, 1,* 87–118.

Hansmann, H., & Kraakman, R. (2001). The End of History for Corporate Law. *The Georgetown Law Journal, 89,* 439–468.

Johnston, D. (1988). *The Roman Law of Trusts.* Oxford, UK: Clarendon.

Johnston, D. R., & Post, D. G. (1997). And How Shall the Net be Governed? A Meditation on the Relative Virtues of Decentralized, Emergent Law . In Kahn, B., & Keller, J. H. (Eds.), *Coordinating the Internet* (pp. 62–91). Cambridge, MA: MIT Press.

Kahn, R. E., & Cerf, V. G. (1999). *What Is The Internet (And What Makes it Work)*. Retrieved April 29, 2009, from http://www.policyscience. net/cerf.pdf

Kang, J. (1998). Information Privacy in Cyberspace Transactions. *Stanford Law Review, 50,* 1193–1294. doi:10.2307/1229286

Keeton, G. W. (1965). *An Introduction to Equity* (6th ed.). London: Pitman.

Lessig, L. (1999). *Code and Other Laws of Cyberspace*. New York: Basic Books.

Lessig, L. (1999). The Law of the Horse: What Cyberlaw Might Teach. *Harvard Law Review, 113,* 501–546. doi:10.2307/1342331

Lessig, L. (1995). The Path of Cyberlaw. *The Yale Law Journal, 104,* 1743–1755. doi:10.2307/797030

Megarry, R. Hon. Sir R. E. (1991). Historical Development. In Fiduciary Duties, Law Society of Upper Canada Special Lectures, 1990 (pp. 1-14). Toronto, ON: De Boo.

Millett, Hon. P. J. (1998). Equity's Place in the Law of Commerce. *The Law Quarterly Review, 114,* 214–227.

Posner, R. A. (2002). *Economic Analysis of Law* (6th ed.). New York: Aspen.

Ribstein, L. E. (2001). Law v. Trust. *Boston University Law Review. Boston University. School of Law, 81,* 553–590.

Rotman, L. I. (2005). *Fiduciary Law*. Toronto, ON: Thomson.

Rotman, L. I. (1996). Fiduciary Doctrine: A Concept in Need of Understanding. *Alberta Law Review, 34,* 821–852.

Rotman, L. I. (1996). The Vulnerable Position of Fiduciary Doctrine in the Supreme Court of Canada. *Manitoba Law Journal, 24,* 60–91.

Vinter, E. A. (1955). *A Treatise on the History and Law of Fiduciary Relationships and Resulting Trust. Cambridge, UK.* W.: Heffer.

Unites States Census Bureau. (2008). *E-Stats: E-Commerce 2006*. Retrieved April 20, 2009, from http://www.census.gov/eos/ www/2006/2006reportfinal.pdf

Weinrib, E. J. (1975). The Fiduciary Obligation. *The University of Toronto Law Journal, 25,* 1–22. doi:10.2307/824874

JURISPRUDENCE

Boardman v. Phipps, 2 A.C. 46 (H.L. 1967).

Cadbury Schweppes Ltd. v. FBI Foods Ltd., 1 S.C.R. 142 (1999).

Digital Equipment Corp. v. Altavista Technology Inc., 960 F. Supp. 456 (D. Mass, 1997).

Frame v. Smith, 42 D.L.R. (4th) 81 (S.C.C. 1988).

Furs Ltd. v. Tomkies, 54 C.L.R. 583 (H.C. Aust. 1936).

Graham v. Mimms, 444 N.E. 2d 549 (Ill. App. Ct. 1982).

Guerin v. The Queen, 13 D.L.R. (4th) 321 (S.C.C. 1984).

Hodgkinson v. Simms, 1 S.C.R. (1994) 177.

Hospital Products Ltd. v. United States Surgical Corp., 55 A.L.R. 417 (H.C. Aust. 1984).

Keech v. Sandford, 25 E.R. 223 (CH. 1726)

LAC Minerals Ltd. v. International Corona Resources Ltd., 61 D.L.R. (4th) 14 (S.C.C. 1989).

Laskin v. Bache & Co., 23 D.L.R. (3d) 385 (Ont. C.A. 1971).

Lloyd's Bank Ltd. v. Bundy, 1 Q.B. 326 (C.A. 1975).

Meinhard v. Salmon, 164 N.E. 545 (N.Y.C.A. 1928).

Reno v. American Civil Liberties Union, 521 U.S. 844 (1997).

Securities & Exchange Commission v. Chenery Corp., 318 U.S. 80 (1943).

Tate v. Williamson, 2 L.R. CH. APP. 55 (1866).

Western Canadian Shopping Centres inc. v. Dutton, 2 S. C.R.,534 (2001).

ENDNOTES

[1]   As Lessig (1999) explains (p. 505, n. 11), "A 'bot' is a computer program that acts as an agent for a user and performs a task, usually remotely, in response to a request."

[2]   Of course, the fourth contestants do not know whether the previous bidders had bid over the value of the prize in question, thereby making it more obvious that they should bid $1. Yet, in such circumstances, the fourth contestants should have then bid only $1 over the highest previous bid to ensure optimal bidding, a factor which was expressly accounted for in the study.

[3]   The study initially suggests (p. 5) that: "… because contestants who use optimal bidding strategies on Contestants' Row are much more likely to win, the sample of contestants spinning the wheel is biased in favor of optimality. One would therefore expect to see better average performance on the Wheel Spin than on Contestants' Row." However, the study concludes (p. 14) that even though winning on Contestants' Row is a prerequisite for competing in the Wheel Spin, which might suggest a selection bias in the sampling of Wheel Spin contestants, such a selection bias is not responsible for the results found. Indeed, the study found that, of the 117 contestants on the Wheel Spin who bid sub-optimally on Contestants' Row, 114 of them (or almost 97.5%) made optimal decisions in the Wheel Spin (which is a higher percentage of optimal decision-making than observed in the contestants who bid optimally on Contestants' Row).

Section 3
Marketing Ethics in E-Business

Chapter 6

E-Business Goes Mobile:
A Fiduciary Framework for Regulating Mobile Location Based Services

Abe Zakhem
Seton Hall University, USA

ABSTRACT

Where others have remarked on the possible fiduciary regulation of e-commerce in general, this chapter makes a more specific and demanding normative claim; notably, that we in fact ought to regulate Mobile Location Based Services (MLBS) along fiduciary lines. Part I describes the limited-access nature of fiduciary relationships and the conditions of peculiar vulnerability and dependence that attract fiduciary obligations. Part II explains the dynamics of user-provider relationships in MLBS environments and argues that the conditions present therein generate fiduciary obligations. Part III describes and addresses a likely criticism; in particular, that the imposition of fiduciary obligations on MLBS providers is morally incompatible with the special fiduciary status rightly and already afforded to equity holders. In response, I argue that those who argue along these lines tend to confuse a manager's nominate function with their strict fiduciary duty to refrain from the opportunistic exploitation of those they serve.

INTRODUCTION

The growth and power of mobile technologies has moved well beyond the realm of simple telecommunication. Current mobile devices are vast storehouses and organizers of information, holding our calendars, to-do lists, photos and music collections, as well as important documents, and databases. Additionally, mobile social network applications

DOI: 10.4018/978-1-61520-615-5.ch006

and Global Positioning System (GPS) enabled devices have spurred the growth of robust mobile communities where users create complex digital identities, make friends, tweet and blog, and collect and share various forms of digital media. Indeed, because of such applications, mobile phones are now the largest technology platform on the planet; and, this is just the beginning. Hardware advances promise "smart" phones that will have the same power as desk top computers. Access to broadband third and fourth-generation networks and the advent of ubiquitous and cloud computing and Mobile 2.0 technologies are among the drivers of significant and rapid technological innovations. With prevailing market conditions driving down costs and promises of over 5 billion worldwide mobile subscribers expected by 2011, the implications for e-business and the opportunities for a variety of mobile location based services (MLBS) are profound (Infonetics Research, 2008). Some of the hotly anticipated services include location-based advertising, mobile social networking platforms, and the development and use of mobile social mapping technologies.

As profound, however, are the concerns about user privacy and, in particular, the potential for the opportunistic exploitation of sensitive user information in MLBS networks (Ardagna, Cremonini, Damiani, De Capitani di Vimercate, and Samarati, 2008). Where others have remarked on the possible fiduciary regulation of e-commerce in general, this chapter makes a more specific and demanding normative claim; notably, that we *ought to* regulate MLBS along fiduciary lines. The first section describes the limited-access nature of fiduciary relationships and the conditions of peculiar vulnerability and dependence that attract fiduciary obligations. The second section explains the dynamics of user-provider relationships in MLBS environments and argues that the conditions present therein generate fiduciary obligations. Finally, the third section of the paper describes and addresses a likely criticism; in particular, that the imposition of fiduciary obligations on MLBS providers is morally incompatible with the special fiduciary status rightly and already afforded to equity holders.

PECULIAR VULNERABILITY AND DEPENDENCE

Given our limitations (e.g., limited time, resources, knowledge or cognitive capacities), we often must rely on others to act on our behalf. Decisions regarding proper medical treatment, legal questions, and retirement investments, for example, are typically best handled by soliciting and following the respective opinions of doctors, lawyers, and mutual fund managers. Yet, benefiting by these sorts of dependencies also requires that we entrust others with limited-access to sensitive information (e.g., our medical history, details of potentially incriminating events, and financial assets and liabilities) and grant limited control over something that we value (e.g., our health, legal status, and capital) for a limited purpose (e.g., medical care, legal defense, or retirement investing). While we certainly hope that those who are granted this limited-access and control will in fact act in our best interests, the fear of opportunistic exploitation looms. We find that just as doctors, lawyers, and money managers are certainly in privileged positions to offer helpful or even quite necessary services, they are likewise in positions where they can take advantage of beneficiary trust.

With this in mind, the conventional imposition of fiduciary obligations has a clear purpose: to control opportunism in limited-access arrangements, under which the professional relationships mentioned above and other similarly structured relationships (e.g., between legal guardians and wards) necessarily fall (Flannigan, 2004). Contracts, legislation, or professional codes can thus recognize the fiduciary status of those entrusted with limited-access to another's property or assets, in which the other party is recognized as the beneficiary of the fiduciary's service. The designation of fiduciary status will then carry the highest expectations for fiduciary honesty, care, and loyalty and stand in sharp contrast with typical market relationships in which all parties are assumed to act out of their own self-interest (*Meinhard v. Salmon*, 1928). Although the bounds of fiduciary obligations and the proper utilization of fiduciary metaphors are highly contested, it is generally expected that fiduciaries are obliged to forgo opportunistic behavior, to obtain no material gain without consent, and to avoid or otherwise disclose potentially conflicting interests with third-parties. Fiduciary obligations are therefore quite "special," in that recognized demands will typically supersede any general duties we may have to others at large. For instance, when exercising discretion over a plan, a mutual fund manager is bound as a fiduciary to place the interests of investors ahead of any concerns about how decisions might impact the performance of their own financial portfolio.

Despite these rather unforgiving obligations, sometimes referred to as punitive, penal, rigid, severe, or draconian, many regard fiduciary obligations as a mere matter of social and legal convention and without significant moral import. John Boatright, for example, maintains that the so called "special" force behind fiduciary duties is simply an extension of the public will and has nothing to do with the moral features of the limited-access relationship itself. On this analysis, fiduciaries are bound to subordinate self-interest in the ways described above merely as a matter of public policy. In a normative sense, argues Boatright, fiduciary obligations ought to be formally recognized only as they demonstrably advance the public good (e.g., by promoting corporate and societal wealth by establishing strong managerial accountability to shareholders) but are otherwise morally insignificant (Boatright, 1994).

Focusing on matters of agency, contract, and public policy, however, Boatright's analysis fails to consider the normative significance of beneficiary vulnerability and dependence in limited-access relationships. In fact, some notable opinions take beneficiary vulnerability and dependence to be the underlying notion behind all fiduciary relationships. Regarding certain equity relationships, Australian High Court Justice Dawson remarked that underlying seemingly disparate cases of fiduciary obligation is the notion that beneficiaries are in "a position of disadvantage or vulnerability" relative to those entrusted to protect or promote their interests (*Hospital Products Ltd v. United States Surgical Corporation*, 1984, at 55). Concerning cases of custodial care, Canada's Supreme Court Justice Wilson described fiduciary relationships as possessing certain essential characteristics:

The fiduciary has scope for the exercise of some unilateral discretion or power and the beneficiary is "peculiarly vulnerable to" or "at the mercy of" the fiduciary holding the discretion or power (*Frame v. Smith*, 1987). Additionally, noted philosopher Robert Goodin concludes that vulnerability and dependence, rather than promises or other self-assumed obligations, "plays the crucial role in generating special responsibilities" and thus serves as the basis for fiduciary obligations in various trust-based relationships (1985, p. 107).

A more rigorous account of that which makes fiduciary relationships so peculiar is as follows. First, fiduciary obligations arise from the relationship itself and not from previous or more general conditions of vulnerability and dependence. As the business ethicist Alexei Marcoux explains, "a lawyer is not a fiduciary to me before I retain his services. However, upon retaining his services, my vulnerability to him gives rise to fiduciary duties on his part" (Marcoux 2003, p. 7). Using another example, an incapacitated person is certainly vulnerable to all sorts of mischief and may in fact depend upon others for his or her very survival. A physician's fiduciary responsible to an incapacitated person, however, only arises when medical treatment is actually administered. Obligations to provide medical services in the first place, which may be required by Good Samaritan laws, are nevertheless beyond the fiduciary scope.

Second, as clearly expressed in Judge Wilson's opinion, fiduciaries are granted a considerable degree of discretionary power over beneficiary resources. This aspect, referred to as "control vulnerability," again arises from the nature of the relationship itself, as beneficiary control is necessarily relinquished for the fiduciary to properly perform his or her nominate function and in doing so administer a limited and perceivably beneficial service (Marcoux 2003). In this way, patients under general anesthesia, for example, can be said to give up control over their respiration and pain management so that physicians can successfully perform the desired medical procedures.

Third, limited-access relationships display what Marcoux refers to as "information vulnerability." Fiduciaries are typically granted privileged access to information concerning beneficiary affairs and in many ways control the "flow of information" to their beneficiaries (Marcoux 2003). Exacerbating the degree of information vulnerability is the fact that a fiduciary's "expert" knowledge is often the result of mastering very technical procedures, practices, and language that cannot be readily deciphered by lay persons. This often leaves the beneficiary wanting of translation and places the fiduciary in a position to manipulate the appearance of relations and interactions in self-serving ways (Flannigan, 2006).

Fourth, exploiting peculiar vulnerability and dependence undermines beneficiary integrity and erodes the social glue that holds together the relationship. Fiduciaries are employed to serve a limited purpose and in order to fulfill this purpose, they must be granted privileged access and control and enough discretionary space to freely perform their nominate function. Distinct from other sorts of relationships, those of a fiduciary nature thus bank on beneficiary trust. Breaches of trust in limited-access relationships are appropriately described as "corrosive" or "parasitic" on the relationship itself as they demonstrably "strangle" our faith in others (Flannigan, 2006, p 212; Marcoux, 2003, p. 7). Additionally, widespread cases of opportunistic exploitation in limited-access relationships can likewise impede the overall social utility of the relationship. Noted legal scholar Robert Flannigan explains that

The (fiduciary) actor is brought within a sphere of purpose. When the actor exploits that purpose, we experience a loss that is different in kind from that associated with breaches in pure exchange relations. The attack has come from within…there is a denial of mutual worth…The (fiduciary) actor has betrayed our purpose – having traded on trust. (2006, p.212)

Concluding that at least some fiduciary obligations are morally substantial regardless of public easily follows from this analysis. More specifically, where conditions of peculiar vulnerability and dependence arise we have sound moral reasons for imposing fiduciary regulations. The basic moral principle at work is expressed thusly: *those entrusted (i.e., granted or otherwise possessing limited control and access) while acting on another's behalf have a moral obligation to pursue the limited purpose of their engagement (i.e., perform their nominate function) without opportunistically exploiting beneficiary vulnerability and dependence.* This is not just a commonly shared intuition, but is in fact grounded in the position of distinct and peculiar advantage that the fiduciary assumes over those interests he or she is explicitly or tacitly entrusted to serve.

As mentioned earlier, the bounds of fiduciary obligations and the proper utilization of fiduciary metaphors are highly contested. Some commentators suggest that a proper understanding of fiduciary obligations requires case-by-case contextualization and the concept therefore resists generalization. Focusing on the moral aspect of fiduciary relationships, however, seems to favor a more generalized reading. Flannigan, for example, holds a strict and generalized account of fiduciary obligation and explains that fiduciaries "are under a duty to forgo self-interest to the extent they have access to the assets of another for a defined or limited purpose… and the circumstances will not matter except to the extent they are offered to establish consent" (2006, p. 214). As such, "fiduciary regulation has a general operation because the generic nature of opportunism requires a generic regulation (Flannigan, 2006, p 221). While this position may find support in conventional legal analytics, it also finds moral support in the analysis laid out above. That is, conditions of "peculiar" vulnerability and dependence generate moral obligations of a fiduciary nature regardless of other contextual factors.

While the compatibility of fiduciary regulation and e-commerce in general has been suggested elsewhere, the first part of this paper sets up a stronger, normative claim: we have sound moral reasons for imposing fiduciary obligations where conditions of "peculiar" vulnerability and dependence obtain. The next section extends this analysis to the case of MLBS management.

PECULIAR VULNERABILITY AND DEPENDENCE IN MOBILE LOCATION BASED SERVICES

In a very general sense, MLBS are information services accessible with mobile devices and which, among other things, make use of the location of the user. Despite the wide variety of MLBS applications, there seems to be a general logic at work. Users access or subscribe to a MLBS through a specific software application. The service provider transmits the data and request for service through a mobile network and subsequently provides a context for the query. Context is generally defined as "any information that can be used to characterize the situation of an entity" (Dey, 2001, p.4). More specifically, and in addition to establishing the user's static and dynamic location, MLBS are being equipped to leverage the extensive information generated and held on mobile devices (e.g., contacts, friends, and social networks, calendars and to-do lists, software applications and widgets, entertainment preferences, and spatio-temporal patterns) to obtain, store, interpret, and ultimately create a much more meaningful contextual model. As the Nokia Research Center technologists have realized, in the mobile service realm "context is king" (Nokia Technology Insights, 2009). The MLBS provider will then deliver content appropriate to the user's established request and context (Pedrasa, Perera, and Seneviratne, 2008; Chen and Kotz, 2000). Based on Harmon and Daim's (2008) survey work and general analysis of location-based services,

the varieties of MLBS include the categories and applications as shown in Table 1.

It is important to note that MLBS typically constitute limited-access arrangements, especially within e-commerce applications (Harmon and Daim, 2008, p.51). Users tend to more positively contribute to their wellbeing or protect certain assets by enlisting MLBS providers. Furthermore, benefiting by MLBS requires that users entrust service providers with limited-access to sensitive information (e.g., through user-defined privacy settings) and grant limited control over something that users value or can significantly impact their wellbeing (e.g., a user's "digital identity") for a limited purpose (e.g., to benefit by the one or more of the publicized MLBS applications listed above).

While users are encouraged to act on the belief that MLBS providers will fulfill the limited purpose of the engagement, the fear of opportunistic exploitation looms. Possible opportunistic abuses include: identity theft, the unauthorized selling of sensitive data to third parties, spamming, unauthorized user profiling and tracking, and data and reality mining (Ardagna et al, 2008). For those unfamiliar with the term, "reality mining" is of particular concern to MLBS service users and involves the study of human interaction vis-à-vis establishing behavior patterns through mobile technologies and/or other sensors, such as radio frequency identification devices (Greenberg, 2008; Green, 2008). Since fiduciary regulations have been traditionally used to prevent and rebuke opportunism in limited-access arrangements, applying the concept to limited-access MLBS is at least compatible with this function.

Mere compatibility, however, does not mean that we *ought to* regulate MLBS in such a manner. While there are several relevant normative and non-normative considerations as to whether or not to impose fiduciary obligations, MLBS do evidence a *peculiar* degree of user vulnerability and dependence commensurate with the analysis laid out in the first section.

Table 1. Varieties of MLBS

MLBS service category	MLBS applications
Navigation	Routing/dynamic navigational guidance Traffic status by location, Weather warnings Points of interest, essential services
Workforce Management	Tracking mobile employees Sales force management Travel information services Personal security Employee networking
Emergency Services	Police, Fire, Medical Roadside assistance
Location Based Advertising	Location and behaviorally-relevant coupon presentation, targeted ads, and promotional messages
Mobile Commerce	Customer identification in a store or a neighborhood E-wallet/Mobile point of service, dynamic pricing Promotional opportunity on the sale of target product
Local Search	Locates nearest restaurants, shows, dinner reservations, concierge services, transportation schedules, points of interest
Social Networking	Friend finder, mobile dating, mobile location experience sharing, user-generated content Geo-blogging, geo-tagged photos and videos

First, a specific form of user vulnerability and dependence newly arises when using MLBS for intended purposes. While mobile device users may be vulnerable to a wide variety of external threats from hackers, stalkers, and other criminal elements, MLBS use opens users up to attacks from within the user-service provider relationship. In mobile location based advertising environments, for example, MLBS providers may collect and store data regarding user location, movement, and purchasing habits with the intent of delivering point-of-service promotions. Accordingly, opportunistic abuses *arising from* the relationship would, as mentioned above, include the unauthorized selling of user information to third parties.

Second, MLBS service providers are granted a considerable degree of discretionary power over user resources. Much of this power has to do with the very nature of contextualization. As mentioned earlier, users turn over control of varying degrees of information (e.g., location, contacts, friends, etc.), which is then contextualized by the service provider in order to deliver appropriate content.

Contextualization itself can take various forms and establishing context is often a monumental task. Chen and Kotz take note of the following contextual factors. Computing context includes technological considerations such as network connectivity, bandwidth, and peripheral devices. User context includes user profiles, location, and social settings. Physical context refers to such things as lighting, noise, traffic, temperature and other features of the surrounding environment. Time and historical context reference computing, user, and physical context across a defined time span (Chen and Kotz, 2000). Contexts can further be established vis-à-vis personalized, passive, and active awareness interactions. Utilizing Chen and Kotz's distinctions, Barkhuus and Dey explain that

Personalization is where applications let the user specify his own settings for how the application should behave in a given situation; passive context-awareness presents updated context or sensor information to the user but lets the user decide how to change the application behavior, where

active-context awareness autonomously changes the application behavior according to the sensed information. (Barkhuus and Dey, 2003, p.150)

Yet, even in cases where more personalized awareness applications are employed, context still needs to be created and interpreted and doing so necessarily requires relinquishing control of private user information. Furthermore, given the inherent difficulty in establishing a meaningful context, designing and accomplishing ease-of-use in MLBS would seem to demand more autonomous forms of contextualization and thereby minimize front-end user interaction.

Third, MLBS users clearly display a considerable degree of information vulnerability. MLBS providers are granted privileged and limited-access to user information and decidedly control the flow of information to and from mobile devices. Despite the multitude of privacy enhancing technologies, the user is often in a position where it is simply impossible for them to detect instances of opportunistic exploitation. Furthermore, the complex and very sophisticated nature of interface, service, and network design requires technical training and knowledge that very few have mastered. Behind even the most intuitive user interfaces, the mostly "invisible" back-end of MLBS platforms and location-based security protocols are truly written in "foreign" languages and using complex protocols and high-level mathematical logarithms. In this way, corrupt motives, or those that betray the limited-access purpose of the relationship, can easily be disguised by a façade of virtue and regularity. These factors, as is the case in many e-commerce applications, place the MLBS provider in a distinct position to manipulate the appearance of relations and interactions in self-serving ways despite the prevalence of technological and existing regulatory controls.

Fourth, exploiting beneficiary vulnerability and dependence undermines beneficiary integrity and erodes the "glue" necessary to that holds together the relationship. Like many e-business transactions, MLBS necessarily bank on user trust

(Harmon and Daim, 2008). MLBS are employed to serve a limited purpose and the outrage when user trust is exploited is notable (Ardagna et al, 2008, p.308). In the same sense as with doctors and their patients, lawyers and money managers and their respective clients, breaches of user trust in MLBS are demonstrably "corrosive" or "parasitic" on the relationship itself. To raise a simple but illustrative question, who would use or continue to use a MLBS where the provider traded off on user trust in order to advance their own interests or the interests of third parties? Furthermore, publicized cases of opportunistic exploitation in location-based services have been shown to hamper wide-spread market adoption and thereby undercut the social utility of MLBS (Harmon and Daim, 2008).

Concluding that MLBS ought to be regulated in a fiduciary fashion thus follows from points one through four. To re-iterate the basic moral principle at work: *those entrusted (i.e., granted limited control and access) to act on another's behalf have an obligation to pursue the limited purpose of their engagement (i.e., perform their nominate function) without opportunistically exploiting beneficiary vulnerability and dependence.* MLBS are limited-access arrangements in which providers offer beneficial services to users. Without explicit indications otherwise, MLBS thus purport to act in the limited interests of users. MLBS relationships likewise place users in positions of peculiar control and information vulnerability and dependence and providers thereby bank on user trust. Opportunistically trading on user trust consequently undermines the moral integrity of the relationship. Even if the link between opportunistic exploitation and overall social utility is not clearly established, we have sound moral reasons for requiring fiduciary regulation of MLBS.

In addition to readings that suggest that a fiduciary regulation of e-commercial transactions is efficient and well placed, the second part of this work makes a stronger argument. From a moral point of view, e-commercial MLBS ought to be subject to fiduciary regulation. This position, of course, will not be without criticism. Accordingly, the next section addresses a powerful and likely criticism; primarily, that extending fiduciary obligations to MLBS service providers conflicts with fiduciary obligations rightly afforded to the equity interests of shareholders and is subsequently conceptually incoherent and/or practically impossible.

CONFLICTING FIDUCIARY OBLIGATIONS?

The argument for imposing fiduciary obligations on MLBS providers advanced in the previous section suggests what some will refer to as a "multi-fiduciary" stakeholder management framework. That is, managers of e-commercial MLBS would hold fiduciary obligations to at least two stakeholder groups, equity holders and MLBS users.

In general, the historical development of a multi-fiduciary view of management proceeds as follows. As previously described, fiduciary obligations command the highest expectations for fiduciary honesty, care, and loyalty and have been particularly attractive for those seeking to define overarching managerial obligations. Thus in one direction, we find that traditional, shareholder-centered positions affirm and then heavily underscore the fiduciary character of manager-shareholder relationships. The intent is to situate the managerial obligation to advance shareholder interests (e.g., driving corporate profitability) as constituting the core of the corporate objective function.

Alexei Marcoux, for example, rightfully points out that shareholder are peculiarly vulnerable to managers in the ways outlined in the first section. The agential aspect of shareholder-manager relationships is undoubtedly limited-access; shareholders turn over limited control of their assets to a firm's management in exchange for a fair return on investment. Shareholders suffer the "special" disadvantage of having their assets in the hands of a management team in possession of all the relevant knowledge, in control of all aspects of their investment, and in control of the flow of information to shareholders. Shareholders are not aware of the day-to-day operations of a firm and must rely on intermittent, sophisticated, and easily manipulated corporate performance reporting. Furthermore, the threat of opportunistic exploitation is parasitic on the manager-shareholder relationship and clearly undercuts the social utility of turning over private capital to corporate managers (Marcoux, 2003). This is in sharp contrast with the nature of managerial relationships with other stakeholders (e.g., employees, suppliers, consumers, or communities, etc.), which are almost fully regulated by contractual means, are transparent with regard to appearance of opportunistic harm, and are readily severable as soon as harm occurs (Marcoux, 2003,).

For some time, however, traditional stockholder views of the firm have lost ground to stakeholder formulations. Thus in another direction, we find fiduciary obligations metaphorically extended to cover "stakeholder" interests, leading to the multi-fiduciary account of stakeholder management theory. From this perspective, managers are, at least in Evan and Freeman's original and highly discussed position, obliged as fiduciary agents to champion the "vital interests" of all stakeholder groups (1988).

The multi-fiduciary stakeholder view, once advanced but ultimately abandoned by Evan and Freeman, has come under fire, leading to what some have referred to as the "stakeholder paradox." Alexei Marcoux, for example, makes the following argument. With respect to the limited-access arrangement between managers and shareholders, acting as a fiduciary requires management to place the interests of the shareholders ahead of their own interests and ahead of the interests of third parties. Multi-fiduciary stakeholder theory claims that managers are fiduciaries to a number of stakeholder groups. Marcoux concludes that multi-fiduciary stakeholder management theory is both conceptually inconsistent and practically impossible for the following reasons:

(1) It is conceptually impossible to simultaneously place the interests of the shareholders ahead of all the others, the interests of employees ahead of all others (including shareholders), the interests of customers ahead of all the others (including shareholders and employees)...etc. (2) It is practically impossible to serve the interests of each of these groups simultaneously. As most everyone recognizes, the interests of shareholders, customers, suppliers, employees, and communities in the management of a firm's assets are conflicting. (Marcoux, 2003, pp. 3-4)

John Hasnas reaches a similar conclusion (1998). On both accounts, multi-fiduciary stakeholder theory is considered morally lacking and ultimately non-fiduciary in character.

Certainly, the claims advanced in this paper do not call for fiduciary obligations between management and stakeholders in general. In a much more narrow sense, *managers ought to bear fiduciary obligations to those stakeholders engaged in limited-access arrangement and subsequently displaying a "peculiar" degree of vulnerability and dependence.* Nevertheless, Marcoux's and Hasnas' arguments must still be treated seriously. If Marcoux and Hasnas are right, it would thus seem conceptually impossible to place the equity interests of shareholders "ahead of" all other interests (including MLBS users) and simultaneously place the privacy interests of MLBS users "ahead of" all other interests (including shareholders). This narrow version of fiduciary obligation would also lead to practical difficulties in that one would expect, for example, the resources devoted to protecting and promoting user privacy in MLBS to at some level conflict with shareholder's desires for increasing returns on their investment.

The problems brought up by Marcoux and Hasnas, however, can be alleviated by making an important distinction between nominate and fiduciary functionality. In general, managers at all levels are more or less responsible for profit

and loss and have the responsibility and authority to administer a variety of activities and processes critical for a firm's success. The nominate functions of specific roles, however, will vary. Chief executive officers tend to preside over short and long term strategic planning and are responsible for the overall direction of corporate governance. As mentioned, however, shareholder centric theorists argue that proper corporate governance involves maximizing profit for shareholders, while stakeholder theorists may call for maximizing value for all stakeholders (Freeman, 2008). Operations managers design and supervise various organizational processes and safety managers are chosen to ensure employee health and safety within these processes. Safety managers may be further nominated (e.g., by OSHA regulations or internal procedures) to champion employee health and safety concerns as taking priority over issues of employee productivity and efficiency. Human resource managers are selected to recruit, screen, and train employees to carry out specified functions and may be further nominated to hold and protect sensitive employee information. Thus corporate nominate functions will involve various types of limited-access arrangements where the interests of some groups are necessarily put "ahead of" the interests of others.

Conventional fiduciary obligations, however, operate independently of the often idiosyncratic and nominate functionality of various corporate and other social roles. Robert Flannigan explains that

Actors, for example, may agree to negotiate contracts for their employers. Or they may agree to hold and manage property for aged beneficiaries. Or they may decide to have children. In each case, the access they acquire by taking on these functions or roles (agent, trustee, parent) is understood to be for the purpose of pursuing the interests of their beneficiaries...Where that understanding exists, where access is limited (but only to the extent that it is limited), we impose a duty to act without self-interest in the course of the nominate func-tion...Conventional fiduciary regulation operates independently as a general regime of obligation. It is distinguishable from idiosyncratic nominate regulation... (and that) fiduciary jurisdiction, traditionally understood, is a relatively narrow form of regulation... concerned exclusively... with the narrow mischief of opportunism on the part of those with limited-access...Conventional fiduciary accountability is also narrow in the sense that it has only a negative operation...The duty is abnegation – a duty not to engage one's self-interest. (2004, p. 7)

Cast in this way, fiduciary obligations hold fast regardless of nominate variation (provided that the limited-access nature of the arrangement withstands) and do not impose the strong sorts of positive obligations inferred by Marcoux and Hasnas.

Consequently, a conventional interpretation of multi-fiduciary stakeholder management will not result in the conceptual and practical difficulties implied by Marcoux and Hasnas. Regarding conceptual inconsistency, fiduciary obligations would command MLBS managers to regard MLBS users and shareholders in exactly the same way: to not divert or exploit the value of user and shareholder assets without consent. This interpretation further eliminates the practical possibility for stakeholder conflict along fiduciary lines, as managers are simply called to perform their nominate function without opportunistically exploiting the limited purpose of their engagement. This is not to say that there will be no stakeholder conflict, say, regarding how much to invest in MLBS security measures, but rather, that conflict of this sort is simply not fiduciary in nature. It seems that conceptual and practical difficulties only appear to arise when fiduciary metaphors are extended far beyond conventional interpretations, as appears to be the case in Marcoux and Hasnas' shareholder-centric theories and in Freeman's original stakeholder position, and become intertwined with and reified in contestable professional roles.

Conventional fiduciary interpretations also have the added advantage of focusing squarely on the problem of opportunism in whatever guise it appears. While some may regard the imposition of fiduciary obligations as harsh, the harshness is "constructive," as the ethical wrong doing when breaching fiduciary trust and exploiting the peculiarly vulnerable and dependent is both socially damaging and morally reprehensible (Flannigan, 2004).

CONCLUSION

We are currently at the beginning of an increasingly mobile computing age and the opportunities for successful e-commercial ventures are rapidly expanding. As opportunities increase, however, so does the potential for the opportunistic exploitation of sensitive user data, particularly in loosely regulated environments. While current regulatory efforts and privacy security protocols have not alleviated consumer concerns, some have suggested regulating e-commercial ventures along fiduciary lines. This paper makes a stronger and more specific point; notably, that we have a sound moral argument, independent of the prevailing political opinion, for imposing fiduciary regulations on MLBS providers. In a world where opportunistic exploitation is pervasive, this would provide us with flexible regulatory means that could be consistently applied to a variety of limited-access arrangements.

REFERENCES

Ardagna, C., Cremonini, M., Damiani, E., De Capitani di Vimercate, S., & Samarati, P. (2008). Privacy-Enhanced Location Services Information. In Acquisti, A., Gritzalis, S., Lambrinoudakis, C., & De Capitani di Vimercati, S. (Eds.), *Digital Privacy: Theory, Technologies, and Practices.* Boca Raton, FL: Auerbach Publications.

Barkhuus, L., & Dey, A. (2003). Is Context-Aware Computing Taking Control away from the User? Three Levels of Interactivity Examined. In *Proceedings of UBIComp 2003: Ubiquitous Computing*. Berlin: Springer Publishing.

Boatright, J. (1994). Fiduciary Duties and the Shareholder-Management Relation: Or, What's So Special about Shareholders? *Business Ethics Quarterly*, *4*, 423–429. doi:10.2307/3857339

Chen, G., & Kotz, D. (2000). *A Survey of Context Aware Mobile Computing Research* (Dartmouth Computer Science Technical Report TR2000-381). Retrieved June 1, 2009, from http://www.cs.dartmouth.edu/~dfk/papers/chen-survey-tr.pdf

Dey, A. K. (2001). Understanding and using Context. *Personal and Ubiquitous Computing*, *5*(1), 4–7. doi:10.1007/s007790170019

Evan, W., & Freeman, R. E. (1988). A Stakeholder Theory of the Modern Corporation: Kantian Capitalism. In Beauchamp, T., & Bowie, B. (Eds.), *Ethical Theory and Business* (3rd ed.). Englewood Cliffs, NJ: Prentice Hall.

Flannigan, R. (2004). The Boundaries of Fiduciary Accountability. *The Canadian Bar Review*, *83*, 35–90.

Flannigan, R. (2006). The Strict Character of Fiduciary Liability. *New Zealand Law Review*, 209-242.

Freeman, R. E. (2008). Managing for Stakeholders. In Zakhem, A., Palmer, D., & Stoll, M. (Eds.), *Stakeholder Management Theory: Essential Readings in Ethical Leadership and Management*. New York: Prometheus Books.

Goodin, R. (1985). *Protecting the Vulnerable*. Chicago: University of Chicago Press.

Green, K. (2008, March/April). TR10: Reality Mining. *MIT Technology Review.com*. Retrieved June 1, 2009, from http://www.technologyreview.com/Infotech/20247/

Greenberg, A. (2008, April 23). What your Cell Phone Knows about You. *Forbes.com*. Retrieved June 1, 2009, from http://www.forbes.com/2008/05/22/reality-mining-cellphone-tech-wire-cx_ag_0523reality.html

Harmon, R., & Daim, T. (2008). Assessing the Future of Location-Based Services: Technologies, Applications, and Strategies. In Unhelkar, B. (Ed.), *Handbook of Research in Mobile Business: Technical, Methodological, and Social Perspectives* (2nd ed.). Hershey, PA: IGI Global.

Hasnas, J. (1998). The Normative Theories of Business Ethics: A Guide for the Perplexed. *Business Ethics Quarterly*, *8*, 19–43. doi:10.2307/3857520

Infonetics Research. (2008). *Mobile Subscribers to Hit 5.2B in 2011*. Retrieved June 1, 2009, from http://www.infonetics.com/pr/2008/ms08.sub.nr.asp

Marcoux, A. (2003). A Fiduciary Argument against Stakeholder Theory. *Business Ethics Quarterly*, *13*(1), 1–24.

Nokia Technological Insights. (2009). *Location, Context, and Mobile Services: The Context*. Retrieved June 1, 2009, from http://research.nokia.com/files/insight/NTI_Location_&_Context-Jan_2009.pdf

Pedrasa, J. R., Perera, E., & Seneviratne, A. (2008). Context Aware Mobility Management. In Unhelkar, B. (Ed.), *In Handbook of Research in Mobile Business: Technical, Methodological, and Social Perspectives* (2nd ed.). Hershey, PA: IGI Global.

Williams, D. H. (2006, October 25th). LBS Development – Determining Privacy Requirements. *Directions Magazine*. Retrieved June 1, 2009, from http://www.directionsmag.com/article.php?article_id=2323

Chapter 7
The Ethical Implications of A/B and Multivariate E–Commerce Optimization Testing

J.J. Sylvia IV
The University of Southern Mississippi, USA

ABSTRACT

A/B and multivariate website optimization may not seem ethically problematic at first blush; however, in this chapter I will consider some of the less obvious elements that have been tested, such as header color, button design, and the style of tabs used for linking to product details. A/B and multivariate testing has shown that these seemingly insignificant changes can increase average order value and decrease abandoned shopping carts, among other results. I will consider these tests through the lens of the major ethical systems of utilitarianism, Kant's respect for person's principle, and virtue ethics, using specific case studies and examples of testing results. I conclude that this type of practice is likely ethically problematic in many uses, as understood through all three ethical systems. Along the way I will be careful to demonstrate how the manipulation resulting from A/B and multivariate testing is different and more problematic than that of advertising in general.

INTRODUCTION

Certainly, many ethical issues related to the creation and use of websites have already been broached. The 2000 Children's Online Privacy Protection Act addresses the privacy of minors (COPPA, 2009). In addition to privacy concerns, copyright issues have also played a large role in the development of the Internet (Montecino, 1996). A more recent issue is accessibility standards for web surfers with disabilities (*United States Access Board*, 2009). Some of these issues have been addressed by acts of Congress, while others have gotten their precedent from law suits, but overall, many of these issues are still in flux and still under debate. I will broach an issue that has not been previously addressed from an ethical or legal view.

Manipulative advertisements have long been the focus of major discussions within business ethics, and this discussion can certainly be ported

DOI: 10.4018/978-1-61520-615-5.ch007

over to advertising via the Internet. However, the Internet also opens up a new avenue for manipulation based on the layout, design, and copy of a particular website. It may be possible that many of the elements on a company's website might actually still fall under the topic of advertising; however, many of the changes that are made on websites due to A/B and multivariate testing clearly should not be understood as elements of advertising. For example, one would not be likely to categorize the color and/or style selection of category tabs as a form of advertising. Therefore, although the discussion may sometimes mirror that of manipulative advertising, a new discussion about implementing the results from A/B and multivariate tests needs to be highlighted.

By and large, companies which advertise their ability to implement multivariate tests on behalf of another company promote the activity in a way such that it appears altruistic or morally exemplar, rather than ethically questionable or problematic. Multivariate testing for websites has been occurring since at least 1995, but these tests have recently gained widespread exposure to a much larger audience of web developers. A/B and multivariate testing is rising in popularity of use and increasing in simplicity of implementation. Despite this increased use, the ethical implications of using such tests have not been raised or explored in either professional literature or on the web in general. For example, is it acceptable that websites conduct this kind of testing? Questions such as this have not yet been broached. In this chapter I strive to set the agenda for the broader topic of the ethical implications of multivariate tests, hoping to open up the issue for further commentary rather than trying to cover every aspect in depth.

BACKGROUND

A/B and multivariate tests on the Internet allow a website to test two or more versions of the same page and measure desired outcomes. A/B split testing allows one to:

randomly divide your visitors into two groups and show each group a different version of a page to determine which version leads to higher conversion, average order value, application completion, or other target. These visitors are then tracked and a report is generated that describes the impact of the A or B page version on this outcome. (Roche, 2004)

Multivariate testing, on the other hand, is:

a process by which more than one component of a website may be tested in a live environment. It can be thought of in simple terms as numerous split tests or A/B tests performed on one page at the same time. (Search Engine Marketing, 2009)

One example of this practice would be a site testing the placement of a search box on the page in order to see if it gets used more frequently depending on whether it is on the top right or left of a page. This type of testing came about long before the Internet and was used for many different purposes; however, the Internet allows for easy use and quick testing of many different design elements at the same time. Results can also be quickly linked to sales related figures such as percentage of shopping carts abandoned or average order size.

Discussion of manipulative advertising has a long history within the realm of business ethics, but this issue falls outside of the typical realm of those debates. The tests I discuss have nothing to do with the products, or brands, or marketing messages themselves. Instead, they focus on elements such as the placement of search boxes, header color, button design, and the style of tabs used. A standard example of manipulative advertising would be a commercial for a fast food company displaying large images of juicy hamburgers that create a feeling of hunger or desire for a hamburger in the viewer, which he or she did not previously have. It is not immediately obvious how web page elements such as search boxes and tabs are similarly manipulative. Multivariate test results can make the connection more obvious, and the results could possibly be linked to specific users or demographics. For example, a particular demographic may be more likely to make a purchase when a headline is displayed in a specific color, and could then be targeted by a dynamically generated website which identifies the demographic of the user and displays the color that has been shown to increase the odds of that person making a purchase.

Privacy is another major ethical concern for e-commerce sites. Technology allows a site to record and store a great deal of information about particular site visitors; however, the goal of multivariate testing is not necessarily to tie the results to any particular user. Instead, overall changes in the actions of the majority of users are of concern, so individual privacy is not the main issue. Yet, multivariate testing on the Internet is still quite young, so although individual privacy is not yet of concern with these tests, this practice may still develop in the future and be cause for even further concern. In general, though, multivariate testing falls outside the scope of some of the largest e-commerce ethical concerns, and very little has been written which directly relates to the practice.

THE ETHICAL ISSUE

The ethical issues surrounding the use of A/B and multivariate website optimization testing are important because these tests are now easily available to all e-commerce sites through Google's free Web Analytics software. One can visit http://www.google.com/analytics/sign_up.html to create an account and have code automatically generated which can be inserted into a webpage. Previously, a great degree of planning and technical expertise was needed to implement such a test, but it has

quickly become both free and easy to implement and thus widely available to any e-commerce site, regardless of its size or budget. This type of testing could have benefited from an ethical analysis at any point, but as the popularity of this type of web development increases, the importance of such an analysis increases. A 2007 Internet Retailer magazine survey of 243 chain retailers found that 23.3% of websites were currently using A/B or multivariate testing (Brohan, 2007). However, this use is certainly on the rise. A 2009 survey of 650 chief marketing officers found that 28% of the marketers planned to deploy multivariate testing this year (*Marketers plan to spend*, 2009). Many companies offer multivariate testing services, but Google alone highlights case studies of companies such as RE/MAX, The Huffington Post, and The American Cancer Society (*Google Analytics*, 2009).

I will argue that this practice leads to manipulative site design aimed at increasing the amount of money spent by consumers on an average order; however, a more charitable interpretation of this testing is certainly available. One could understand the tests as aiming to help consumers best achieve what it is they would like to achieve. For instance, the placement of the search box might be important because a consumer might not be as likely to locate it in one spot rather than another. If the consumer ultimately would like to make a purchase, but simply is not able to because he or she overlooks the search box, it would certainly make sense to test locations of the search box for visibility.

Companies which offer multivariate testing often promote the service as beneficial to not only the e-commerce company, but also the consumers using the site:

Here at CSN Stores, customers are our highest priority and multivariate testing lets us tune our sites so that they are highly intuitive for the largest number of customers," noted Chuck Casto, Vice President of Corporate Communications for CSN

Stores. "By continuously testing and targeting, we are able to enhance the CSN online shopping experience for consumers. (Online Businesses Doing, 2009)

An interpretation such as this suggests that site optimization simply makes it easier for a consumer to complete some task he or she was already trying to complete. Certainly there are situations in which this really is the case, but clearly not all tests are aimed at such a lofty goal. Instead, the emphasis is often on *creating* a conversion or sale.

On a surface reading, this does not seem any more problematic than a brick and mortar store putting candy bars next to the checkout or playing carefully selected music throughout the store. However, multivariate testing via the Internet allows for much faster and more accurate results than other media. One is able to draw direct statistical conclusions from the changes made on a webpage to changes in sales figures. Why does the design of a button or the color of a heading affect the likelihood of a person completing a purchase? As of yet, not many tests have been very concerned with this 'why' question because the focus is on the results. Without a solid answer to the 'why' question, though, the practice of multivariate testing can be understood as an extremely manipulative practice. By implementing the results of these tests, web developers can be understood as using minute, manipulative tactics in order to increase the amount of money spent by consumers. Michael Phillips defines manipulative advertising as "advertising that tries to favorably alter consumers' perceptions of the advertised product by appeals to factors other than the product's physical attributes and functional performance," (1994). I believe this definition can be adjusted for manipulative web site design in the following way: manipulative design is design which tries to increase the likelihood of a customer purchase or the amount of a purchase by altering design elements unrelated to information about a product's physical attributes or functional performance,

using design changes that have been statistically shown to lead to such increases.

At least part of the answer to the 'why' question will likely lie with psychological responses. For example, particular colors are often hypothesized to create a particular mood or emotion in the person experiencing that color. The employees of Google have done research on the effect of the shade of blue used in links:

Google Mail uses a very slightly different blue for links than the main search page. Its engineers wondered: would that change the ratio of click-throughs? Is there an "ideal" blue that encourages clicks? To find out, incoming users were randomly assigned between 40 different shades of links – from blue-with-green-ish to blue-with-blue-ish. It turned out blue-ness encouraged clicks more than green-ness. Who would have guessed? And who would have cared? Google, of course, which wants to get people clicking around the net. (Arthur, 2009)

For a detailed critical review of empirical research on the affects of color, reference Whitfield, T., & Wiltshire, T. (1990).

In addition to color, there is perhaps a similar reaction to other elements in web page layout and design. Consider example results from ShopNBC.com:

The five tested elements were the location and size of the Add to Cart button; headlines identifying cross sells; styles of the tab linking users to product detail; color, size and style of clearance and limited time pricing offers; and highlighting of payment options for qualified buyers.

ShopNBC found that variations in a single element with the others remaining unchanged produced a relatively small effect in improving results. The right combination of options drawn from all five elements, however, produced the 16% lift in aver-

age order value, besides increasing the number of shoppers who initiated the checkout process by 4%. (Multivariate Testing Produces, 2008)

Results such as this are not atypical. The changes produced only a small result in the number of consumers actually initializing a checkout. Instead, what increased was the average order value. The main problem was not that many customers were unable to figure out how to add items to their cart or checkout. The changes simply influenced them to buy *more*. Again, no one seems to care about *why* the style of a tab or the color of a clearance offer increases average orders – the pragmatic results are enough. Manipulation of the consumer, that is, making the customer more likely to buy a product or products through layout and design changes unrelated to the details of the merchandise being sold, is often viewed as the desirable end goal rather than a potentially ethically problematic practice. Companies are actively being encouraged to engage in these practices in order to increase profits. This encouragement has thus far lacked any ethical reflection on the potential problems associated with such practices.

Contrast this with the oft-considered ethical issues of manipulative advertising. One issue with advertising is that it is said to *create* desires in consumers that did not exist before. The process is relatively straightforward and easy to understand. I see an advertisement with a giant picture of a hamburger and this makes me desire a hamburger, even if I had not been feeling particularly hungry before. Whether or not this practice is unethical can be debated in a somewhat straightforward manner. On the other hand, it is not nearly as clear what is going on as the result of multivariate tests. How could the style of tabs and the size of a checkout button possibly influence one to purchase more?

Perhaps a better analog to an ethical issue is the use of color in physical stores and logo designs. Many restaurants prominently feature the color red in their stores and logos because the color has been shown to induce hunger in those seeing

it (Stoll, 2004). For a similar reason, check-out lines in grocery stores are lined with candy in the hopes of inducing an impulse purchase. What is different about online multivariate testing is the ability to test several layouts at the exact same time while continuously monitoring information. If a particular change produces the best results only on weekends, for example, while another produces the best results on weekdays, little effort is required to display the optimal page layouts on different days, at least compared to the effort of reorganizing a grocery store to have different displays on weekends than weekdays. Thus, A/B and multivariate testing issues lie somewhere between the ethical debates relating to advertising and marketing.

The problem is that there has to be some type of design on a site and an easy to navigate, functional interface is certainly desirable. The difficult question that remains is: what is the difference between acceptable good design practices and manipulation? The answer may be intent. A site which is created with the goal of being user-friendly and aesthetically pleasing seems laudable; however, once design changes are made solely on the basis of how much extra money these changes can pull out of consumers, a line is perhaps crossed. Instead of the intention being to create a usable site, the intention shifts to doing whatever is necessary to make more money, and this includes mental manipulation of consumers. If seeing a headline in red rather than blue can be shown to statistically generate more revenue, then by all means the site will post their headlines in red.

Another aspect of the difficulty in dealing with the issue is the phenomenological experience on both sides. As someone shopping on the Internet, I do not directly experience subtle elements such as colors or styles pulling me toward some purchasing decision. On the other end, I have also worked as a web developer; although I never used multivariate testing specifically, one of the goals was to design the site in such a way that it would maximize profits. For example, adding a feature which allows for magnification of an image in order to better see the details seems like a good idea that will help consumers decide whether or not they like a particular product. It may even be the case that in seeing the detail, they decide this is a product that they do not want. The phenomenological experience of making a change such as this and making a change in the tab style or button color is very similar. Although the actions in these two situations are similar, the goals are completely different. In the former example, the goal is to help the consumer gain more information about a product and enable them to better determine whether or not he or she really desires to make a purchase. In the latter case, the goal is to increase the chance that the consumer will want to make a purchase or buy more items, regardless of whether or not he or she really desires to make the purchase. For a web developer who must necessarily be worried about increasing profits, taking the time to understand and reflect on the differences in these somewhat similar situations may be quite difficult.

Further, with corporate focus heavily emphasizing bottom lines and profit increases, web developers may feel pressure to make any change that is going to increase sales, regardless of the reason or manipulation involved. On one hand, these design tweaks aimed at manipulation might be easily ignored in a free market economy where *"caveat emptor"* is still a predominate sentiment. Certainly it seems unlikely such actions would ever be deemed illegal, yet it seems pertinent to consider whether such actions might be unethical. In trying to analyze the ethical implications of A/B and multivariate testing, I will be making several basic assumptions about which I would like to be clear up front. I will assume that based on the statistical results of multivariate tests, it is the actual changes to site design that are driving the increase in sales and not some outside force or circumstance that was not able to be measured. I believe enough case studies have been completed to show that these sometimes minute

changes can make a difference in sales. I will consider whether there may be ethical problems with such an approach through three ethical systems: utilitarianism, Immanuel Kant's deontology, and virtue ethics. I have selected these systems because they are the most common ethical views typically discussed in relation to business ethics in general. Within each, I will assume the validity of the ethical system being discussed, in order that criticisms of a particular ethical system should not be lodged against the ethical analysis made through that particular system.

Utilitarianism

Utilitarianism is a system based on a Greatest Happiness Principle which states that "actions are right in proportion as they tend to promote happiness, wrong as they tend to produce the reverse of happiness," (Mill, 2004, p.6). Further, John Stuart Mill defines happiness to be pleasure and the absence of pain. This system of ethics was founded by Jeremy Bentham but greatly expanded by John Stuart Mill. Although many variations of utilitarianism have since arisen, I will be considering this issue mostly from the perspective of hedonistic act utilitarianism, meaning the consequences of each individual act should be considered in making a moral judgment.

A utilitarian must ask if the overall good is served by manipulating consumers into buying more. Certainly, on the side of the company this does seem to be toward the good. Similar utilitarian arguments have been made supporting the use of manipulative advertising, such as the statement by William H. Genge: "Where does the money go? The answer is: It provides jobs and livelihoods for hundreds of thousands of people – not only in the advertising and communications sector but for all the people employed by fast-food companies and, indeed, all marketing organizations" (1985, pp. 58-59). Genge is arguing that advertising money is not wasted because the advertising causes consumers to make more purchases, and these purchases keep the economy moving and growing, allowing for more people to be employed – thus more people make more money and presumably the overall good is increased.

A similar argument could be applied to the manipulative changes made to web pages due to multivariate testing. Although the consumer may be manipulated to purchase more than he or she would have on his own, this is to the positive because the extra money allows the company to continue growing and expanding, possibly even hiring more employees as they do so and thus increasing wealth for many people. Sales allow for the economy to continue moving. What of the consumer, though?

The negative impact on the consumer does not *prima facie* seem to be very large. He or she was presumably already seeking to make some kind of purchase, and although the amount of the purchase has increased, that increase by itself is not typically significant. Average orders are increasing in the range of ten to twenty percent not two or three hundred percent. Although there may be a higher likelihood of buyer's remorse, the negative impact hardly seems to be problematically large. On this reading, many people see an increase in utility as the economy grows, while a consumer sees a small decrease in utility. Perhaps multivariate testing is not problematic from a utilitarian perspective.

Along those lines, arguments have been advanced that advertising and purchasing products actually increase overall happiness. In summarizing such an argument by Theodore Levitt, Michael Phillips says that "we do not merely buy a physical product, but also a set of positive feelings connected with it by advertising. If his [Levitt's] argument is sound, those feelings give us extra utility above and beyond the utility we get from the product's performance of its functions" (1994, p. 41). Although multivariate testing does not create exactly the same type of positive feelings as advertising, it must in some sense create a desire to purchase more, therefore

the consumer might be understood as being able to fulfill a further desire.

The impact of these arguments arises because of the capitalistic assumption that the economy should always continue to grow and that making purchases really does make a person happier. Yet, this basic assumption is rarely questioned within the advertising community. It does seem clear that purchases can and do increase the happiness of a consumer. We have all witnessed and likely experienced the euphoria of making a large purchase and the joy that comes with it – whether it be a new car or a new computer. However, that euphoria very often wears off quite fast. Capitalism provides the answer to this problem as well though, because once the euphoria wears away and the computer starts running more slowly than we would like, we can just buy yet another new computer which brings back that feeling of euphoria in a never-ending cycle. Presumably that cycle of buying and the related euphoria is the good life, and offers increased utility all around. Consumers buy more and more, growing the economy and jobs, and thus everyone can afford to continue buying more and more, bouncing from one euphoric purchase to another.

This is a good story, but I am not entirely convinced that it is the whole story. Yes, that euphoria exists; yes we can keep achieving it by purchasing more and more, but when this becomes the singular focus – when businesses are given the okay to manipulate others to make purchases, many can feel that something is missing. Consider John Stuart Mill's distinction between the quantity and quality of pleasure. In opposition to Jeremy Bentham's strict utility calculations, Mill believed that some pleasures are better than others. If one accepts Mill's distinction, it then becomes an important question to ask what the quality of pleasure derived from making purchases is and whether an alternative option exists that would provide a higher quality of pleasure. This is an open question that has yet to be settled, but I believe a very strong argument could be made that the pleasure from the euphoria of purchases is of a much lower quality than other activities such as having fun with friends and family. Although the empirical data is not completely cut and dry, research does seem to show that once basic needs are met, increased wealth does not seem to increase happiness (Meyers, 2004). Mill himself believed that one needed to appeal to a "competent judge" in order to determine an issue such as this. In other words, only a person capable of fully experiencing both types of pleasures could truly determine which is the higher pleasure (Mill, 2004, p. 15).

The utilitarian case against multivariate testing is further complicated by the use of such tests by philanthropic organizations. The American Cancer Society used Google Analytics to improve many faucets of their web campaign, from search engine advertising copy to homepage layout to email marketing. Alexander Negash, the user experience manager for the American Cancer Society says, "Google Analytics is literally helping in the fight against cancer" (*Google Analytics – Case Studies, 2009*). Even if one understood the American Cancer Society as using multivariate testing with the strictest of intentions to increase the amount of money spent by their visitors – which is clearly not the case here – a utilitarian may still have trouble criticizing such an action because of the overall good being done. Further, the visitor is not being manipulated into being a materialistic consumer – if anything, one would have to understand them as being manipulated into being an altruistic giver.

Should a utilitarian support implementing the manipulative changes suggested by multivariate testing? This remains a bit of an open question, because the results of the actions are somewhat difficult to fully deduce. The typical story is that purchases do increase utility overall, but that story has recently been called into question more and more. It is certainly a possibility that the huge emphasis on purchasing and consumerism is actually not good for a person, but perhaps the better argument is that there are alternative options that are of a higher quality of pleasure. For example, one might gain pleasure, and thus utility, from buying a new pair of shoes every week; however, a higher quality of pleasure might be found instead in reading a book which has been checked out from one's local library. Although pleasure is to be found in both activities, many would argue that becoming engrossed in a book is of a much higher quality pleasure than buying new shoes. If this can be safely asserted, then a utilitarian should oppose manipulative site design aimed solely at increasing conversion rates on a product like shoes.

To take this analysis one step further, it should be noted that Mill acknowledges the difficulty in applying act utilitarianism to every decision which needs to be made. For that reason he supports a type of rule utilitarianism which says that one should do those actions which, when followed, typically tend to produce the greatest good for the greatest number, even if they might not in every particular case. Looking at multivariate testing from this perspective, one might argue that the practice of making decisions based on emotional manipulation, while beneficial in certain cases, as a rule makes for a less rational and self-disciplined society. Overall, rule utilitarianism could offer a more substantial criticism of such a manipulative practice, similar to virtue ethics which is analyzed later.

Deontology

In his deontological ethical system, Immanuel Kant proposed that one should always treat others never merely as a means, but always also as an end (Kant, 1993, p. 30). In other words, in order to treat a person ethically one must respect his or her autonomy. Drawing from literature on manipulative advertising, one can see that the case has been made that advertising undermines the autonomy of consumers. This same type of argument might be lodged against the implementation of results from multivariate testing. If designing or laying out a page in a certain manner consistently causes consumers to spend greater amounts of money, perhaps this is an assault on autonomy.

Even within ethical debates on advertising, this point may be disputed, and I believe it may be more difficult to make the same case for multivariate testing. For starters, Andrew Sneddon says one can consider the phenomenological experience of being a consumer: "it typically does not feel like one is being jerked around and parted from one's money like a puppet" (2001, p. 15). I am sympathetic with that reading of manipulative advertising, and even more so when it comes to multivariate testing. It is almost difficult to fathom that when one is shopping on an e-commerce site, something other than one's autonomy is driving the purchase decision.

Advertising's manipulation of autonomy at least has a somewhat simple story behind it: advertisements create or enhance desires and offer products as easy methods of gratification. Almost anyone can consider this view and find it at least within the realm of feasibility that this may actually happen. We all have desires, and certainly different experiences can cause desires to arise. Looking at a picture of a large hamburger can certainly make a person hungry and create a desire for a hamburger that did not exist before.

The effects of multivariate testing do not offer a story that is understood quite so simply. At best one might argue that a psychology of aesthetics is being employed in which the layout, color, and size of various elements is able to manipulate one in a way that they want to or are more comfortable with spending more money. However, this story is not as easily intuited. I can easily think of the time I have seen a hamburger and felt hungry, but there is not a similar experience I can point to where I experienced a red headline and suddenly felt an urge to spend more money.

If the phenomenological experience does not reflect it and the common sense story is hard to interpret, is there really any manipulation going at all? Roger Crisp offers a discussion of manipulation and offers an understanding which might be able to make sense of what is happening: "A more convincing account of behaviour control would be to claim that it occurs when a person causes another person to act for reasons which the other person could not accept as good or justifiable reasons for the action" (1987 p. 416). Through this interpretation, we might consider something to be a case of manipulation if it causes action due to reasons the person being manipulated would not find justifiable. This offers a clear way of understanding the manipulation going on in the case of multivariate testing. If I were to know that I was going to spend extra money simply because of the aesthetic design of a particular page that had been fine-tuned for just such a purpose, I would not find that to be a justifiable reason for my increased purchased. Therefore, according to Crisp, I could conclude that I had been manipulated into spending more money. This manipulation undermines my autonomy and therefore the website designers could be understood as acting unethically under Kant's respect for person's principle.

Yet, one further problem to this type of argument remains: every web site must have some type of design to it. Even something as simple as using black text on a white background is done by deliberate choice when creating a web page. How can one differentiate between good design principles and unethical manipulation of consumers? An answer to this type of question is not entirely clear.

A possible answer to this problem would be to suggest that any use of A/B or multivariate testing is problematic in that it aims to manipulate consumers. Unfortunately, reality is not that cut and dry. As I have mentioned, there are many cases where legitimate ease of use questions might be resolved by multivariate testing. Perhaps an information site is experiencing a large number of visitors leaving after spending a few seconds on the main page. This may be because the search box is not well placed or is too small to be noticed. A multivariate test may help bring this out, and allow a site visitor to search for the information they wanted to find. It would be much more difficult to make the case that any type of unethical manipulation is going on in this multivariate test, so the answer must be more complex than claiming that all multivariate tests are unethical.

The intentions of the web developer may be the strongest criterion from which to make an evaluation of the ethical position. What exactly is the goal of the multivariate test? What is being measured? If the main criterion is the increase of income, this may be a warning sign that the intentions of the web developer are unethical.

Aside from multivariate tests, other methods exist for measuring outcomes which may actually be more useful if the sole criterion is not increased profit. Focus groups can use a website or a particular page. In addition to the verbal feedback that they are able to give, their actions on the page and even their eye movement can be tracked. This data can best answer questions about what the person was trying to achieve, whether or not they had trouble, and also what could be done to improve the webpage. In a focus group, the testing is more clearly focused on helping a consumer achieve something they were already attempting to do, rather than searching for any combination of factors that will increase sales. The biggest problem with focus groups as opposed to multivariate testing is the much greater expense and effort involved in setting up a focus group as opposed to a multivariate test.

Drawing the line which demarcates a particular multivariate test as unethical under a deontological system using the respect for person's principle is a difficult task, but I argue that at least in some cases, the intentions of the web developer clearly makes the implementation of the results of such tests an unethical action. Using these results is a form of manipulating consumers.

Virtue Ethics

Whereas deontology is a rule-based system and utilitarianism focuses on the consequences of ac-

tions, virtue ethics considers a person's character to be of chief importance. An Aristotelian interpretation of this would suggest that the best characters are those that experience emotions that are the means of two extremes. For example, a virtuous person acts courageously, rather than rashly or cowardly. Virtue ethics is often thought to offer one of the best arguments against manipulative advertising (Phillips, 1994). In the video *Advertising and the End of the World,* Sut Jhally notes that one of the main purposes of advertisements is to tell us that we can become happy through the consumption of objects (1998). Neil Postman and Steve Powers reiterate this message:

Boredom, anxiety, rejection, fear, envy, sloth – in TV commercials there are remedies for each of these, and more. The remedies are called Scope, Comet, Toyota, Bufferin, Alka-Seltzer, and Budweiser. They take the place of good works, restraint, piety, awe, humility, and transcendence. On TV commercials, moral deficiencies as we customarily think of them do not really exist. A commercial for Alka-Seltzer, for example, does not teach you to avoid overeating. Gluttony is perfectly acceptable, maybe even desirable. The point of the commercial is that your gluttony is no problem. Alka-Seltzer will handle it. (Postman and Powers, p. 125)

These commercials emphasize removing the symptoms of moral deficiencies, rather than working on improving the virtues. Cleary this practice is objectionable from a virtue ethics standpoint. The problem with multivariate testing is similar. By implementing the results of the multivariate tests, a web developer could be understood as aiming to overcome virtues such as moderation, self-control, and self-discipline. Often the underlying question in a multivariate test is what one can do in order to encourage consumers to spend more money.

As with manipulative advertising, the virtue ethics argument against implementing results of multivariate tests is pretty straight forward and shows the actions as clearly objectionable, because they hamper the development of virtues in the consumer. However, Virginia Held also offers another perspective from which to criticize the actions from a virtue ethics perspective: they undermine intellectual and artistic integrity (1984, 64-66). Although her writing focuses on advertisements, the criticisms may be ported over. The web developer and/or designer is in a real sense creating a work of art through the web page. Focusing on conversions and increased sales rather than the design itself is lacking in artistic integrity Furthermore, intellectual integrity is also undermined.

The biggest problem with putting forth a virtue ethics critique such as this is the possibility of a response that implementing the results of multivariate tests actually highlight virtues. As previously mentioned, most of the literature on such tests frames them in a way that virtues such as altruism could easily be highlighted. A tendency could develop in which one views the situation as being merely relative: some people see such actions as manipulative while others see them as altruistic. Such a reading of the situation is likely, because one can certainly find examples of cases where multivariate testing is being used in a manner that is altruistic. The problem which may elude one who is not paying close attention to the issue is that not all uses of multivariate testing are equal.

Once again an issue of demarcation arises. Which uses of multivariate testing highlight positive virtues and which are problematic? If the discussion and reflection on such a question is left entirely up to those businesses doing the testing, recent history seems to suggest that all such testing will championed as virtuous, but this is clearly not correct. To get a handle on the battle of language really going on, consider this report on the benefits of multivariate testing:

When it comes to segmenting, a large dating site found that those visiting its page during normal business hours were receptive to a main photo depicting a happy couple, while those visiting in the evening were much more receptive to a photo of three provocative- looking blond women. The psychographic implications of that particular finding opened a treasure trove of marketable data for the company. (Quant is King, 2007)

Consider the language being used here. The visitors were 'more receptive' and the findings opened a 'treasure trove' of data. With this type of language it sounds as if the lives of everyone involved in the process are being improved. The company has access to important and valuable data and the visitors are being helped out by images with which the associate more pleasantly. But the alternative reading is that this valuable data is being used by companies to manipulate consumers. Segmentation of visitors to different times of the day allows for even greater manipulation. The "psychographic" implications on this reading are not a treasure trove of data, but rather a worryingly large amount of information that is being used to manipulate visitors into continuing to use the site and likely purchase a membership or view more advertising. Is the visitor actually benefiting? From a virtue ethics perspective that question is actually irrelevant. This type of manipulation implements data in a way that strives to weaken self-control and moderation from the visitors. Billing this practice as mutually beneficial is certainly not showcasing intellectual integrity.

Issues, Controversies, Problems

Perhaps the greatest challenge when considering A/B and multivariate tests is trying to get businesses to understand the subtle ethical considerations involved. First, most business are going to focus on the very pragmatic issue of increasing sales above and beyond all other considerations. However, even if a business is attempting to behave ethically, the ethical issues involved in multivariate testing can be easily written off in light of activities which appear much more blatantly unethical. Google, whose unofficial company motto is "do no evil," is the company that offers free multivariate testing for any website.

As demonstrated earlier through the words of Chuck Casto, Vice President of Corporate Communications for CSN Stores, one can easily summarize multivariate testing in a way that makes the actions appear to be altruistic rather than manipulative. Through this interpretation websites are simply trying to help customers achieve the actions they already want to perform. Because this is likely to be true in some cases, it is very difficult to determine where to draw the line: which tests lead to results that are actually helpful and which lead to the manipulation of customers. Currently there is no clear answer to this question. It is extremely unlikely that any company is going to completely eliminate multivariate testing from their web development strategy – the activity is actually becoming more popular than ever.

If any progress is going to be made in spreading the notion that this practice is unethical, clearer guidelines need to be developed which can differentiate between those uses which are ethically problematic and those which are not. A deontological system seems to be the best suited for developing such guidelines, but others may be possible. Deontology may be fruitful because a designer should be able to step back and examine the intentions of changes he or she is making to a particular page. Future work on this issue may be able to better explore the practical issue of ethical boundaries for the use of multivariate testing. One strategy that has been adopted widely in cases where testing on people occurs is informed consent. Currently, it is difficult or impossible for a typical consumer to tell if a website is conducting A/B or multivariate testing. It would at least a step in the right direction if a website were to post a notice informing a visitor that the testing was occurring. This would alleviate some of the ethical problems, but others would remain: a

deontologist could still claim, for example, that the website was using the consumer merely as a means to increased profit.

Solutions and Recommendations

Multivariate testing seems more popular than focus groups, likely because it is both easier and cheaper to conduct as well as being more pragmatic. As with manipulative advertising, multivariate testing seems to be a practice that is clearly legal, but ethically questionable. Manipulative advertising has become so widespread that it is extremely unlikely there will ever be a reduction or elimination of such ads. Multivariate testing for websites is still a relatively new practice, so there may be a greater hope of intervening at this point in time.

If a professional code of conduct for multivariate testing could be developed, this may help stem the tide of its unethical use. Many companies offer multivariate testing as a third party service, so if they were to adopt a code of conduct, this would greatly reduce the number of companies participating in such a use, however, it would not prevent any particular company from undertaking their own internal multivariate testing.

FUTURE RESEARCH DIRECTIONS

This discussion would benefit greatly from empirical research into the question of why certain changes influence consumers to change their habits in the way that they do. The question of why may not be important for some ethical theories such as deontology, which is based on intentions, but those answers may be important for evaluation within other ethical frameworks.

Very little literature currently exists on the ethics of A/B or multivariate testing. However, the ethical issue is similar to that of manipulative advertising, therefore a broad understanding of the issues associated with manipulative advertising would stimulate the understanding and discussion of the ethical issues that arise from the manipulation that occurs through multivariate testing.

CONCLUSION

Previously the discussion of multivariate testing has been left up to corporations and other businesses who charge to implement such tests. Little or no criticism of the practice has arisen. A/B and multivariate tests can be understood as problematic, at least in many uses, through three prominent ethical theories. However, if companies ever decided to include such tests with individualized data on consumers, the manipulation occurring would be even worse. Some segmenting is already done, as with the dating web site that draws a distinction between visitors during the day and visitors at night. But it is certainly possible to narrow those distinctions even further. Perhaps concerns about online privacy have attracted enough attention that this might not happen, but even without this connection, A/B and multivariate tests should be considered problematic.

Due to possible praiseworthy uses of such tests, some companies may not even realize that there are, or take time to reflect on, these ethical problems with such tests. Focusing on such an issue, even if it seems minor in light of other ethical concerns of e-commerce, may at least spread the notion that such tests may be used unethically so that those companies concerned with behaving ethically can stop to consider their own use of such tests.

It seems that there are clear cases of uses that are ethically praiseworthy and clear cases of uses that are ethically blameworthy, but in the middle will exist many uses which cannot be so easily categorized. Further effort at evaluating such middle cases through various ethical theories may be beneficial to all. At the very least, a better understanding of why such changes cause consumers to increase purchases could be beneficial. Such answers could be incorporated into current media awareness campaigns, such as that by the Center for Media Literacy. Understanding the process and being aware of the causes and implications of A/B and multivariate testing would be a major step forward.

REFERENCES

Arthur, C. (2009, July 8). *Google's Marissa Mayer on the importance of real-time search*. Retrieved September 7, 2009, from http://www.guardian.co.uk/technology/2009/jul/08/google-search-marissa-mayer

Brohan, M. (2007). *Internet Retailer Survey: Form and function*. Retrieved September 7, 2009, from http://www.internetretailer.com/article.asp?id=23262

COPPA - Children's Online Privacy Protection. (n.d.). Retrieved September 11, 2009, from http://www.coppa.org/comply.htm

Crisp, R. (1987). Persuasive Advertising, Autonomy, and the Creation of Desire. *Journal of Business Ethics*, *6*(5), 413–418. doi:10.1007/BF00382898

Genge, W. (1985). Ads stimulate the economy. *Business and Society Review*, *1*(55), 58–59.

Google. (n.d.). *Google Analytics - Case Studies - American Cancer Society*. Retrieved June 13, 2009, from http://www.google.com/analytics/case_study_acs.html

Google. (n.d.). *Google Analytics*. Retrieved September 7, 2009, from http://www.google.com/analytics/index.html

Held, V. (1984). Advertising and program content. *Business & Professional Ethics Journal*, *3*(3/4), 61–76.

Internetretailer.com. (2008, May 29). Multivariate testing produces winning product pages for ShopNBC.com & InternetRetailer.com, *Daily News*. Retrieved May 28, 2009, from http://www.internetretailer.com/dailyNews.asp?id=26545

Internetretailer.com. (2009, March 17). *Marketers plan to spend more online and aim to measure results better*. Retrieved September 7, 2009, from http://www.internetretailer.com/dailyNews.asp?id=29781

Jhally, S. (Director). (1998). Advertising and the End of the World [Documentary]. USA: Media Education Foundation.

Kant, I. (1993). *Grounding for the Metaphysics of Morals With on a Supposed Right to Lie Because of Philanthropic Concerns*. Indianapolis, IN: Hackett Pub Co Inc.

Meyers, D. (n.d.). The Secret to Happiness, *YES! Magazine*. Retrieved June 9, 2009, from http://www.yesmagazine.org/article.asp?ID=866

Mill, J. S. (2004). *Utilitarianism*. Boston: Public Domain Books. doi:10.1522/cla.mij.uti

Montecino, V. (1996). *Copyright and the Internet*. Retrieved September 11, 2009, from http://mason.gmu.edu/~montecin/copyright-internet.htm

Phillips, M. (1994). The Inconclusive Ethical Case Against Manipulative Advertising. *Business & Professional Ethics Journal, 13*(4), 31–64.

Postman, N., & Powers, S. (1992). How to Watch TV News. Boston: Penguin (Non-Classics).

Roche, M. (2004, July 26). *Scientific Web Site Optimization using AB Split Testing, Multi Variable Testing, and The Taguchi Method | WebProNews*. Retrieved June 12, 2009, from http://www.webpronews.com/topnews/2004/07/26/scientific-web-site-optimization-using-ab-split-testing-multi-variable-testing-and-the-taguchi-method

Search Engine Marketing (SEM) Glossary. (n.d.). Retrieved September 7, 2009, from http://www.anvilmediainc.com/search-engine-marketing-glossary.html sitespect.com (n.d.). *Online Businesses Doing More With Less In An Uncertain Economy Via Web Optimization Technology*. Retrieved May 20, 2009, from http://www.sitespect.com/news-online-business-121608.shtml

Sneddon, A. (2001). Advertising and Deep Autonomy. *Journal of Business Ethics, 33*(1), 15–28. doi:10.1023/A:1011929725518

Stoll, K. (2004, August 18). *Color your hunger*. Retrieved September 13, 2009, from http://findarticles.com/p/articles/mi_qn4179/is_20040818/ai_n11815804/

Technology Weekly. (2007, May 30). *Quant is King*. Retrieved June 13, 2009, from http://technology-weekly.mad.co.uk/Main/Home/Articlex/6f452b7cb4db47a8afc00ff9b8fb9752/Quant-is King.html

United States Access Board. (n.d.). Retrieved September 11, 2009, from http://www.access-board.gov/

Whitfield, T., & Wiltshire, T. (1990). Color psychology: a critical review. *Genetic, Social, and General Psychology Monographs, 116*(4), 385–411.

Chapter 8
Ethics in E–Marketing:
A Marketing Mix Perspective

Erkan Özdemir
Uludag University, Turkey

ABSTRACT

Some of the ethical issues experienced in traditional marketing practices are encountered in those of e-marketing as well. However, e-marketing practices raise specific new and different ethical issues as well. For instance, new forms of dynamic pricing, spam email advertising, and the use of tracking cookies for commercial purposes have all raised ethical issues. These issues can be examined from the perspective of such components of the traditional marketing mix as product, price, place and promotion, which are under the control of marketing executives. As such, an awareness of the ethical issues in e-marketing under the control of marketing executives is central to the realization of an ethical climate in e-businesses. The aim of this chapter is to present a critical analysis of ethical issues in e-commerce in relation to the marketing mix. The topics discussed within this framework will be enlightening for the ethical decision making process and practices of e-marketing.

INTRODUCTION

As use of the Internet becomes more widespread, businesses are taking advantage of this media to reach prospective customers and offer them products and services. While some businesses conduct part of their activities on the Internet, others carry out nearly all of their business transactions over the Internet. Therefore, e-marketing currently constitutes a significant part of marketing activities of businesses (Krishnamurthy, 2006). Indeed, it is no longer a choice, but in fact has become a necessity for businesses to engage in e-commerce. The Internet now represents a market place that exists in tandem with more traditional markets. This alternative shopping environment is drawing many businesses onto the online world and, as such, the competition on the Internet is inevitably increasing (Kim & Kim, 2004). However along with new opportunities for commerce, the Internet, as a new

DOI: 10.4018/978-1-61520-615-5.ch008

business environment, also brings along new opportunities for unethical behavior. Further, the global aspect of the internet makes it difficult to implement legal codes, since e-commerce spans national jurisdictions. As such, it becomes all the more important for businesses to establish ethical codes on issues and to train their employees in ethical decision making.

Ethical issues typically discussed in relation to in e-business include information security, privacy, and intellectual property (Caudill & Murphy, 2000; Foxman & Kilcoyne, 1993; Franzak et al., 2001; Kelly & Rowland, 2000; Maury & Kleiner, 2002; Milne, 2000; Rodin, 2001; and Stead & Gilbert, 2001). In addition to these commonly discussed topics, the competitive nature of e-business also raises a number of ethical issues specific to marketing. Indeed, the business function in which unethical practices are most likely to be recurrent in e-businesses is that of marketing. Marketing, the most important outward looking department of businesses, unlike the other business functions, mediates between consumers, shareholders, suppliers and other stakeholders. This external role of marketing and stakeholder pressures cause unethical practices to emerge more in marketing (Chonko, 1995). Therefore, the managers most likely to face ethical dilemmas are those involved in marketing. The same is true in the context of e-businesses as well. It is because businesses that transfer their own practices onto the internet have chances to reach more online consumers that intense competition continues to be the case in the medium of internet as well. Factors such as the features unique to e-commerce, the intense competition on internet, the insufficiency of the regulations regarding the internet, information security, and privacy also contribute to the rise of the ethical dilemmas of e-marketing executives who take on the role of customer representatives within the business.

Marketers can only build mutual valuable relationships with their customers through the process of confidence-based cooperation. E-businesses

that create and nurture trust may make it more likely for online customers to visit their web sites again (Dayal et al., 1999). Some of the studies done on this subject also justify this particular suggestion. For instance, in their study, Roman and Cuestas (2008) found that in comparison to traditional shopping, when online consumers perceived higher risks, they were less willing to do online purchasing. Another interesting result in their study is related to the fact that the experienced internet users were better at distinguishing "ethical" web sites in comparison to inexperienced ones. In another study, Roman (2007) found that security, privacy, non-deception and fulfillment/reliability were the powerful predictors of satisfaction and trust of online consumers. Therefore, the establishment of trust is essential to successful e-businesses.

The aim of this chapter is to examine the ethical issues encountered by e-marketers in e-businesses with regards to the elements of the marketing mix and to offer some solutions and suggestions to these ethical issues in e-marketing. In order to realize this aim, the nature of ethics, e-marketing ethics, stakeholder relations, and ethical theories are reviewed first. Then, the ethical issues in e-marketing are discussed in terms of the elements of the marketing mix. Finally, in the concluding section, some suggestions regarding how to deal with ethical issues in e-marketing are offered.

BACKGROUND: ETHICS AND E-MARKETING, STAKEHOLDERS, AND ETHICAL THEORIES

Ethics and E-Marketing

The aim of marketing in business is to appeal to the wants and needs of potential customers in promoting the products of a business. In order to realize this aim, the e-marketing executives, first of all choose their target markets, and then plan the marketing mix consisting of product, price, place and promotion. The marketing mix can be defined as the set of controllable marketing variables that consists of the blend of these four strategy elements meeting the wants and needs of the target market (Boone & Kurtz, 1995; Kotler & Armstrong, 1991). The marketing mix, one of the basic concepts of modern marketing, was for the first time defined as the four Ps by Jerome McCarthy (1964). During the same period of time, Borden (1964) designated twelve elements of marketing mix consisting of sub-mixes within each P. The marketing mix classification made by McCarthy is still pertinent today despite many advances in marketing thought and conceptualization (Yudelson, 1999). Simplicity of use and understanding has helped this classification be widely used and appreciated by executives and academicians (Dominici, 2009). In addition to the fact that marketing mix is a useful classification in study and analysis, the combination of the variables has a significant impact on the level of success of the marketing function.

In the years during which the marketing mix was first created, physical products, physical distribution and mass communication were dominant. However, the dynamics of the medium of e-business today has also affected the elements of the traditional marketing mix. For instance, the online sales potential of various product categories has emerged, and the issues of online brand management and brand loyalty have gained importance. Regarding the issue of pricing as the issues of price transparency and price adjustment become more important, it has become a must for the price expectation and value perceptions in the online market to be learnt. The fact that the structure of intermediaries has changed in the issue of distribution has turned out to be one of the most important dilemmas faced by e-businesses. E-commerce has caused a reduction in intermediaries. One of the most important effects of e-commerce has turned out to be the promotional element of the marketing mix. Online advertisements and communication especially have become outstanding element of

e-marketing activities (Kımıloğlu, 2004). Due to the impact of e-commerce on the elements of marketing mix and the features unique to online environment, many additions have been made to the elements of marketing mix by various authors, existing terms have been replaced by different names, or different marketing mix elements have been established. For instance, in his study, Robins (2000) introduced the benefits of buying online from the buyers' perspective as 8I (Inexpensive, Interactive, Involving, Information-rich, Instantaneous, Intimate, Individual and Intelligent) and presented this 8I as the new personal, electronic marketing mix. Kalyanam and McIntyre (2002) presented the elements of marketing mix three-dimensionally and developed a different e-marketing mix model in the form of $4P+P^2$ (Personalization and Privacy) $+ C^2$ (Customer Service and Community) $+ S^3$ (Site, Security, Sales and promotions). Constantinides (2002), on the other hand, proposed a different model of web marketing mix in the form of 4S (Scope, Site, Synergy, and System). Harridge-March (2004) examined electronic marketing by using the 7P model (in addition to 4P, process, physical evidence and people were added) and stated that he proposed an important structure to the executives for the assessment of e-marketing operationalization of this framework. Chen (2006), on the other hand, discussed the e-marketing mix as 8P (the other Ps added to the existing 4P: Personalization, Payment, Precision, and Push and pull). As is clearly seen, no single comprehensive e-marketing mix has been successfully created to replace the traditional marketing mix. Considering the existing state of affairs today, as noted by Dominici (2009), the basic structure of 4P still constitutes the core of operative decisions with some extension and adjustment. The traditional marketing mix is composed of the variables controlled by the marketers. Therefore, the decisions and practices of marketers regarding the elements of marketing mix may illustrate the ethical issues likely to emerge in the online environment. In this study, the e-marketing mix is examined in accordance with the traditional 4P classification. At this stage, it will be beneficial to explicate such concepts as business ethics and marketing ethics in the assessment of ethical issues in e-marketing.

Ethics can be defined as an "inquiry into the nature and grounds of morality where morality means moral judgments, standards, and rules of conduct" (Tsalikis & Fritzsche, 1989, p. 696). Business ethics, as an area of applied ethics, can thus be described as the totality of the ethical principles and rules determining which business practices are morally acceptable and which are not (Schoell et al., 1993). Marketing ethics, as a sub-field of business ethics, concerns the application of ethical evaluations to marketing decisions (Smith, 1993) or as the application of ethical codes or systems in the field of marketing (Gaski, 1999). Marketing executives are often faced with decisions that give rise to ethical dilemmas that can be related to the elements of the marketing mix of product, price, place and promotion (Rallapalli et al., 2000). As businesses move more of their efforts to the Internet, the implementation of e-marketing applications enhance the ethical issues faced by marketing executives. This is because of the unique technological environment of e-commerce. In order to understand better the ethical approach that will be adopted toward marketing practices here, we must first consider the various stakeholders that are involved in marketing endeavors.

Stakeholders

The marketing decisions of marketing executives are affected by many internal and external people, groups and organizations with different wants and needs. Therefore, it is very important to know the stakeholders and their influences in assessing whether marketing decisions are ethical. Stakeholder approach postulates that those who have an interest in or are affected by a business have a stake in its decisions. In the context of e-

marketing, the stakeholders include the organization itself, customers, rivals, the government, and, cyber society more generally (Caudill & Murphy, 2000). According to marketing executives, the most important stakeholders are the customers of a company (Vitell et al., 2000). It is important to remember the customers are primary since they make the purchase of the products and services offered by the company and therefore allow them to stay in business (Schlegelmilch, 1998). According to Maury and Kleiner (2002), ethically questionable practices of businesses can directly affect the consumers, other businesses (such as competitors) and investors. The wants and needs of each stakeholder vary. For instance, consumers want their rights of privacy not to be exploited and expect trust in business transactions (Caudill & Murphy, 2000). Likewise, groups such as the suppliers that have business ties with the businesses and investors are entitled to rights such as those to information security and confidentiality. Paying attention to the rights of all stakeholders can also be useful in terms of enhancing trust in businesses. The fact that trust is placed in the centre of the strategies of e-businesses in the online environment has an important role in customer loyalty for e-businesses (Urban et al., 2000). The ethical environment of the business is also affected by the attitude and practices of the business staff (Fudge & Schlacter, 1999). For instance, the cognitive morality development of marketing executives and employees will have an important role to play in shaping up the ethical aspect of the decisions to be made.

The ethical issues traditionally found in marketing emerge as a result of the relationship of the marketing executives with the parties in the exchange process. The failure of the parties concerned to fulfill the duties and responsibilities lead to the emergence of ethical issue (Lund, 2000). Similarly, the failure of the parties concerned to fulfill the duties and responsibilities during the e-marketing process also lead to ethical conflicts. In this respect, being aware of the duties and responsibilities of the stakeholders and fulfilling them will minimize the emergence of ethical issues. For instance, marketers with the responsibility of meeting pre-determined sales quotas may resort to such unethical e-marketing tactics as spamming, creating difficult-to-close pop-ups windows, and misinforming customers with deceiving online advertisements. Similarly, confidential personal information of the customers may be misused in efforts to increase business. Therefore, fulfilling the responsibility towards the customers naturally raises important ethical questions.

Ethical Theories

As marketing practices can affect the well being of a number of different stakeholders, marketing ethics should provide marketing executives with guidance in resolving moral quandaries. Ultimately, the goal of moral theory is to provide pragmatic guidance. In making marketing decisions, marketing executives will, consciously and unconsciously, make use of and apply different ethical standards. Ethical theories are designed to provide guidance on the use of these standards, and have traditionally been divided into teleological and deontological types of theories. Deontological approaches to ethical theory stress the intentions underlying behavior, rather than the results of a behavior (Ferrell & Gresham, 1985). Teleological approaches, on the other hand, focus on the final effects of human actions and maintain that it is the outcomes of behavior that determines whether a behavior is ethical or not (Hair & Clark, 2007). In other words, it is clearly observed that while the key issue in deontological approaches is the intrinsic righteousness of a behavior, the key issue in teleological approaches, on the other hand, is the amount of good or bad represented in the consequences of the behaviors (Hunt & Vitel, 1986). Marketing executives with a deontological approach consider the perceived ethical problem and perceived alternatives according to the deontological approach to eventually reach

ethical judgments based on the perceived intrinsic qualities of the actions. Marketing executives with a teleological approach, on the other hand, consider the perceived ethical problem and perceived alternatives by focusing on the outcomes. Therefore, the ethical judgments of teleological marketers will be made in accordance with the consequences of the behavior and action. In focusing on different aspects of the actions, both approaches can be useful in considering the ethics of marketing practices.

Marketers may also use virtue theory, an important element of ethical thinking, as an approach aimed at coping with ethical problems. "An ethics of virtue assumes that being human entails living in community and developing certain virtues or skills required for a humane life with others. By trial and error, human reason arrives at certain core virtue for community living – such character traits as honesty, truthfulness, compassion, loyalty, and justice" (Williams & Murphy, 1990, p.23). The difference of this approach from the other two approaches is seen in the evaluation of whether this behavior is ethical. While the deontological approach focuses on intentions underlying behavior, and the teleological approach focuses on the results of a behavior, the virtue approach emphasizes the character of the moral agent. The decisions and practices of e-marketers regarding marketing mix may emanate from this approach as well. Therefore, in the rest of the study, this approach will also be considered as the foundation in the explication of ethical considerations towards marketing mix.

ETHICAL ISSUES IN E-MARKETING FROM THE PERSPECTIVE OF THE MARKETING MIX

Clearly, the Internet helps businesses overcome the limitations arising from differences of time and space (Herschel & Andrews, 1997). When compared to more traditional business mediums, increased flexibility in creating information, interactive technologies, and the combination of modalities of television, print, and radio in a single medium represents some of the technological advantages offered in e-commerce (Cook & Coupey, 1998). However, many consumers are still suspicious about the functional mechanisms of electronic commerce; its process, effects, as well as the quality of many of the products offered online. The most important reason that online consumers have to be reluctant about online shopping is the lack of trust, which is very widespread amongst both businesses and consumers on the web. Lack of trust is a long term barrier preventing consumers from making comprehensive use of the e-business potential (Kraeuter, 2002; Hoffman et al., 1999). For instance, Hoffman et al. (1999) found that almost the 95% of the consumers refused to provide their personal details to commercial web sites. The consumers either withheld their personal information or provided incorrect information.

The ethical issues likely to be encountered by e-marketers in the practice of e-marketing arise from the elements of the traditional marketing mix of product, price, placement and promotion. By examining the ethical elements of each of these traditional aspects of the marketing mix, we can gain a clearer understanding of the nature of the ethical issues involved in e-marketing. A failure to be sensitive to the ethical issues involved in any of the elements of the marketing mix may have an impact upon all marketing efforts. Given the necessity of trust in e-commerce, such failures will not only be unethical, but will also lead to unsustainable business practices in the long run. Thus, marketing executives need to be particularly sensitive to the ethical components of the marketing mix in e-commerce, if their businesses are to be successful in the long term.

Ethical Issues of Product

Many of the ethical issues encountered by marketers in traditional marketing are also found in e-marketing. In considering the introduction of new products while the basic variable focused on during product development is the economical one, ethical issues arise as well. When such issues are ignored faulty or unsafe products may be put on the market and these products may eventually bring damage to consumers (Morgan, 1993). It is commonly observed that engineering problems, for instance, can raise ethical issues related to product safety (Meel & Saat, 2002). The same sorts of problems related to the offering of unsafe products may also be true for products offered on the internet. Indeed, because of the international scope of the internet and the lack of uniform legal regulations for internet commerce, ethical issues may arise more easily. The problems often seen in traditional marketing such as deceptive information on product labels (Chonko, 1995), environmental problems introduced through product use and disposal (Menezes, 1993), and the failure to fulfill guarantee conditions regarding products are also common ethical issue likely to be encountered in e-marketing.

Other problems related to the product variable as a result of internet marketing include advertising and selling tobacco, alcohol, bullet, stun guns, pornography, online gambling and prescription-only pharmaceutical products on the internet. The lack of regulations regarding the internet, the ease of internet access in general and the ability of children to access the internet cause all these problems to emerge. Therefore, concerns can be raised over many types of products with socially suspicious value sold on the internet. For instance, there are 4.2 million pornographic websites (12% of total websites) in the world. Likewise, 2.5 billion (8% of total emails) pornographic emails are sent daily, 1.5 billion (35% of all downloads) monthly pornographic downloads (peer-to-peer) are made, and $4.9 billion worth of internet pornography sales occur annually (Internet Pornography Statistics). Many people monitor internet use for children because it allows easy access to pornography, enables sales of products aimed at children, and encourages gambling (Rao & Quester). The ethical issues are particularly important in the case of children, who may lack the ability to properly evaluate and foresee the risks of such problematic internet products. Sites designed specifically for children, even when they do not involve pornography or more blatantly objectionable material, can also raise ethical issues. Inappropriate site content and terminology, information gathering, sharing practices and marketing practices are the main areas of ethical issues (Austin & Reed, 1999).

Ethical Issues in Pricing

Pricing is one of the most difficult matters to analyze from an ethical point of view. Ideally, we might think that the price of the product or service should be equal or proportional to the benefit provided by it to consumers (Kehoe, 1985). The ethical issues related to pricing in textbooks on marketing ethics typically include non-price price increases, misleading price reduction, price advertisements which can be misleading or considered as deceitful, price-fixing practices that affect the structure of competition, predatory pricing which aims to gain monopolistic positions, discriminatory pricing, pricing applications of products according to the products' unit or quantity basis, and the practice of misleading pricing methods (Chonko, 1995). While some of these issues comprising areas of ethical issues in traditional marketing are true of e-marketing practices as well, some others are not used in e-marketing. For instance, fixed prices are often not used for sales on the internet. Instead, many marketers use techniques of real-time pricing such as online auctioning and dynamic pricing. Price adjustments, enabled by sophisticated software are made online in real time based on demand and product inventory level. As a result of this, businesses may offer different prices for the products/services sold on the internet. However, in that case, different consumers may have to pay different prices for the same product or service. This, consequently, constitutes a degree

of discrimination on the mere basis of internet accessibility. Internet pricing also has the potential of determining real-time individual prices. Actually, all these new internet pricing techniques raise concerns of privacy and other ethical issues (Rao & Quester, 2006).

One of the greatest benefits of the internet for consumers is that it allows them to be able to compare the prices of products or services based on their own criteria. Internet sites such as pricegrabber.com, shopping.com, price.com, and kelkoo.com are designed to provide price comparisons and make the job of price comparison all the more easier for consumers, giving consumers the chance to purchase the most appropriate product for the most reasonable price. However, according to Rao & Quester (2006), this price transparency constitutes downward price pressure on other retailers. In fact, this downward price pressure taken together with general profit pressures may force businesses to be inclined to other unethical practices in e-commerce.

Ethical Issues of Place

The internet is a type of electronic medium enabling the electronic distribution of products/services. The increasing competitive pressures brought about through e-commerce are forcing companies to seek out multiple channels of distribution (Rao & Quester, 2006). The ethical issues in traditional marketing related to place arise due to the power relationship in the channels and due to the fact that the businesses with this power use it to force intermediaries in the channel to carry out unethical practices (Laczniak & Murphy, 1993). For instance, a manufacturing business that does not comply with the required standards of products, but is that is more powerful in comparison to the retailers in the channel, may force the retailing businesses to exercise the practice of these unethical activities. This causes these unethical practices to spread throughout the distribution channel. However, as an important channel between the producers and the consumers, the internet alters traditional power relationships, providing for some mitigation against the ability of companies to force intermediaries to engage in unethical behavior. Nevertheless, new forms of ethical issues can confront e-marketers once again. For instance, direct selling of products by producers through the establishment of company web sites can lead to the elimination of intermediaries (retailers, wholesalers, outside sales reps) and thus have a serious economic impact on traditional business partners. In this regard, some businesses sell their products/services online, thus directly competing with their own intermediaries, jeopardizing the viability of those intermediaries businesses. This raises ethical issues of loyalty and the obligations of businesses toward their long term intermediaries (Stead & Gilbert, 2001).

The internet is fast developing as a key player in the relationship between producers and consumers and is commonly used as an important medium in the supply chain (Chaudhury et al., 2001). However, as seen, the use of the internet as an efficient channel of distribution enhances the problems of parallel imports causing harm to the authorized distributors and causing diversion of goods from the legal supply chains. Not surprisingly perhaps, the cost of the products of grey market being sold on the internet now reaches into the billions of dollars (Rao & Quester, 2006).

Ethical Issues in Promotion

New information technologies can offer opportunities to marketers in areas such as improved market segmentation and target marketing. At the same time, marketing executives are faced with potential ethical dilemmas resulting from the potential invasion of consumer privacy made possible with these technological practices (Foxman & Kilcoyne, 1993). In fact, the ethical issues here involve both technological and personal aspects. While the technical problems are related to the technological potentials of the process,

personal problems are related to the behavior of individuals in relation to these technologies. For instance, inappropriate distribution of information by an individual, or inappropriate responses to information, are examples of personal actions (Langford, 1996).

The most popular commercial use of web sites is for advertising their businesses and/or products and services. The primary aim of businesses in setting up their own web sites generally is to advertise the products and services they offer (Chaudhury et al., 2001). For instance, in a study in which they examined the web-based strategies of the Top 100 U.S. Retailers, Griffith and Krampf (1998) found that most of the retailers used their web sites for advertising, public relations and customer service access. Advertising has an important social role and also has high degree visibility and pervasiveness. Therefore, the number of those people reached and affected by the advertisements is quite high. For these reasons, advertising is the subject of criticism and controversy (Ergin & Ozdemir, 2007).

The short-term thinking of marketing executives can lead businesses towards short term advertising campaigns and planning that has the potential to push them towards unethical forms of promotion. The marketing executives in businesses in which short-term thinking is dominant focus on making short-term profits. This, on the other hand, causes marketers to implement aggressive actions such as pop-ups, deceiving banners, hyperlinks and other forms of intrusive mechanisms which impinge on personal privacy. This danger may prevent the businesses from adopting a proactive ethical attitude towards the consumers within their e-marketing strategies (Gauzente & Ranchhod, 2001).

Marketers are less limited, by regulation or cost, in online advertising products with potentially negative effects than traditional marketers. Besides, the internet is not a typical broadcast medium like radio and television, since the audience interacts with the medium and they can decide the amount of time of exposure that they want to the web site content. Therefore, the time-based restrictions of the traditional medium are not true for the internet. This causes advertising-related problems to emerge. For instance, while we see restrictions of alcohol and tobacco in terms of where and when they are advertised offline, it is not obvious whether these products are restricted in online advertising (Rao & Quester, 2006).

While new ethical issues in promotion are raised in e-commerce, most of the unethical cases in traditional advertising mediums are encountered in the internet medium as well. For instance, Hyman et al. (1994) stated 33 issues that they determined in relation to advertising ethics and indicated in order of importance. In the top three of these issues were the use of advertising deception, advertising to children, and cigarette and tobacco advertising. Alcohol advertisements, negative political advertisements and sexual stereotyping in advertisement constitute other significant ethical issues. These issues related to advertising ethics are also true for the online advertising as well. The ethical evaluation of marketing practices in e-marketing will include treating issues such as deceptive practices against vulnerable consumers, children and the elderly, the marketing of defective products, and the invasion of consumer privacy. Certainly, empirical studies done on the subject of e-marketing suggest that such concerns are warranted (Mattsson & Rendtorff, 2006). For instance, Rao and Quester (2006) studied the consumer's opinion of ethical priorities amongst 190 sophomore commerce students and at the end of the study they ranked the e-marketing ethical issues according to their means. Consequently, they found that in 5 of the first 10 statements out of 36 statements of the e-marketing ethical issues, children were targeted with e-marketing practices. These five ethical issues targeting children include using sexual content and nudity on websites accessible to children to increase website traffic, collecting consumer information from children independently, using violence on

websites accessible to children to increase website traffic, including links with unfiltered websites that children may access, offering incentives such as free games to children for providing personal information. This result demonstrates that the use of e-marketing practices targeting children creates a serious concern for consumers. The growth in children's access to the internet today has led to the emergence of thousands of child-oriented web sites. Consequently, the interest in ethical issues on this subject has also increased. As a matter of fact, most of the web sites targeting children are heavily laden with commercial promotions. Although some web sites do not specifically target children, they are easily accessed by children. This eventually causes children to be exposed to advertising for many products that children do not use, and which could be inappropriate for them. For instance, Nairn and Dew (2007) found in their study that three quarters of the online advertisements they observed were banner advertisements, and that the majority of the products advertised were targeted for a general audience.

Although technology offers the convenience of shopping without physically leaving the comfort of one's home through the internet, it also has less desirable results such as the reduction of individual privacy and the decrease of trust in business (Peace et al., 2002). It is because e-businesses gather information initially for promotional purposes and information that will help in marketing decisions and their applications that they may also impinge upon the privacy rights of users. The issue of privacy is also crucial for data mining, in which information is gathered in an attempt to better understand consumers' preferences and analyze them for future marketing purposes (Olson, 2008). All of this data provides businesses with opportunities to advertise their products and sell them on the internet in innovative ways (Milne, 2000). In e-marketing, different techniques are used to collect information about consumers and those who visit the web site of businesses. These techniques may lead to the invasion of the right of privacy and thus to ethical issues in e-marketing. Some of these uses of information include the gathering of descriptive personal information such as name of the customer, his/her address, or credit card number which the customer gives voluntarily while making a purchase on the web site of the firm. This information can also be used for promotional purposes related to the consumer in the future.

A second issue concerns the self-divulgence of information in order to access web sites. Accordingly, some web sites require their visitors to register to their web sites in order to have access to their web sites. Likewise, user surveys and online contest are other tools used to collect descriptive personal information. Third, free merchandise given out by companies is often used as another way of information gathering. Fourth, e-businesses compile anonymous profiles of data based on visiting consumers. This information can consist of the operating systems, country of origin, Internet Protocol (IP) addresses and the internet providers of users (Kelly & Rowland, 2000).

Fifth, companies make use of cookies to track data. Cookies are small data structures used by the web sites or servers to store and retrieve information on the user's side of the internet connection. They are sent by the host web site or server and reside in the personal computer of the user (Charters, 2002). Each time the user visits a web site, the web site can open the cookie. This process is also known as "profiling" or "data-mining" (Stead & Gilbert, 2001). For instance, when consumers register to a web site, a "cookie" is registered on their PC and this cookie enables the seller to monitor the purchases of the consumer as well as the web sites they visit (Laczniak & Murphy, 2006). Today the web sites of some businesses ask for permission in order to able to send cookies and make a pledge that consumers' personal information will be kept confidential and that they will use the cookie to enhance the speed of access (Franzak et al., 2001). In addition, many businesses, in an attempt to enhance the trust of

consumers in web sites using cookies, only allow first party cookies (the ones put in by the web site visited) and put limitations on those cookies placed by an outside firm (Laczniak & Murphy, 2006). Finally, monitoring online news groups and chat rooms is also another method of information gathering used by businesses in e-commerce (Kelly & Rowland, 2000).

One of the least satisfying dimensions of the development of the e-commerce environment is the use of spam advertising. Spam e-mail consists of the use of unsolicited mass e-mailings to consumer e-mail accounts for advertising purposes. The most irritating form of spam includes advertisements for easy (high cost) financing, gambling sites, pornographic material and diet supplements (Laczniak & Murphy, 2006). The concerns caused by spam e-mail can include the overloading of consumer e-mail accounts, the use of false e-mail identities, the inclusion of inappropriate mail content, and the use of misleading enticements and fraud (Rao & Quester, 2006). In the worst case, sending spam e-mail can be used as an offensive or retaliatory move to shut down the PC or the server the spam is sent to or interfering with the business of companies or other entities targeted. In some cases, such attacks a lead to the loss of crucial information of the targeted user (Franzak et al., 2001).

Another ethical issue regarding e-commerce promotions is the use of pop-up advertising. When an internet user accesses a web site, pop-ups are separate windows that appear automatically in the browser of an internet user. These pop-ups that are generally used for promotion purposes that allow the users to access other web sites by clicking the appearing windows. Although pop-ups are usually easy to close, some of the intrinsic problems of pop-ups arise due to the non-availability of the function of closing down the window, recurrent appearance of those windows and their ability to lock up the screen of the PC (Palmer, 2005). Aggressive actions such as pop-ups, deceptive banners, hyperlinks and other mechanisms seen in some e-businesses also constitute other ethical issues in e-marketing (Gauzente & Ranchhod, 2001). Finally, the use of ethically controversial subjects such as sex, nudity and racist language in advertising messages in order to attract attention and interest are used in e-marketing activities as well.

CONCLUSION

Marketers can easily adopt the new technologies offered in e-commerce since they enable efficiency of exchange, enhance productivity, cost less and offer great conveniences. However, the new technologies may involve some ethically questionable practices and can have unintended negative effects. Moreover, since technology, as pointed out by Palmer (2005), alters the nature of the kind of relationship between businesses and consumers, marketing executives need to pay particular attention to these possibilities and their ethical implications before adopting such technological transformations. It is simply because ethics is crucial to establish and maintain a long term relationship with consumers that businesses should beware of engaging in potentially problematic practices in e-commerce. Therefore, in order for a commercial web site to be able to function successfully from an ethical point of view, e-marketers should be able to understand how the ethical perceptions of the consumers are shaped (Roman & Cuestas, 2008).

Paying close attention to the ethical use of web-based technologies will constitute a differentiating power for businesses (Gauzente & Ranchhod, 2001). In fact, what lies at the roots of this power is trust. Trust forms the essence of ethical relationships. Therefore, e-businesses should establish a relationship based on trust both with the consumers and employees and should further develop it (Peace et al., 2002). However, as pointed out by Bush et al. (2000), lack of regulation on the internet may cause frequent ethical abuses.

In addition, as noted by Camenisch (1991), the tension between the difficulties of economic life in today's competitive markets and ethics affects many executives and businesses. In order to be able to overcome these difficulties, it is an important duty of marketing administration to develop a culture encouraging ethical behavior and preventing unethical behavior (Hunt & Parraga, 1993). In this respect, the creation of an ethical climate should be an essential aspect of e-businesses.

It is possible to define the ethical climate of a business in terms of the business's typical practices and procedures. Written ethics policies or codes of conduct, ethics training programs, and the installation of an ethics officer or ombudsperson to assist employees in resolving ethical dilemmas constitute the manifestations of a positive ethical climate. In addition, in order to improve the ethical conduct of e-businesses, consumer welfare should be focused upon, accountability should be emphasized, and the effect(s) of decisions on other relevant stakeholders should be analyzed well. In an attempt to achieve this, the effects of the decisions to be made not only on the investors and top executives, but also on all stakeholders, including consumers, employees, lenders, communities, governments and competitors should be considered as well. Adopting participation in decision making, paying more attention to the legal arrangements and analyzing the ethical issues within a larger perspective are other significant practices to be implemented in developing an ethical climate (Sama & Shoaf, 2002). All these endeavors should be realized while taking into account the dynamics unique to e-marketing, and should be tracked to ensure personal and organizational development. As pointed out by Bartlett and Preston (2000), by establishing a positive ethical stance, businesses can obtain such benefits as higher sales through the development of a positive public image, increased employee commitment for the firms' objectives, higher job satisfaction, and increased efficiencies through developing internal and external relationships.

In this chapter, the ethical issues in e-marketing were examined with regards to the elements of marketing mix for professionals working in e-businesses and particularly in the e-marketing departments of e-businesses, as well as for academic researchers interested in these issues. This study aimed to contribute to the ethical e-marketing decision making process of professionals working in the field of e-marketing by making them, and others, more informed of the ethical aspects of e-marketing and by placing those issues within the traditional understanding of marketing represented by the notion of the marketing mix.

REFERENCES

Austin, M. J., & Reed, M. L. (1999). Targeting children online: Internet advertising ethics issues. *Journal of Consumer Marketing, 16*(6), 590–602. doi:10.1108/07363769910297579

Bartlett, A., & Preston, D. (2000). Can ethical behaviour really exist in business? Journal of Business Ethics, 23(2), 199–209.Boone L. E., &. Kurtz, D. L. (1995). Contemporary marketing, 8. Ed., Fort Worth, TX: The Dryden Press.

Borden, N. H. (1964). The concept of the marketing mix. *Journal of Advertising Research, 24*(4), 7–12.

Bush, V. D., Venable, B. T., & Bush, A. J. (2000). Ethics and marketing on the internet: practitioners' perceptions of societal, industry and company concerns. *Journal of Business Ethics, 23*(3), 237–248. doi:10.1023/A:1006202107464

Camenisch, P. F. (1991). Marketing ethics: Some dimensions of the challenge. *Journal of Business Ethics, 10*(4), 245–248. doi:10.1007/BF00382961

Caudill, E. M., & Murphy, P. E. (2000). Consumer online privacy: Legal and ethical issues. *Journal of Public Policy & Marketing, 19*(1), 7–19. doi:10.1509/jppm.19.1.7.16951

Charters, D. (2002). Electronic monitoring and privacy issues in business-marketing: The ethics of the doubleclick experience. *Journal of Business Ethics, 35*(4), 243–254. doi:10.1023/A:1013824909970

Chaudhury, A., Mallick, D. N., & Rao, H. R. (2001). Web channels in e-commerce. *Communications of the ACM, 44*(1), 99–104. doi:10.1145/357489.357515

Chen, C. Y. (2006). The comparison of structure differences between internet marketing and traditional marketing. *International Journal of Management and Enterprise Development, 3*(4), 397–417.

Chonko, L. B. (1995). *Ethical decision making in marketing*. Thousand Oaks, CA: Sage Publications.

Constantinides, E. (2002). The 4S web-marketing mix model. *Electronic Commerce Research and Applications, 1*(1), 57–76. doi:10.1016/S1567-4223(02)00006-6

Cook, D. L., & Coupey, E. (1998). Consumer behavior and unresolved regulatory issues in electronic marketing. *Journal of Business Research, 41*(3), 231–238. doi:10.1016/S0148-2963(97)00066-0

Dayal, S., Landesberg, H., & Zeisser, M. (1999). How to build trust online. *MM, Fall,* 64–69.

Dominici, G. (2009). From marketing mix to e-marketing mix: A literature overview and classification. *International Journal of Business and Management, 4*(9), 17–24.

Ergin, E. A., & Özdemir, H. (2007). Advertising ethics: A field study on Turkish consumers. *The Journal of Applied Business Research, 23*(4), 17–26.

Ferrell, O. C., & Gresham, L. (1985). A contingency framework for understanding ethical decision making in marketing. *Journal of Marketing, 49*(3), 87–96. doi:10.2307/1251618

Foxman, E. R., & Kilcoyne, P. (1993). Information technology, marketing practice, and consumer privacy: Ethical issues. *Journal of Public Policy & Marketing, 12*(1), 106–119.

Franzak, F., Pitta, D., & Fritsche, S. (2001). Online relationships and the consumer's right to privacy. *Journal of Consumer Marketing, 18*(7), 631–641. doi:10.1108/EUM0000000006256

Fudge, R. S., & Schlacter, J. L. (1999). Motivating employees to act ethically: An expectancy theory approach. *Journal of Business Ethics, 18*(3), 295–304. doi:10.1023/A:1005801022353

Gaski, J. F. (1999). Does marketing ethics really have anything to say? – A critical inventory of the literature. *Journal of Business Ethics, 18*(3), 315–334. doi:10.1023/A:1017190829683

Gauzente, C., & Ranchhod, A. (2001). Ethical marketing for competitive advantage on the internet. *Academy of Marketing Science Review, 5*(4), 1–7.

Griffith, D. A., & Krampf, R. F. (1998). An examination of the web-based strategies of the top 100 U.S. retailers. *Journal of Marketing Theory and Practice, 6*(3), 12–23.

Hair, N., & Clark, M. (2007). The ethical dilemmas and challenges of ethnographic research in electronic communities. *International Journal of Market Research, 49*(6), 781–800.

Harridge-March, S. (2004). Electronic marketing, the new kid on the block. *Marketing Intelligence & Planning, 22*(3), 297–309. doi:10.1108/02634500410536885

Herschel, R. I., & Andrews, P. H. (1997). Ethical implications of technological advances on business communication. *Journal of Business Communication, 34*(2), 160–170. doi:10.1177/002194369703400203

Hoffman, D. L., Novak, T. P., & Peralta, M. (1999). Building consumer trust online. *Communications of the ACM, 41*(4), 80–85. doi:10.1145/299157.299175

Hunt, S. D., & Parraga, A. Z. V. (1993). Organizational consequences, marketing ethics and sales force supervision. *JMR, Journal of Marketing Research, 30*(1), 78–90. doi:10.2307/3172515

Hunt, S. D., & Vitell, S. J. (1986). A general theory of marketing ethics. *Journal of Macromarketing, 6*(1), 5–16. doi:10.1177/027614678600600103

Hyman, M. R., Tansey, R., & Clark, J. W. (1994). Research on advertising ethics: Past, present, and future. *Journal of Advertising, 23*(3), 5–15.

Kalyanam, K., & McIntyre, S. (2002). The E-marketing mix: A contribution of the e-tailing wars. *Academy of Marketing Science Journal, 30*(4), 487–499. doi:10.1177/009207002236924

Kehoe, W. J. (1985). Ethics, price fixing, and the management of price strategy. In Laczniak, G. R., & Murphy, P. E. (Eds.), *Marketing ethics - Guidelines for manager* (pp. 71–83). Lexington, KY: Lexington Books.

Kelly, E. P., & Rowland, H. C. (2000). Ethical and online privacy issues in electronic commerce. *Business Horizons, 43*(3), 3–12. doi:10.1016/S0007-6813(00)89195-8

Kim, E. Y., & Kim, Y. K. (2004). Predicting online purchase intentions for clothing products. *European Journal of Marketing, 38*(7), 883–897. doi:10.1108/03090560410539302

Kımıloğlu, H. (2004). The "E-Literature": A framework for understanding the accumulated knowledge about internet marketing. Academy of Marketing Science Review, 6, 1-36.Kotler, P., & Armstrong, G. (1991). Principles of marketing, 5th Ed., Upper Saddle River, NJ: Prentice-Hall.

Kraeuter, S. G. (2002). The role of consumers' trust in online-shopping. *Journal of Business Ethics, 39*(1/2), 43–50. doi:10.1023/A:1016323815802

Krishnamurthy, S. (2006). Introducing e-markplan: A practical methodology to plan e-marketing activities. *Business Horizons, 49*(1), 51–60. doi:10.1016/j.bushor.2005.05.008

Laczniak, G. R., & Murphy, P. E. (1993). *Ethical marketing decisions: The Higher Road*. Boston: Allyn and Bacon.

Laczniak, G. R., & Murphy, P. E. (2006). Marketing, consumers and technology: Perspectives for enhancing ethical transactions. *Business Ethics Quarterly, 16*(3), 313–321.

Langford, D. (1996). Ethics and the internet: Appropriate behavior in electronic communication. *Ethics & Behavior, 6*(2), 91–106. doi:10.1207/s15327019eb0602_2

Lund, D. B. (2000). An emprical examination of marketing professional's ethical behavior in differing situations. *Journal of Business Ethics, 24*(4), 331–342. doi:10.1023/A:1006005823045

Mattsson, J., & Rendtorff, J. D. (2006). E-marketing ethics: A theory of value priorities. *International Journal of Internet Marketing and Advertising, 3*(1), 35–47.

Maury, M. D., & Kleiner, D. S. (2002). E-commerce, ethical commerce? *Journal of Business Ethics, 36*(1/2), 21–31. doi:10.1023/A:1014274301815

McCarthy, E. J. (1964). *Basic marketing: A managerial approach* (2nd ed.). Boston: Irwin.

Meel, M., & Saat, M. (2002). Ethical life cycle of an innovation. *Journal of Business Ethics, 39*(1/2), 21–27. doi:10.1023/A:1016319714894

Menezes, M. A. J. (1993). Ethical issues in product policy. In Smith, N. C., & Quelch, J. A. (Eds.), *Ethics in marketing* (pp. 283–301). Boston: Irwin.

Milne, G. R. (2000). Privacy and ethical issues in database/interactive marketing and public policy: A research framework and overview of the special issue. *Journal of Public Policy & Marketing, 19*(1), 1–6. doi:10.1509/jppm.19.1.1.16934

Morgan, F. W. (1993). Incorporating a consumer safety perspective into the product development process. In Smith, N. C., & Quelch, J. A. (Eds.), *Ethics in marketing* (pp. 350–358). Boston: Irwin.

Nairn, A., & Dew, A. (2007). Pop-ups, pop-unders, banners and buttons: The ethics of online advertising to primary school children. *Journal of Direct. Data and Digital Marketing Practice, 9*(1), 30–46. doi:10.1057/palgrave.dddmp.4350076

Olson, D. L. (2008). Ethical aspects of web log data mining. *International Journal of Information Technology and Management, 7*(2), 190–200. doi:10.1504/IJITM.2008.016605

Palmer, D. E. (2005). Pop-ups, cookies, and spam: Toward a deeper analysis of the ethical significance of internet marketing practices. *Journal of Business Ethics, 58*(1-3), 271–280. doi:10.1007/s10551-005-1421-8

Parental Control Software. (n.d.). *Internet Pornography Statistics*, Retrieved September 27, 2009, from http://www.parental-control-software-top5.com/internet-statistics.html

Peace, A. G., Weber, J., Hartzel, K. S., & Nightingale, J. (2002). Ethical issues in e-business: A proposal for creating the ebusiness principles. *Business and Society Review, 107*(1), 41–60. doi:10.1111/0045-3609.00126

Rallapalli, K. C., Vitell, S. J., & Szeinbach, S. (2000). Marketers' norms and personal values: An empirical study of marketing professionals. *Journal of Business Ethics, 24*(1), 65–75. doi:10.1023/A:1006068130157

Rao, S., & Quester, P. (2006). Ethical marketing in the internet era: A research agenda. *International Journal of Internet Marketing and Advertising, 3*(1), 19–34.

Robins, F. (2000). The E-marketing mix. *The Marketing Review, 1*(2), 249–274. doi:10.1362/1469347002529134

Rodin, T. (2001). The privacy paradox: E-commerce and personal information on the internet. *Business & Professional Ethics Journal, 20*(3-4), 145–170.

Roman, S. (2007). The ethics of online retailing: A scale development and validation from the consumers' perspective. *Journal of Business Ethics*, *72*(2), 131–148. doi:10.1007/s10551-006-9161-y

Roman, S., & Cuestas, P. J. (2008). The perceptions of consumers regarding online retailers' ethics and their relationship with consumers' general internet expertise and word of mouth: A preliminary analysis. *Journal of Business Ethics*, *83*(4), 641–656. doi:10.1007/s10551-007-9645-4

Sama, L. M., & Shoaf, V. (2002). Ethics on the web: Applying moral decision-making to the new media. *Journal of Business Ethics*, *36*(1/2), 93–103. doi:10.1023/A:1014296128397

Schlegelmilch, B. B. (1998). *Marketing ethics: An international perspective*. London: International Thomson Business Press.

Schoell, W. F., Dessler, G., & Reinecke, J. A. (1993). *Introduction to Business*. Boston: Alyn and Bacon.

Smith, N. C. (1993). Ethics and the marketing manager. In Smith, N. C., & Quelch, J. A. (Eds.), *Ethics in marketing* (pp. 3–34). Boston: Irwin.

Stead, B. A., & Gilbert, J. (2001). Ethical issues in electronic commerce. *Journal of Business Ethics*, *34*(2), 75–85. doi:10.1023/A:1012266020988

Tsalikis, J., & Fritzsche, D. (1989). Business ethics: A literature review with a focus on marketing ethics. *Journal of Business Ethics*, *8*(9), 695–743. doi:10.1007/BF00384207

Urban, G. L., Sultan, F., & Qualls, W. J. (2000). Placing trust at the center of your internet strategy. *Sloan Management Review*, *42*(1), 39–48.

Vitell, S. J., Dickerson, E. B., & Festervand, T. A. (2000). Ethical problems, conflicts and beliefs of small business professionals. *Journal of Business Ethics*, *28*(1), 15–24. doi:10.1023/A:1006217129077

Williams, O. F., & Murphy, P. E. (1990). The ethics of virtue: A moral theory for marketing. *Journal of Macromarketing*, *10*(1), 19–29. doi:10.1177/027614679001000103

Yudelson, J. (1999). Adapting McCarthy's four P's for the twenty-first century. *Journal of Marketing Education*, *21*(1), 60–67. doi:10.1177/0273475399211008

Chapter 9
Moral Guidelines for Marketing Good Corporate Conduct Online

Mary Lyn Stoll
University of Southern Indiana, USA

ABSTRACT

Corporate social responsibility (CSR) is highly valuable for transnational corporations, but entails special requirements of heightened honesty in the marketing of CSR as compared to other goods and services. Companies need help in finding appropriate venues for advertising CSR. The Internet is an ideal medium for advertising CSR because it affords a global reach and greater space than the confines of standard advertising venues. However, using the Internet also poses special challenges in terms of perceived epistemic criteria for truth in a company's online presence. This chapter highlights both the problems and benefits of marketing good corporate conduct online and provides moral guidelines for marketers of good corporate conduct.

INTRODUCTION

Most companies now include corporate social responsibility as a part of their stated goals in business practice. Whether as simple as a corporate code of conduct or as complicated as including social responsibility in a company's fundamental structure, corporate social responsibility (CSR) is now par for the course. This is in part due to Sarbanese-Oxley and changes in the federal sentencing guidelines (Stoll, 2008). This heightened concern with CSR has

also grown because consumers and investors have become more willing to hold transnational corporations morally accountable for their actions. The rise of nongovernmental watchdog organizations that have gone global along with the companies they track has further helped to make social accountability crucial to business practice. However, making consumers aware of moral guidelines at work in the creation and distribution of goods and services is importantly different from standard corporate attempts to sell products. Traditional advertising and public relations practices that may serve well

DOI: 10.4018/978-1-61520-615-5.ch009

in marketing goods and services are often inappropriate in marketing good corporate conduct.

This chapter explains both why there are more stringent guidelines for marketing good corporate conduct and how companies may best inform the public while still operating within the limits of what is morally acceptable conduct. The Internet is an important part of marketing corporate social responsibility initiatives for a number of reasons, but there are also special limitations and problems associated with providing information about good corporate conduct online.

BACKGROUND

Many of the issues faced by those charged with communicating corporate social responsibility initiatives are the same as those faced by individuals advertising goods and services more generally. When it comes to advertising, there are already a number of ethics codes in place. The Better Business Bureau Code of Advertising, the Australian Advertiser Code of Ethics, the British Codes of Advertising Sales Promotion, and the Canadian Code of Advertising Standards share the following key principles. First, it is essential to recognize that advertisers must meet responsibilities to consumers, local communities, and society at large. Secondly, advertising should adhere to standards of decency, honesty, and truth. This, of course, entails that advertisers ought to avoid misrepresentation and outright deception in ads. Advertisers also ought to respect a sense of fair play with other market competitors. Finally, advertisers must consider how their behavior affects the advertising industry as a whole (Spence and van Heekeren, 2005).

Despite these codes many advertisers clearly diverge from the requirements of honesty and avoidance of misrepresentation. Consumers facing the glut of beer and automobile ads promising a hot date still know that beer and nice cars won't actually guarantee delivery of the promised blond

bombshells in the advertisements. Puffery is common practice in advertising. Puffery refers to "exaggerated claims, comments, commendations, or hyperbole, and in its most common usage, puffery is based on subjective views and opinions" (Spence and Van Heekeren, 2005, p. 46). The public is fully aware that puffery is common. According to an article in *Adweek*, 74% of Americans strongly or somewhat strongly believe that advertisers regularly and deliberately stretch the truth (Spence and Van Heekeren, 2005). So long as the positions endorsed in ads are presented as subjective opinion rather than as objective rationally defensible claims, the Federal Trade Commission tends to let this sort of misrepresentation slide.

It could be argued that so long as consumers understand that the claims made in advertisements are exaggerated it is no more a case of outright lying than it is an outright lie for an actor in *Hamlet* to pretend to be a Danish prince when he is, in fact, a middle class man from Los Angeles. Artistry is never a matter of perfect representation of reality and the public knows that advertising is as much an art form as it is an attempt to provide information to the public about a company's products. Given this context, puffery is likely not problematic so long as one is not targeting marketing efforts towards children or to those who are mentally incompetent due to age or disease. With a minimally rational target audience, puffery is not necessarily deeply problematic, since context allows a rational agent to discern fairly easily the actual likely results of purchasing a product even if ads are unduly hyperbolic in their expression of purported benefits.

Puffery in advertising good corporate conduct, however, is much more problematic. If one is duped by a beer advertisement into thinking that Budweiser really will improve one's romantic life, the harm done is minimal. If a company knowingly advertises good corporate conduct in an attempt to reap the financial benefits of being perceived as having morally good corporate character when in fact the company does not deserve that kind of moral praise or support, morality itself is treated as a mere means to the end of profit. That is a much more serious moral offense. Not only would such a company be guilty of lying, the company would also be guilty of treating morality, itself of the utmost possible value, as being of less value than profit. Money, however, is a mere means that is good only for the ends it allows one to achieve. This kind of moral mistake undermines something of great moral value in order to achieve something of limited or perhaps even negative moral value. Phillip Morris, for instance, ostensibly spent more money advertising its corporate giving to Meals on Wheels and its investments in the company's Supplier Diversity Program than it did on actually supporting the programs in question. Given that the company was engaging in the campaign in response to moral censure for having targeted ads to children for its addictive products, it is clear that the measures were motivated by profit rather than by a sense of moral duty. So Philip Morris was trying to reap the financial benefits of morally good conduct without actually delivering the goods (Stoll, 2002). This is dishonest and deeply morally suspect.

In another case, British Petroleum (BP), donated four million dollars to the National Fish and Wildlife Foundation in partial compensation for a huge spill of over two hundred million gallons of oil, the largest spill recorded on Alaska's North Slope; the spill was foreseeable and avoidable, as BP was fully aware that its pipes were corroded and needed repair. Making reparations and paying a hefty twelve million dollar criminal fine was the least they could do. But then the company included its attempts at compensation and reparation in its 2007 Sustainability Report posted online (MacDonald, 2008). Again, failure to indicate that the donation was actually in reparation for past harm makes the company seem as if it deserves greater moral praise than it does. While omitting information is par for the course in advertising other products like shampoo in a thirty second ad, information crucial to judging the morality of

corporate behavior can not be treated so flippantly and disingenuously; it was entirely possible to provide adequate context in the report without any added cost, but BP simply opted not to do so.

This sort of dishonest advertising of good corporate conduct also provides a disincentive to any other company which hopes to become more competitive while adhering to CSR. Sometimes doing the right thing costs more, but a company could offset that cost if more customers were willing to purchase its products because of the added value of its products having been produced and marketed in a morally desirable fashion. If companies that do not engage in good conduct can still reap the benefits of seeming to have acted in accordance with moral requirements, then companies that truly embrace CSR may well be driven out of the market. Thus, it is especially important to ensure that consumers have an accurate and clear understanding of whether or not a company has acted in a morally responsible fashion. (For a more in depth account of this argument see Stoll 2002). Given that companies have a heightened duty to be honest and forthright in communicating good corporate conduct, the internet, unlike many other standard advertising venues, is often a good place to begin if a company hopes to provide clear and accurate information about its CSR initiatives.

SPECIAL ISSUES IN MARKETING GOOD CORPORATE CONDUCT ONLINE

Benefits of Marketing Good Corporate Conduct Online

The most obvious benefit of marketing online is that the Internet is a cost effective way of providing a great deal of information, especially through a corporate website. While some companies opt to let the public know about their good deeds in television advertisements or print ads, both of these forms of advertising entail severe content limits, making it difficult to communicate with the kind of clarity that careful moral judgment requires. Online communications on a company's website are not as limited and can be returned to for repeat viewing to ensure clarity, unlike many television advertisements. Bennett (2008), for instance, contends that the public is already used to looking online to find out what companies are doing: "(o)rdinary people are making your business their business as democratized media and information have made corporations more accessible—and accountable. Anyone with internet access can find out just about anything they want about your company"(p. 19). In support of his claim that consumers really are concerned with the morality of corporate conduct, Bennett notes that according to at least one survey, over 1/3 of respondents in the United States and the United Kingdom indicated that they had actively searched for information about corporate reputation and corporate ethics in the months preceding the survey. In France, over half of the respondents had sought to learn more about corporate conduct. If the findings in this survey are correct, then online outreach makes sense. Individuals are already looking actively for information about corporate conduct; companies that make that information available online are just making it easier for the public to find out how the company views its own conduct (Bennett 2008).

Second, the Internet allows greater opportunity for the public to provide interactive feedback via emails, online forums, or even interaction in talk rooms where both critics and corporate representatives might be able to discuss potential moral problems with corporate conduct. Open dialogue could lead to changes for both companies and critics. A company, for instance, may only realize that its conduct is unacceptable after receiving online feedback. Alternatively, corporate critics may occasionally change their minds after company representatives clarify the reasoning behind corporate policy. Respondents in the aforementioned survey indicated that it was important for companies to maintain an open dialogue with consumers; 80% of United States respondents, 78% of United Kingdom respondents, and 92% of French respondents stated as much. Not only does the public expect companies to maintain dialogue, they also often see it as a matter of personal duty to ensure that they, as members of the public, censure unethical corporate behavior. The majority of survey respondents in the United Kingdom and France indicated that they were paying closer attention to corporate conduct along with two thirds of American survey participants. Eight out of ten surveyed in the United States and France, and nearly seven out of ten surveyed in the United Kingdom said they felt a personal responsibility to censure unethical companies by refusing to purchase their products (Bennett, 2008). Going online to communicate good corporate conduct not only gives a company more space to explain their efforts, it also encourages companies to continually reevaluate policy initiatives in light of public feedback. Like any moral agent, a company can benefit from the perspective provided by feedback from those affected by its actions. A company can also rethink how it presents information on CSR policies in light of common misconceptions communicated by those with whom the company interacts online. This will help to ensure that the heightened clarity conditions for marketing CSR are met.

Third, the internet could allow companies to link outside analysis of corporate conduct directly to the company's website for a more objective account. Thus, a company need not rely on mere assertions of its good conduct; it can point to independent outside observers who verify that the company is meeting its commitment to CSR. Using outside analysis to verify claims is especially beneficial in attempts to communicate good corporate conduct. Most moral agents are leery of individuals who feel the need to tell everyone about their good character, since this may seem self serving. Such behavior may appear to indicate that the motivation for good conduct is popularity rather than a respect for the moral value of others. A profit driven company is importantly different from ordinary individuals in that a company's role responsibilities require that it make a profit to survive. So it may be that in order to do what is right a company must also make sure that its consumers and investors understand the value that their commitment to moral conduct adds to the goods and services it provides. Touting good conduct may be essential to a company actually being able to engage in maximally desirable moral conduct. But the average person may not think through all of the subtle differences between appropriate criteria for judging the moral conduct and character of corporate as opposed to individual moral agents. Providing links to outside verification of good conduct online allows a company both to ensure that others are aware of the moral value added to its products by the standards to which the company adheres, while at the same time not offending those who think it untoward for any moral agent (corporate or individual) to brag about their good conduct.

The Drawbacks of Marketing Good Corporate Conduct Online: Difficulties with Establishing Trust and Truth Online

Despite the numerous benefits of marketing good corporate conduct online, the practice does have important drawbacks. Many persons in business might think that explaining CSR to the public is just like any other part of business, and that, in this respect, just as a corporate report might be put online, so should CSR audits. Marc Gonzalves, Corporate Affairs Manager of the Billiton Mining Group, for instance, says that "undertaking Corporate Social Responsibility Programs will be the same as having to print annual corporate reports. It is what business is about" (Kapelus, 2002, p. 279). Communicating CSR online, however, is different from other aspects of public relations and advertising in a number of ways. First, there

is the heightened requirement for clarity in providing information necessary to moral judgment. Second, the Internet itself may present special drawbacks for communicating this specific sort of information.

Problems with Trust in Online Relationships

When communicating information on the Internet, the medium by which the message is conveyed affects both the message and its efficacy. Trust is importantly different in online interactions. Not only is the relationship of trust mediated because individuals are dealing with institutions that have designated and skilled media communications professionals, but the relationship is further mediated by technology when it occurs online. Consider the following example. Suppose that an individual, Mr. A, is subletting an apartment in the United States from a woman, Ms. B, who has returned to India for the summer. The rent check submitted by Mr. A to Ms. B never finds its way to the landlord during the first month of his stay. Mr. A must then deal with a long series of email interactions with Ms. B in order to sort things out. She believes that the error was due to the bank or to Mr. A not having accurately filled out the check. Mr. A firmly believes that he made no such error and suspects that Ms. B is short on cash and making excuses. If the two could interact face to face, they may have less difficulty trusting one another. Each could judge facial cues, or use his or her comprehension of body language and tone of voice to discern motivation and guilt. It is not just trained psychologists who can understand the tell tale cues of a liar who covers his mouth or refuses eye contact. But neither Mr. A nor Ms. B can make use of this skill set in determining who is telling the truth as the two are separated by an ocean and can only interact online.

Hubert Dreyfus has analyzed the problems that online interaction can create for human abilities to understand and trust one another at length. Dreyfus (2009) warns that:

(W)hen we enter cyberspace and leave behind our emotional, intuitive, situated, vulnerable, embodied selves, and thereby gain a remarkable new freedom never before available to human beings, we might at the same time, necessarily lose some of our crucial capacities: our ability to make sense of things so as to distinguish the relevant from irrelevant, our sense of the seriousness of success and failure that is necessary for learning, and our need to get a maximum grip on the world that gives our sense of the reality of things. (pp. 6-7)

The ability to understand oneself and the world is compromised in crucial ways in online interaction. Not only will individuals miss out on important body language cues, but it is often difficult to maintain a sense of seriousness and reality in online interactions. Without the physical presence of the other person in front of one, how vulnerable can either party be? Without that vulnerability how seriously can one take the situation? For those who think that online interactions really are not that different from face to face interaction, consider how much easier it often seems to confront someone online rather than in person. Public reactions to texted or emailed break up messages are not merely fodder for the tabloids. They indicate how problematic, and sometimes cowardly, opting for technology mediated communication can be. Furthermore, consider how much time online is spent gaming or playing at different identities. One can easily lie about one's looks, age, or even gender when on Facebook or in an online dating forum. In online role playing games, one could inhabit a world with others appearing to them as wood nymphs, dragons, or sorcerers. While this kind of creativity and experimentation can have valuable results, it also means that there may be a sense of unreality attached to online communicative interaction and

that trust online is established both differently as well as more cautiously.

A company explaining its attempts at social responsibility online could be analogous to that dream date or the company could be more like the jerk next door merely pretending to be something he is not. Without the background context of meaning, habits, and skills that face to face interaction provides, the truths discovered online may often rightly be regarded with greater skepticism. In the United Kingdom, online ads were responsible for one third of complaints to the Advertising Standards Authority (ASA). A full 90% of the advertisements that were subject to complaints occurred on corporate campaign websites that the ASA does not regulate (Carter, 2007). While it may be that companies are just more likely to make false claims about their sustainability or their products when it is legally permissible to do so (even if it violates their own ethics codes), it may also be the case that corporate decision makers simply do not take online communications as seriously. This may also help to explain why the ASA regulates other sorts of corporate communication but not the claims made on corporate websites; the ASA may not take online communications as seriously either.

It might be easier for company officials to feel that information posted online is somehow less real, less risky, and need not be taken as seriously as company reports printed out and kept physically present in desks and on shelves, or less serious than moral promises made to another's face. It is perhaps no surprise that documentarians and news reporters so often want to confront the CEO's of companies involved in ethics violations and scandals face to face. Not only does it make for a more entertaining story, it makes both the infraction and guilt more difficult to deny. Removing room for denial allows the moral wrong to be felt more deeply and seriously than mere numbers on a page in a report ever could. Part of the ability to judge moral character and conduct comes from these kinds of face to face interactions and the

CEO is often the closest thing one can find to the face of a corporate institution. To blame the CEO alone is often unfair, but the desire to speak face to face about moral infractions is likely in part due to an array of emotional and embodied habits of judgment that human beings are often unaware are at work in determining their moral choices and judgments. Without face to face interaction, it is easy to feel that the infraction is dismissed and the requisite examination of conscience never undertaken.

Obviously it is impractical for an institution to answer every moral question with a face to face interview with the company's CEO. Communication for a company will of necessity be mediated. Given the costs of making information available to all relevant stakeholders, the Internet is often the best option. But the aforementioned kinds of problems endemic to online communication between individuals also follow companies trying to communicate corporate social responsibility online. While companies must communicate via their emissaries, an online discussion of corporate policy deprives an individual of body language cues concerning whether or not the individual involved actually believes the claims he or she is making. While establishing trust between institutions and individuals is always strained, communications via online media can be even worse. This situation is further complicated by the fact that the individual searching for information about corporate social responsibility has no idea whether or not his emails and online chat messages are being received by an individual and replied to by a human being making a careful judgment call. It may well be the case that his emails and chat messages are returned by a computer program designed to generate replies automatically. The critic of corporate conduct may be all too used to being forced online to provide feedback knowing full well that this is merely an attempt to ensure that his grievances will never be heard or at the very least delayed indefinitely. If online communications really are conducted so as to ensure dialogue,

this might not be a problem. But if companies regularly refuse to allow customers interaction with customer representatives in any way except online, customers and critics will begin to suspect that the Internet is merely yet another wall erected between companies and accountability. Making online chat rooms or emails of company officials available, but never paying attention to that feedback would then be a dishonest attempt to make it seem like a company was engaged in a careful examination of conscience when in fact no such action had occurred. This problem is not unique to corporate communications online as Mr. A and Ms. B could attest insofar as each fears the other is not hearing the moral critique advanced across the distance of email. But it is still a danger that is endemic to corporate attempts to communicate CSR online that must be understood and dealt with accordingly.

Problems in Establishing Truth Online

The Internet also poses problems in that the sheer amount of information provided can make it nearly impossible to gain real comprehension. One could literally search for more information endlessly. What is gained in breadth could be lost in precision and clarity. Listing an array of detailed charts with no concise explanation of their meaning will not adequately convey CSR initiatives. While a two minute ad is too brief, a three hundred page report awash with detail but lacking context is equally inadequate. Companies that are honestly committed to CSR must also be weary of using the sheer girth of the Internet to make claims seem true rather than providing reliable evidence. If company Q responds to critics by merely inundating the web with claims to the contrary (rather than by providing evidence that these criticisms are unfounded or remedying the situation), it may well be the case that those researching the matter might never even find the opposing viewpoint. But this is akin to a child who convinces everyone else (and perhaps even herself) that she is innocent of having hit her brother merely because she said she was innocent so very many times. While knowledge is not formed merely by repetition, the vastness of the web may incline those seeking more information about a company towards this sort of epistemological vice. An unethical company would exploit that opportunity rather than take the critique seriously. Dean (2008) echoes precisely this worry:

Today, the circulation of content in the dense, intensive networks of global communications relieves top-level actors (corporate, institutional, and governmental) from the obligation to respond. Rather than responding to messages sent by activists and critics, they counter with their own contributions to the circulating flow of communications hoping that sufficient volume (whether in terms of number of contributions or the spectacular nature of a contribution) will give their contributions domination or stickiness…The proliferation of distribution, acceleration, and intensification of communicative access and opportunity far from enhancing…resistance results in precisely the opposite. (p. 102)

The glut of information available on the Internet that is not contextualized deprives information of meaning. Without a meaningful way to determine what really matters, shock, spectacle, repetition, and newness are given unwarranted power to determine what counts as truth for online researchers (Dean, 2008). This undermining of meaning and truth online, however, makes it difficult for companies whose actions really do match their online accounts of corporate social responsibility to effectively communicate this, especially when their less responsible competitors are willing to trade on repetition and spectacle to convince consumers and critics to opt for their products and services. The trust that a responsible business deserves may never take root in a context where

truth is not especially relevant to determining the moral beliefs of investors and consumers.

Although it was not a case of marketing CSR online, the Sony corporation once paid for a fake blog, a 'flog,' to be produced in which two men tried to convince their families to get them a Sony PS 2 for Christmas (Beard, 2007). This kind of activity makes it difficult for anyone to trust that blogs are what they claim to be. Imagine if Sony had paid bloggers to talk about how morally upright the company was. If the blog was sensational enough, it might get more viewers than a more truthful appraisal on a more morally responsible competitor's website. Even if the flog was premised upon a lie, it might be difficult to discern as much given the difficulty of establishing truth online and the extent to which web surfers prefer the sensational to the evidentially verifiable. Walmart presents another even more worrisome example. In response to critics who charged that Walmart does not treat workers with the respect they deserve, Walmart set up a flog called "Wal-Marting Across America" in which Walmart paid two professionals to take a 2800 mile road trip across America parking their RV in Walmart lots. The flog, however, nowhere mentioned that its writers were paid to create the flog. A second subsidiary flog of "Wal-Marting Across America" was called "Paid Critics" and was designed to expose Walmart critics (Fernando, 2007). Instead of engaging in dialogue with critics or remedying the situation, in this instance, the company opted to trade on the epistemic pitfalls of online information gathering to make it seem as though outside observers were coming to the company's defense, when in fact, the flog was funded by the company itself. While Walmart can and ought to defend itself against criticisms it believes to be false, doing so in such a deceptive fashion hardly speaks well to the company's overall moral character. This kind of behavior would undermine companies marketing CSR online who provide links to outside analyses. If competitors provide links that seem independent, but are not, this undermines trust for both the responsible and irresponsible companies. But without trust, a company following CSR can't compete and continue to be morally responsible if morally good behavior is more expensive. Thus flogs are doubly morally wrong. First, they are inherently deceptive to an unacceptable degree. Second, they undermine the conditions that make corporate social responsibility in general a viable business strategy.

To further complicate matters, outside of the confines of corporate communications the Internet is awash with anonymous commentators and bloggers; this alone could make the intended audience believe that CSR initiatives communicated online are also just so much talk with no action to back it up. Castelfranchi and Tan note that several experts in online communication have found that individuals are more likely to break commitments made online that those made face to face (2001). Given this sort of experience, corporate critics may rightly be wearier of trusting promises made online by anyone, whether by an individual or a corporation. Trust may be further compromised by worries that just as the Internet provides consumers and critics with more information about companies, it also provides companies with a greater ability to engage in surveillance of critics visiting their sites. The interactivity of the net could become a double edged sword. If everyone who complains about a disconnect between CSR reports and actual corporate behavior is required to communicate that complaint by email then has her inbox inundated with unwanted emails for sales, few will register those complaints.

Finally, even if CSR marketers could get past all of the aforementioned problems in communicating CSR online, complete with links to outside independent observers who verify their claims, the reliability of those outside observers may also be put into question. McChesney argues that the press, traditionally a means by which corporate wrongdoing was exposed when government oversight

was lax, is no longer able to serve that function well given the current political economy of media in the United States (2008). As newspapers and newsrooms become just one tiny part of giant multimedia conglomerates, the news becomes increasingly underfunded. McChesney worries that even at its best, so called objective journalism was equally an effort to make it seem that conservative media outlet owners did not exert undue control over the news. This was achieved by trusting official sources, like government, university professors, and official business spokespersons. These sorts of strategies can skew news content. Surveys show that 40-70% of the news is actually taken directly from press releases issued by public relations (PR) experts. Without the money to pay for a team of investigative reporters, it becomes ever easier to substitute the slick prepackaged PR in the place of independently verified news stories. If one is tempted to think academics alone can provide objectivity, it is important to remember the ever increasing pressure that universities face to bring in outside grant money, which is itself usually financed by the very same institutions the academy is ostensibly charged with overseeing. Furthermore, since no media outlet can survive without advertising dollars, McChesney is concerned that the press will not do an adequate job of reporting corporate wrongdoing. As evidence, he notes how despite the huge growth in business coverage over the last decade, the press did not report Enron's misdealings until 2002. McChesney speculates that this is in part because Enron actively courted both the *New York Times* and Viacom. Enron had major business ventures with both media companies and paid several prominent journalists $50-$100,000 as consultants. Lest one object that perhaps publicly funded news providers should have been immune, McChesney notes that Enron was also an underwriter for a six part PBS series on globalization (2008).

Even if journalists had the funding and could avoid conflicts of interest in reporting wrongdoing of potential advertisers, in text advertising could further warp their intended message. Fox, for instance, embeds hyperlinks to ads in various words in its online content. Fox contends that this is acceptable because reporters do not know which word will by hyperlinked in advance, but it is easy to see how these hyperlinks could be used to blunt or undermine whatever journalistic pieces were critical of advertisers (Beard 2007). If a reporter critiques food safety for Tyson chicken, but Tyson has an embedded hyperlink touting how safe it is, the news report is immediately undermined before it can even be read in full.

Finally, online communication must contend with the problem that the net itself is not entirely neutral. Companies can find a multitude of ways to pay to ensure that their perspective on their conduct and their products is linked more gratuitously throughout the web. Even Wikipedia could play into the hands of companies whose potential critics are living in third world countries where English is not dominant. While the worst effects of sweatshops and refusal to adhere to environmental safety standards are often abroad, victims who write and publish stories documenting harms in their indigenous languages may not be heard on Wikipedia since Wikipedia tends to prefer references written in English (Garfunkel 2008).

Solutions and Recommendations

Given the dearth of challenges facing anyone who attempts to market corporate social responsibility at all, it may seem impossible to do so online. Explaining to online consumers in a morally acceptable fashion how and why a company's products are of greater value because moral standards were met during production and marketing raises a number of challenges. While the challenges are real, the hurdles are not insurmountable. With respect to the worry that human beings are simply not skilled in making moral evaluations and establishing trust in online contexts, it is important to recognize that the Internet is a fairly

new phenomenon. It is also important, however, to remember that individuals and institutions constantly change and evolve. Even though it is difficult to form trust in online interaction, many people do, and many companies are clearly profitable online. If humans were not able to make decent decisions at all online, society as a whole would be far less able to engage in the amount of E-commerce already extent. But an awareness of how usual skill sets put in play during face to face interactions are undermined in Internet contexts is also important. For this reason, communicating corporate commitment to social responsibility can not be left entirely to website managers. People will trust a company that is represented in their community. Face to face interaction with company leadership may be requisite at various points if claims of moral praise will ever be fully trusted. A manager who shows up to the charity auction is doing more than fostering warm fuzzy feelings in the community at large; she is also putting herself out there in the community to hear moral critique and feedback. The vulnerability of face to face interaction will at points be necessary to establish the trust requisite for believing that a company really is morally responsible rather than merely playing to what seems popular. Marketing corporate social responsibility is complex and must be multifaceted. This shows that online marketing of CSR is one part of a much bigger puzzle, but it is necessary nonetheless. A corporate website is often the one place a company can afford to significantly elaborate on policies, providing a key part of the kind of contextualized information necessary for consumers to make informed moral judgments.

To the objection that online interactions are not taken seriously, it should be noted that online commercial transactions are taken seriously. For instance, while there was much initial mistrust of online banking, most bank costumers now participate in the practice. While it is important to recognize the temptation to shirk accountability more in online contexts, that does not mean the online medium of communication undermines all attempts at seriousness. Even Dreyfus, the philosopher who launched the critique of seriousness in online interactions, argues that people committed to a cause offline will see the Internet as one tool among many. Dean also notes that according to a survey of 159 producers of blogs online, 60% said that participation in online political forums lead them to be involved in at least one political gathering or protest. Since becoming more active online, 29% reported being more active in political acts and 63% reported the same level of commitment (Dean, 2009). While these numbers refer to individual bloggers, they do indicate that although many take commitments online less seriously than promises made face to face, Internet involvement is not without efficacy. People involved online also tend to be involved offline. Seriousness is not completely lost simply because individuals engage online as well.

A lack of seriousness in online claims of corporate social responsibility is probably an indication that the company lacks seriousness in meeting the commitment to CSR in general, not just online. Although specifically referring to the commitment to democracy and justice in society more generally, the following comments from Dean (2009) apply equally as well to corporate social responsibility: "the technology is offering new standards, platforms, and ways of expression. So we can study them and learn to better utilize them, but real change goes deeper and it comes from somewhere else" (p. 125). If a company is not committed to moral standards in its business practice, then that is beyond the scope of this chapter. This chapter is concerned with providing moral guidelines in marketing CSR online for companies that truly do embrace CSR.

A company that does not bother to ensure that complaints are read by anyone or reach the relevant corporate leaders is not a company committed to social responsibility, since an accountable moral agent is one that takes moral critique seriously. Nike, for instance, early on its attempt to deal with critics who frowned upon their use of sweat shop labor, invited Dartmouth graduates to tour some of its Vietnamese and Indonesian facilities for three weeks and paid for their trip. Nike also posted a virtual tour of some facilities in Viet Nam on its website along with a report made by the students. At the same time, however, Nike was aware that an audit by Ernst & Young of one of its subcontracted Vietnamese factories showed that it did not have potable water and that toxic chemicals onsite were up to 177 times allowable safety limits at another Vietnamese factory producing its products (Bell de Tiennes & Lewis, 2005). The problem in this case is not that Nike used the Internet to post the virtual tour. The moral problem in this situation is that Nike didn't take corporate social responsibility seriously. While select testimonials are fine for advertising shoes as products, there are higher standards for moral conduct. Just because one has behaved admirably in some cases, does not mean that one's behavior on the whole is morally acceptable. The fact that an individual did not steal from his neighbors 49 out of 50 times does not make him any less of a thief when he steals a car on his 50th opportunity. By the same token, even if 99.9% of its factories were in compliance, the few that were not meant that people's lives were put at risk so Nike could make money. Even one foreseeable death due to violation of safety standards is one too many. Here the moral mistake was applying the usual standards of advertising, when marketing CSR puts a greater burden of honesty and clarity on the part of the company. The same problem applies to Walmart's flogs. When it comes to meeting moral criticism head on, a handful of subjective testimonials (especially when they are financed by corporate officials) are morally inadequate. The requirements for honesty are much higher when it comes to communicating information crucial to determining moral praiseworthiness or blameworthiness for conduct. Companies that use the interactivity of the net to fill the complainer's inbox with ads also miss the point. While gluts of ads might work in selling shoes, they are not appropriate as a reply to moral censure. Interactivity should foster mutual comprehension and critical reflection, not silencing of questioning.

Objections concerning the difficulty of establishing an objective account of whether or not a company deserves moral praise and investment premised upon that praise are, however, more serious concerns. The glut of information available on the web makes it especially important that attempts to communicate corporate social responsibility online are clear. It can be tempting to just post every social audit with every single bar graph and loads of confusing acronyms when a company has all the space in the world on its website. But information needs to be digestible. Few consumers have the time or skills to go through every inch of detail. If the goal is to be socially responsible, transparency is important, but true transparency requires clarity and context. If a company is committed to CSR and wants the public to know it, they can not hide behind piles of figures and confusing details. A summary highlighting key points and directing those who want to know more to the appropriate locations on the site is what is needed. For this reason, it is better to charge CSR communication to individuals who understand that the requirements are different from standard advertising and PR. Companies can look to firms that specialize in communicating good corporate conduct rather than relying on professionals skilled primarily in puffery. They can also rely upon outside audits and partner with nonprofits to ensure that their examination of conscience is truly fair and objective.

It is also important to realize that links to outside analyses may themselves be brought into question. The press is not always the most neutral source, nor is the government or the academy. This problem is endemic to all epistemic journeys that involve politically and economically powerful entities. Yet despite this general problem, large institutions do seem occasionally to have been able to change for the better over time. Churches go through reform, governments are overthrown, and corrupt companies can and do get found out. If this was not the case, Enron would be a poster child of corporate virtue to this day. While it is important to realize the challenges to providing a truly objective account of corporate commitment to CSR, it is also important to realize that a reasonable case can be made. If a company can reference multiple outside observers, such as the press, non-profits or university researchers, who attest to their virtuous conduct, then that makes a stronger case for their corporate virtues or their vices. Marketers of CSR must avoid the Nike mistake of selective evidence from biased sources.

For companies that truly do hope to embrace standards of corporate social responsibility, it is also important to recognize that the task of marketing good corporate conduct may be more efficient with systemic changes. Truly morally good and socially responsible companies may find it in their best interests to support key government reforms that could make their continued existence more viable. The government could help by requiring social audits by independent firms for repeat offenders. Government could also work to encourage truly independent reporting and social critique by supporting measures. These measures could include better funding for government oversight agencies, speedier and more complete returns of Freedom of Information Act requests concerning companies involved with government contracts or campaign donations, revamping of FCC guidelines to ensure an independent and profitable press, acting on proposed Internet neutrality requirements already outlined by the FCC, and adequate funding of higher education to ensure that social critique is not co-opted by corporate grant funding. A virtuous moral agent can not exist in a vacuum. The same is true of virtuous corporations; they also need a social context that allows them to flourish.

FUTURE RESEARCH DIRECTIONS

While this paper has suggested a number of policy changes to encourage truly morally responsible advertising of good corporate conduct, it remains to be seen which of these suggested changes will be economically and politically viable. In the future, it will be important to document the effectiveness of various measures at communicating good corporate conduct. Virtuous companies can solicit feedback from those who visit their websites to determine which communicative strategies work best. When it comes to providing meaningful context for key claims, feedback from the public is probably the best way to ensure that the message is communicated with the requisite moral clarity. While this paper has suggested that government ought to do more to encourage objective analyses of corporate conduct, industry groups that work with international nonprofits might prove effective as well. Press cooperatives paid for by subscribers might also be useful in monitoring the flow of information, and we should explore other alternative means of funding journalistic watchdogs of business to avoid conflicts of interest. Finally, further empirical information concerning how online users access and process information about corporate commitment to CSR would be helpful in determining the most effective ways to establish trust and to disseminate information in a responsible manner.

CONCLUSION

While it is clearly challenging and difficult for companies committed to producing goods and services in a fashion that is socially responsible to be successful, the task is not impossible. Using the Internet to help consumers to understand and trust that a company is truly committed to corporate social responsibility is one very important tool in ensuring that socially responsible companies can compete and thrive. Online spaces allow companies an affordable means by which to provide detailed information and links to objective analyses by outside parties affirming claims of good corporate conduct. Online spaces also provide the opportunity to engage in dialogue with potential corporate critics concerning both the clarity of corporate attempts to communicate efforts towards social responsibility and the validity of those claims. Companies do need to be careful to watch out for the ways in which online communications present challenges to creating trust with consumers and critics, as well as challenges in establishing the truth and seriousness of corporate commitment to social responsibility. Corporate actors who take advantage of the special problems associated with communicating online can make it exceedingly difficult for morally responsible companies to have their message efficaciously heard. But with an adequate awareness of both the problems and benefits associated with online attempts to communicate good corporate conduct, socially responsible companies can find ways to be competitive in the marketplace and truly to do well by doing good.

REFERENCES

Beard, F. (2007). The ethicality of in-text advertising. *Journal of Mass Media Ethics*, *22*(4), 356–359.

Bel de Tienne, K., & Lewis, L. W. (2005). The pragmatic and ethical barriers to corporate social responsibility disclosure: The Nike case. *Journal of Business Ethics*, *60*(4), 359–376. doi:10.1007/s10551-005-0869-x

Bennett, A. (2008). Consumers are watching you: Ignore the role of the consumer in corporate governance at your own risk. *Advertising Age*, 19.

Carter, M. (2007). Internet Advertising: Video/ethics: Online ads must clean up their act. *The Guardian Supplement*, 6.

Castelfranchi, C. (Ed.) & Tan, Y. H. (Ed.). (2001). Trust and deception in virtual societies. New York: Springer.

Dean, J. (2008). Communicative Capitalism: Circulation and Foreclosure of Politics. In Boler, M. (Ed.), *Digital democracy and media: Tactics in hard times* (pp. 101–121). Cambridge, MA: The MIT Press.

Dreyfus, H. L. (2009). *On the internet* (2nd ed.). New York: Routledge.

Fernando, A. (2007). Transparency under attack. *Communication World, 24*(2), 9–11.

Garfunkel, S. L. (2008). Wikipedia and the meaning of truth: Why the online encyclopedia's epistemology should worry those who care about traditional notions of accuracy. *Technology Review, 111*(6), 84–86.

Kapelus, P. (2002). Mining, corporate social responsibility, and the community: The case of Rio Tinto, Richard's Bay Minerals, & Mbonambi. *Journal of Business Ethics, 39*(3), 275–296. doi:10.1023/A:1016570929359

MacDonald, C. (2008). *Green, Inc.: An environmental insider reveals how a good cause has gone bad. Guilford.* CT: The Lyons Press.

McChesney. R. W. (2008). The political economy of media: Enduring issues, emerging dilemmas. New York: Monthly Review of Foundation.

Spence, E. (Ed.), & Van Heekeren, B. (Ed.). (2005). Advertising ethics. Upper Saddle, NJ: Prentice Hall.

Stoll, M. L. (2002). The ethics of marketing good corporate conduct. *Journal of Business Ethics, 41*(1), 121–129. doi:10.1023/A:1021306407656

Stoll, M. L. (2008). Backlash hits business ethics: Finding effective strategies for communicating the importance of corporate social responsibility. *Journal of Business Ethics, 78*(1-2), 17–24. doi:10.1007/s10551-006-9311-2

Section 4
Privacy and Property Rights Online

Chapter 10

Privacy Revisited:
From Lady Godiva's Peeping Tom to Facebook's Beacon Program

Kirsten Martin
The Catholic University of America, USA

ABSTRACT

The underlying concept of privacy has not changed for centuries, but our approach to acknowledging privacy in our transactions, exchanges, and relationships must be revisited as our technological environment – what we can do with information – has evolved. The goal of this chapter is to focus on the debate over the definition of privacy as it is required for other debates and has direct implications to how we recognize, test, and justify privacy in scholarship and practice. I argue privacy is best viewed as the ability of an individual to control information within a negotiated zone. I illustrate this view of privacy through an analysis of Facebook's Beacon program and place the case in the context of both privacy violations and successful business strategies. I find privacy zones are illuminating for situations from 10th century England to current social networking programs and are useful in identifying mutually beneficial solutions among stakeholders.

INTRODUCTION

After months of Lady Godiva's lobbying for tax reform for their village of Coventry, her husband Leofic, Earl of Mercia, exasperatingly challenged Lady Godiva to ride naked through the town center before all the people in order to get her wished tax relief. Lady Godiva immediately contacted the great magistrates of the city and informed them of Leofic's challenge. Given the dire economic status of the town, these community leaders agreed to have all citizens of Coventry return to their homes and remain behind closed doors so as to not lay eyes upon Lady Godiva during her ride through the center of town. At noon on the appointed day, Lady Godiva let down her hair which covered her in a semi-modest degree, mounted her horse and, accompanied by two knights, rode through town. As was agreed upon, the roads were clear and the market was eerily quiet from the absence of barter

DOI: 10.4018/978-1-61520-615-5.ch010

and negotiations. Suddenly, Lady Godiva was notified of an errant young man peeping through a window by her horse's neigh. The knights soon realized that the young man was Tom the Tailor who was immediately struck blind as some sort of divine punishment but not before he was able to turn and tell his tale to others in the house.[1]

While the notion of privacy has been a focus of concern for centuries, recent developments have changed how we approach acknowledging privacy. Peeping Tom violated privacy norms by peering through a window to view a disrobed Lady Godiva riding through town, whereas current privacy concerns are more likely to arise in a virtual world such as SecondLife rather than in a township, or by compromised by a cyber-voyeur, hacker, on computer program rather than by an individual peering through a window. Two related trends have necessitated a revisiting of how we acknowledge privacy. *First, information is increasingly the basis for an organization's value proposition and, for some, the entire business model.* While business has always relied upon consumer information to complete transactions, more of that information is stored or being repurposed for behavioral marketing or custom recommendations. All organizations have stakeholders such as employees, suppliers, governing bodies, and communities who share information in increasingly important ways.

Second, our information is both greased and sticky. By becoming separated from us, our information is 'greased' in the words of technologist James H. Moor (1997). Information can slip quickly from protected, legitimate storage to an unprotected, illegitimate individual through the click of a mouse. At the same time, I argue that information is 'sticky' in that organizations can link and aggregate information which was previously separated and compartmentalized. As such, our information is increasingly in permanent records which are searchable rather than only observable (Lessig, 1998, 1999). Where department stores previously recorded customers' preferences on

index cards to notify them of upcoming sales, grocery chains now store previous and current purchases to customize coupons and Amazon. com monitors browsing and purchases to suggest additional products. Where Lady Godiva briefly rode through town at noon on an appointed day, Facebook retains information to be linked to, searched, revealed, or repurposed for an indeterminate amount of time. Perhaps the type of information requested and stored is similar, but the format of the information is different in important ways.

These shifts make the information both *valuable* and *vulnerable* to organizations and individuals: valuable by facilitating customized services and lowering transaction costs and vulnerable by being identifiable and searchable thereby creating a larger target for illegal hacking or unintended violations of privacy. As I will argue, these trends point to the control view of privacy as being increasingly useful and applicable: when our information was 'attached' to us as in the case of Lady Godiva in the beginning of this chapter, the ongoing control of information was not important to our conception of privacy as we controlled our information by controlling who has access to ourselves. This is not the case now. Now, our information is not contained by physical barriers but rather negotiated through virtual barriers: the equivalent to asking a friend to turn their back while you change or asking your child to cover their ears while you talk to another adult.[2]

The goal of this chapter is to develop an understanding of privacy for business ethics as a common concern among stakeholders. To do so, I focus on the debate over the definition of privacy as it is required for other debates and has direct implications to how we recognize, test, and justify privacy in scholarship and in practice. I argue that privacy is best viewed as the ability of an individual to control information within a negotiated zone or space. I illustrate this view of privacy through the case of Facebook's Beacon program and analyze the case in the context of privacy zones and contemporary examples of both privacy violations and successful strategies to protect stakeholders' privacy. While the tactics we employ to protect privacy have evolved, previous work on privacy need not be rendered moot. STS scholar Deborah Johnson refers to this as the genus-species account of moral issues: we may find multiple and evolving applications of privacy while not necessarily changing the genus of privacy (Johnson, 2006).

Privacy scholarship breaks into a series of overlapping and interconnected debates seeking to identify harms, justify the acknowledgment or existence of privacy, or parse privacy rights. Yet these debates, such as the justification, rights, cohesion, and substitution debates, rely upon a definition of privacy: the scope of privacy and/or privacy violations is a fundamental assumption to all subsequent debates. In other words, an argument seeking to justify the acknowledgement of privacy is predicated on a shared definition of the scope of privacy; a discussion about minimal privacy rights secured in the law similarly necessitates a definition of privacy. Thus, in this chapter I focus on the scope or definition debate as it pervades other theoretical debates, provides the basis for research assumptions, and has direct implications for how we recognize, test, and justify privacy in scholarship and in practice. I do not dismiss or ignore these ongoing debates; rather, these debates serve to highlight the benefits and harms, the stakeholders involved, and the pervasiveness of privacy concerns. I discuss these alternative debates at the conclusion of the analysis in light of the arguments of this chapter. In addition, I steer away from conventional approaches which seek to identify privacy minimums by examining privacy as if the parties involved seek mutually beneficial solutions similar to Lady Godiva and the magistrates. As such, I use pragmatic scholarship from legal scholars such as Rosen (2001) and Scolove (2006) for their philosophical reasoning rather than their prescriptions for laws and regulations.

BACKGROUND

The concept of privacy can appear nebulous as teenagers now post all of their activities on Facebook only to declare their lives private, and managers bemoan employee surveillance while implementing behavioral marketing for their online customers. As I illustrate below, seemingly inconsistent strategies, research, or behavior may be grounded in a narrow understanding of privacy. Here I outline two approaches to the scope of privacy—the *control* and *restricted access* versions of privacy—and introduce the mechanism of information or privacy zones as useful to conceptualize privacy within business ethics. As seen in many nuanced arguments for and against each view, how we define privacy has implications not only for our theoretical justifications of privacy, but also in how we conduct research on expectations of privacy and how we navigate privacy in practice.

Privacy as Restricted Access

The restricted access view of privacy holds that privacy is "fundamentally about protection from intrusion and information gathering by others" (Travani and Moor, 2001, p. 6p. and, therefore, requires a degree of inaccessibility of individuals and their information from the senses of others (Allen, 1988; Reiman, 1995; Brin, 1998). According to this view, privacy is protected when information is hidden, and privacy violations occur when information or an individual is revealed. This view of privacy is easy to detect and model as information is either hidden or not hidden, revealed or not revealed. As an added benefit, once the information is revealed, privacy norms do not apply since the information is no longer private and the individual can hold no expectation of privacy.

An immediate reaction is to argue that, according to the restricted access view, individuals are always private since one is always inaccessible to others to some degree (Elgesem, 1999) or, in a related vein, that individuals are never completely inaccessible and therefore never completely private. However, these arguments would hold for both the control and restricted access version of privacy (i.e. one can never be in complete control, ergo, one can never be in total privacy) and are therefore not compelling or helpful in differentiating the two. However, the restricted access view's problems are most clearly illustrated through the belly dancer and prisoner accounts. These problems require some to take what I consider extreme measures to make the case for the restricted access version of privacy.

Privacy without Restricted Access: The Belly Dancer Account

The first problem with the restricted access version of privacy is that we frequently give access to information or reveal ourselves while still retaining an expectation of privacy; restricted access is not necessary for a notion of privacy. We have private conversations, act in ways that we expect to remain private ('what happens in Vegas, stays in Vegas'), or ride naked through a village marketplace as in the case of Lady Godiva. I refer to this as the belly dancer problem: when a belly dancer performs, she reveals herself and gives others access to information about her belly; yet she retains an expectation of privacy in that her belly is not available for public access or surveillance once she decides the dance is over. Further, she is able to negotiate rules, justifiably in my estimation, that no image of her belly can be recorded and leave the privacy zone she has created through a social contract with her guests. Similarly, when Lady Godiva gives her husband access to her body, the case is not a privacy violation nor would he be correct in assuming that the information about her body is public based solely on his access to it.

The line of reasoning which has the belly dancer (or Lady Godiva) *without* any expecta-

tion of privacy relies upon the idea that every dissemination of information is a loss of privacy. This argument is logically extended to conclude that those who argue for privacy must be hiding something (Scolove, 2007) or are hypocritical (Wasserstrom, 1978). Accordingly, economists develop a choice in theory and in research—a false choice, I believe—of releasing information versus protecting information (Acquisti, 2002; Acquisti and Grossklag, 2004). The focus on restricting access becomes a problem of keeping information secret (Scolove, 2007) thus leading to the determination that privacy is inefficient due to fraud and misrepresentations (Posner, 1981) and supports discussions such as "The Economics of Privacy as Secrecy" (Hermalin and Katz, 2005) and the positioning of the rights of society to know information about individuals (Singleton, 1998) for the good of the community (Brin, 1998). As demonstrated by its advocates, such a view of privacy lacks practicality; a definition of privacy is not sustainable or useful if recognizing it necessitates the breakdown of a community or a market or positioning the claims of the individual as subsumed to the needs of an efficient market or a good society.

Restricted Access without Privacy: The Prisoner Problem

Second, an individual can have access restricted without realizing a private situation. Consider a prisoner in a cell who is behind locked, solid doors. The restricted access view of privacy would have the prisoner in a private situation with the guard holding the key since access is technically restricted. Yet, prisoners and prisons are traditionally used to illustrate the harms of a lack of privacy (e.g. Foucoult, 1977; Reiman, 1995); a definition of privacy which implies prisoners are in a private situation becomes theoretically inconsistent. Two extreme and unsustainable reactions attempt to solve the prisoner problem. First, privacy could be relational—as in the prisoner is in a private

Figure 1. Zone of privacy

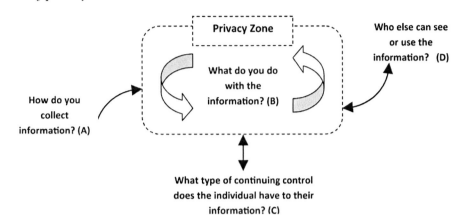

situation with respect to certain people but just not the guards (Allen, 1998). Second, the threat of privacy violations could have a similar if not identical effect to actual privacy violations (Reiman, 1995); however, threats of privacy violations come when an individual does not have control over access to themselves or their information and we quickly fall into the control version of privacy as described below.

Perhaps at the most basic level, most individuals do not subscribe to this version of privacy where every dissemination of information is a loss of privacy.[3] Individuals have an expectation of privacy when they have handed over information as in the case of medical records, vacations in Vegas, or having a conversation over coffee. Thus, research on privacy within business mistakenly attributes individuals as relinquishing privacy when sharing information.

Privacy as the Control of Information

Alternatively, the *control* view of privacy builds on the ability of individuals to determine when, how, and to what extent information is communicated and used. Privacy is not merely the absence of information accessed by others, it is the control we have over our information: the manner in which information about an individual is accessed, dis-

seminated, searched, and communicated to others (Westin, 1967; Elgesem, 1999; DeCrew, 1997; Fried, 1984). As stated perhaps most famously by Westin, "privacy is the claim of individuals, groups, or institutions to determine for themselves when, how, and to what extent information about them is communicated to others" (1967, pg. 7).

One benefit of the control version of privacy is that information can be transferred or revealed to another party without losing the notion of privacy. Even as information becomes separated from us, e.g., is logged into a hospital system or is posted within a defined community within a social networking site, we still retain an expectation of privacy according to this view. In practice, with the advent of information becoming both separated and searchable, control of information no longer equals holding the keys to the locked room or closing the blinds. We have virtual rooms or zones (Moor, 1997; Scolove, 2006). Individuals let information into certain zones and are able to control through agreements when it leaves that zone. As illustrated in Figure 1, individuals are better able to protect their privacy if they know (a) the boundaries of privacy, (b) the uses of information within those boundaries (c) their ability to control and monitor the use of information, and (d) under what conditions information enters and leaves the zone. Lady Godiva created such a space in her agreement with the citizens to turn their backs so as to create a virtual privacy zone within the middle of the village market where her information would remain private. Peeping Tom pierced the veil of that zone by opening the window to view Lady Godiva and sharing what he saw with others.

Some scholars writing on privacy have taken a similar approach to recognizing private areas (Simmel, 1950; Samarajiva, 1997), spaces, (Scolove, 2006; Jiang, Hone, and Landay, 2002), spheres (Rosen, 2001), situations (Moor, 1997), or zones (Moor, 1997). For most, the space or sphere surrounds an individual—either the target of the information or an individual holding the information—and the goal of inquiry is centered on identifying harm of the target or the responsibility for the 'data holder' (e.g., Scolove, 2006). The use of information zones or spaces within computing shifts the use of zones from centering on the individual to a virtual space which can exist without an individual—an important notion with information being housed in negotiated zones which may have multiple parties involved (such as the citizens of a village in the case of Lady Godiva) or multiple stakeholders impacted by that information. Two interesting dissents from the control approach to privacy attempt to demonstrate that information control is not necessary or sufficient for privacy yet inadvertently serve to further highlight the utility of the privacy zone metaphor.

Control without Privacy

The cases where individuals may reveal particular information, such as medical records (Travani and Moor, 2001), or reveal all information (Schoeman, 1984) are positioned as exemplary of individuals who can maintain control of information without an expectation of privacy; this counter argument attempts to maintain that individuals control who has access to information yet cannot simultaneously claim the information is private. However, in the former case, control is maintained through

a surrogate or second party who oversees the zone of privacy agreed to by the individual in much the same manner as I maintain security of my house by contracting with a security company. I still expect security even if I ask someone else to help maintain it, and an individual still can expect privacy when handing over information to a doctor, a lawyer, a close friend, or even a good waiter. In the case of the latter, we may not agree with someone who reveals 'too much,' but that does not necessitate a loss of privacy over other information or forgo the possibility of negotiating the terms of a zone of privacy. It should be said that no one really knows if someone has revealed *everything*—only the individual knows what everything is and if they have revealed it (barring some sort of invasive scanning or interrogation technique, which would be a violation of control and, ipso facto, of privacy by any standards). Rather, this particular dissent becomes a paternalistic argument as to what type and quantity of information is appropriate to share and with whom.

Privacy without Control

A second type of dissent from the control view of privacy attempts to identify situations where individuals are in a private situation without control over themselves or their information. For example, we might consider cases such as an individual on a desert island (Schoeman, 1984) or the impossibility of actually retaining complete control over information (Moor, 1997) to demonstrate that control of information cannot be necessary for privacy. However, individuals do maintain their privacy, if they are in control of their existence on a secluded island, by selecting their zone of privacy. If I fly to a desert island, and can return when I please, I am in a private situation; if I go to a secluded room and return when I please, I can also expect to be in a private situation. Note the similarity to the prisoner problem above: both have individuals secluded with varying degree of restricted access and are therefore considered in a private situation. According to the control approach, the prisoner described above would not be in a private situation when the guard held the key but would be in a private situation if the *prisoner* held the key—a similar analysis to the desert island dilemma (Schoeman, 1984). Counter to these dissents, and key to the control view of privacy, if an individual consents to any information flow–either access by a third party or proactive revelation by the individual—there is no privacy violation (Scolove, 2006, pg. 484); we may not like the type or amount of information revealed, but the individual does not necessarily give up their claim for privacy.

Summary

The differences between the control and the access view of privacy are important as they shape what we see as violations, what we do in practice, and how we conduct research. First, the control view of privacy, particularly with the use of privacy or information zones, fits with our common conceptions of privacy: we regularly give access to information while still retaining an expectation of privacy. We share information with friends, colleagues, and businesses with an expectation that the information will stay within a commonly agreed upon, zone. In other words, one can claim an access view of privacy—such as when AT&T announced that they would use consumer information as they saw fit as the information was 'theirs'—however, stakeholders are free to disagree (AT&T revises privacy policy, 2006). This is a descriptive argument for the control view and is exemplified in the many examples (and failed counter examples) such as the belly-dancer problem and the prisoner problem.

However, more importantly, the control view of privacy supports the values we use to justify acknowledging privacy. Within the debate as to how to justify privacy, privacy is supported by identifying the many intrinsic values—such as autonomy (Johnson, 2001; Moor, 1994, 1997)

and dignity (Rosen, 2001)—and instrumental benefits of adequately maintaining a notion of privacy—such as for self creative enterprise (Rosen, 2001), to develop a unique personality (Smith, 1979), to try new ideas (Rosen, 2001), to have intimacy (Elgesem, 1999), to develop relationships (Rachels, 1975), and in support of a free society (Reiman, 1975). I argue that these values are supported when individuals have control over their information and not when their information is merely hidden or inaccessible. One cannot develop a unique personality or try new ideas while making themselves inaccessible. Similarly, one main avenue to developing different levels of intimacy and types of relationships is by sharing information with different individuals (Elgesem, 1999; Rachels, 1975). A form of privacy which only ensures that individuals make themselves and their information inaccessible does not support these benefits or values.

In addition, individuals can be harmed through violations of privacy—such as a loss of freedom through controlled behavior and choice (Reiman, 1995), incurring an insult to dignity (Reiman, 1975; Rosen, 2001), being misrepresented or judged out of context (Rosen, 2001, pg. 21), and even psycho-political metamorphosis (Reiman, 1995, p. 42) where individuals act/think conventionally and lose "inner personal core that is a source of criticism of convention, of creativity, of rebellion and renewal." Yet the access-view of privacy would have an individual, say a prisoner, who is the prototypical example for the harms of privacy violations, in a private situation. These harms are incurred as an individual loses control of their information and not when they grant access to this information. We may have a fear or run the risk of being misrepresented or insulted when we share information, as when talking to a friend or performing as a belly dancer, but that risk exists due to the lack of control over that information and not from granting access.

As illustrated in the Facebook example below, the access view of privacy stops at initial data collection (a)—once information has been revealed, it is no longer considered private (see Figure 1). This greatly limits the responsibility of organizations to maintain stakeholders' privacy as their privacy obligation ends at the point of revelation. The control view suggests that individuals are concerned about not only (a) how information is gathered, but also (b) the purpose and use of the information, (c) their control of the information, (d) third party access to the information (see Figure 1). I turn now to illustrate the application of privacy zones in a particular example: Facebook's Beacon program.

EXAMPLE: FACEBOOK AND BEACON

In November 2007, the social networking site, Facebook, offered a free tool to online partners called Beacon to track the activity of Facebook members on partner sites and proactively broadcast such off-Facebook activities to their designated Facebook 'friends' (Jesdanum, 2007). Beacon worked by being embedded in a partner's web site, such as Blockbuster, The New York Times, or Overstock.com, and gathering behavioral information which was then sent to Facebook to process as either an alert to designated friends or as an item to ignore. Considered at the forefront of online advertising, Beacon was hailed as a mechanism to target potential customers based on their social network through a friend's implied recommendation.

However, Facebook members had difficulty understanding and navigating their role in Beacon (Nakashima, 2007). A notice to opt-out of Beacon broadcasts appeared in a small window and disappeared without Facebook users taking any action, thereby leaving Beacon broadcasts to be sent to the user's identified friends by default. Further, Facebook users were not given the ability to reject all sharing, and the notification window appeared (and disappeared) every time the user entered a partner site (Perez, 2007a; Jesdanun, 2007).

That Facebook faced and continues to face privacy issues is not in question here: stakeholders of Facebook and the Beacon program raised privacy concerns through blogs and newspaper articles. However, views of privacy which rely upon legal regulations or actionable harms could miss these privacy concerns as (1) these violations were not illegal and (2) all information was revealed by individuals on Facebook or on the partner site. Many famous privacy violations emanate from the improper handling of information that has been previously revealed; and Facebook's Beacon program gathered information from retailers' web sites where consumers assumed it would stay.

While much of the focus has been on Beacon as a 'Peeping Tom,' the following analysis of this case illustrates that (1) Facebook's most egregious mistake centered on the ability of users to be notified and make decisions as to the repurposing of their social network and (2) the partners who willingly implemented Beacon may have been the most underreported offenders.

The Role of Facebook

The initial collection of information and third party access to information, the classic definition of a privacy violation in the access-view of privacy, were not issues for Facebook. In fact, Facebook members voluntarily revealed their information within different Facebook 'zones' with a great degree of notification (Facebook). Rather, Facebook's transgression was in not allowing their members to control the use of their networking information.

First, the initial purpose of members revealing their information on Facebook, and subsequently linking friends to that information, was to share the information in order to build relationships; yet, Beacon leveraged information, contacts, and social networks in order to market third party products. The secondary use of their information as a vehicle for marketing distribution was too far afield from the primary, agreed upon use of the members' data. Second, the role of users was severely limited in the ongoing use of their information due to Facebook's anemic approach to alerting users of Beacon (Nakashima, 2007), the decision to use an opt-out strategy,[4] and the absence of an option to turn off Beacon permanently. In effect, Facebook did not allow members to maintain a relationship with their information.

Fortunately, Facebook's remedy focused on fixing their members' ability to control their information. Within days of the uproar reaching the mainstream media, Facebook decided to give users the option of permanently turning off Beacon for their 55 million users and apologized for their mistakes (Liedtke, 2007). Where previously consent was assumed, Facebook would now ask members to opt in to the service and would not automatically store information from third-party partner sites. In addition, Facebook members are now asked to allow the broadcast of their activity before their off-Facebook activity is sent to their friends thus shifting to an opt-in notification. Even with improvements, Facebook members still were not offered the ability to easily disable the Beacon service. Further complicating the situation and infuriating online communities, users were not informed that data on their activities was *always* flowing back to Facebook nor given the option to block that information from arriving at Facebook; Facebook merely promised to disregard or not use certain information. In fact, if a Facebook member ever decides to have their computer 'remember' their login information, Facebook could tie activities from third-party sites even if the user was logged off Facebook or had opted out of the Beacon broadcast.

The Role of Partners

Little attention was paid to the Facebook partners who voluntarily implemented Beacon. These partners, such as Blockbuster, Sony Online En-

terntainment, eBay, The New York Times, and IAC, took a wide range of approaches to the adoption of this surveillance technology (Perez and Gohring, 2007). As became apparent through journalistic inquiries and the persistence of the online community, Beacon captured detailed data along with the IP addresses of *all visitors* on a partner site—Facebook members and non-Facebook members—and determined whether or not to store and broadcast the information once the tracking information was sent back to Facebook (Liedtke, 2007). Partners controlled if and how the Beacon program would work for them. As advertised on Facebook's website, a partner could "add 3 lines of code and reach millions of users" (Beacon).

Similar to our analysis of Facebook's role, the behavioral and purchasing information on the partner site was freely provided by the user as they shopped. However, users did not expect their purchasing information on overstock.com, for example, to be sent to their Facebook friends: media outlets repeated the story of a woman whose engagement surprise was ruined when she was notified of her boyfriend's purchase of a diamond ring on overstock.com (Nakashima, 2007). Even after modifications, Facebook decided not to modify Beacon's ability to "indiscriminately track actions of all individuals on partner sites which implemented beacon" (Perez, 2007c). This facet of Beacon was referred to as 'broad user tracking' as Beacon captured "addresses of web pages visited, IP addresses, and the actions taken on the site" of nonmembers and members of Facebook and deletes the data upon receiving it (Perez, 2007b). The partners' decisions to repurpose behavioral and purchasing information for marketing and to provide that information to Facebook (a third party) violated the users' expectations of privacy by reusing the information provided and sending the information outside the privacy zone.

Not all partners implemented Beacon without changes. Overstock.com stated, "We have a specific threshold that the program needs to meet, in

Table 1. Privacy violations with Facebook's Beacon program

	Facebook	Partners
a. Collection of information	Facebook member provided links to friends.	User provided behavioral and purchasing information by navigating the partner's web site.
b. Use of information	Links repurposed for marketing by Facebook. Original purpose for members to communicate with friends.	Behavioral information repurposed for marketing by partner site. User used as 'spokesperson' for products and services without their consent.
c. Continued relationship with information	Members initially not adequately notified nor given the ability to opt out of the program.	Users not given the opportunity to (a) not have their information collected or (b) not have their information reused.
d. Third party access to information	No third party received information on member-friend links.	Facebook given access to behavioral information. Member's friends received broadcasted information on browsing and purchasing behavior.

terms of privacy, before we'll be turning it back on" (Perez and Gohring, 2007). Others opted to trust Facebook to delete the information they sent back via Beacon thus still allowing user information to leave the partner's zone. Kongregate, an online gaming site, used the program to track games people play but not other activities on their site. In a similarly nuanced installation, Six Apart asks their users to opt-in and only inserts the script for the Beacon program at that point. In other words, Six Apart turns off Beacon for users who ask not to be included so that their information is never collected or sent to Facebook. Ebay also uses Beacon in a limited fashion by applying the program to sellers only and using an opt-in strategy.

Much has been publicized about both voluntary and involuntary penetrations of information zones through third-party access and surveillance. GE Money lost a digital tape with customer information (Morning Edition); an intruder gained access to a computer that contained customer information of millions of TJX shoppers (Dash, 2007); AOL researchers released three months' worth of users query logs to public which contained personally identifiable information (Hafner, 2006). The decision of partners to release behavioral information is a similar violation of privacy and should be categorized with similar breaches of privacy in business.

Beacon Summary

If we relied upon the access-view of privacy, information which was provided by users (on partner sites) or members (on Facebook) could not be considered private. Yet, the blogging community, members, users, the press, and all other stakeholders of Beacon viewed the Beacon program as violating privacy. As summarized in Table 1, Facebook's mistakes in the initial release of Beacon amount to improper notification of members as to the repurposing of their Facebook relationships. However, the role of some partners was much larger. Partners reused behavioral information for marketing purposes and sent the information to multiple third parties.

Just as Peeping Tom broke the negotiated zone to view Lady Godiva, these partners took information from a negotiated zone and provided it to others in a similar manner. Facebook may have been the public face for this story as the developers of the program, for which they bear some responsibility for the design and use of that program. However, the partners' decisions as to if and how to implement Beacon violated the privacy of Facebook and non-Facebook members and, thus, were the greater violators of privacy. These partners, as the guardians of behavioral information, became the equivalent of Lady Godiva's knights turning and telling the tale of Godiva's ride through the marketplace.

FUTURE RESEARCH DIRECTIONS

In this chapter, I break from conventional approaches looking for privacy minimums and examined privacy as if the parties involved seek mutually beneficial solutions similar to Lady Godiva and the magistrates. Specifically, I argued that privacy is best viewed as the ability of an individual to *control* their information within a particular privacy or information zone and illustrated this view of privacy through the example of Facebook's Beacon program and analyzing the role of multiple stakeholders in violating and securing privacy zones. I turn now to consider the implications to privacy scholarship and practice.

Implications to Theory: Alternative Debates

Within privacy scholarship, scholars have identified a series of problems to solve, debates to enter, and ways to justify privacy (See Table 2). Where the cohesion debate attempts to categorize privacy harms with varying degrees of commonality versus a mere "hodgepodge" of infractions (Rosen, 2001; Alderman and Kennedy, 1995), the rights debate posits the individual against society—or, as Warren and Brandeis famously stated, "against the world" (1890, p. 10)—to identify minimal rights to be protected. In particular, I have leveraged Scolove's (2007) deft handling of the interminable rights debate. Scolove argues privacy is not merely instrumental to society or only intrinsic to the individual but a concept with benefits and harms to all—the individual as a member of society and society as a collection of individuals. Such is the approach taken in this chapter by assuming privacy to be of a mutual concern of individuals

and organizations in the particular case rather than a contentious argument where party 'wins' and another loses.'

The debate over how to justify privacy by identifying the many benefits of recognizing and harms in violating privacy is leveraged in assuming that privacy is important and valuable to different parties. The justification or value for these concerns, it is assumed, has been taken into account by those stakeholders and by Facebook in attempting to navigate the concerns. Rather than taking an antagonistic stance to debating whose 'rights' trump the other, Facebook attempted to find mutually agreeable strategies which met the concerns of multiple parties. Both debates are useful endeavors when identifying commonality among court cases (Prosser, 1960) or moral minimums for future regulation, but these debates may be misplaced given the approach taken in this chapter to highlight opportunities and vulnerabilities.

Alternatively, the substitution debate attempts to build up or substitute privacy with an amalgamation of individual and societal rights, which equate to the right to privacy; this argument attempts to demonstrate that we do not really have a right to privacy, we just call it privacy. For example, rather than a right to privacy, an individual has a right to autonomy with the exchange of information being governed by intellectual property (Samuelson, 2000; Singleton, 1998), information asymmetry (Jiang, Hong, and Landay, 2002), and social contract norms – the latter three concepts being summarized as privacy as fair information privacy (Bennett, 1992). Facebook respected information asymmetry in some circumstances (such as the initial notification of privacy rules and tactics) and not in others (such as the ongoing notification and opt-out policy of Beacon). Facebook respected the control of information without the need for claiming property rights and continued to negotiate their social contract

Table 2. Implications to current privacy debates

	Focus of inquiry	**Examples**	**As contributing to this chapter**	**As impacted by this chapter**
Scope Debate	How do we define privacy to use in theory, research, and practice?	Moor (1994); Schoeman (1984)	Focus of chapter.	Privacy is defined as the control of information inside a particular zone.
Justification Debate	What are the benefits to protecting privacy and the harms of privacy violations?	Reiman (1995); Rosen (2001); Elgessem (1999); Rachels (1975)	Privacy is a highly valued concept with many concerned parties. Supports privacy as a mutual concern rather than a point of contention.	Privacy as controlling information supports the intrinsic and instrumental values we associate with privacy (whereas, privacy as restricted access does not).
Rights Debate	How should a right to privacy be protected? Whose interests are we protecting?	Moor (1997); Johnson (2001) Warren and Brandeis, (1890); Brin (1998)	Approach to privacy as an interest of both individual and society. Privacy is something a community acknowledges to the benefit of the individual and the community.	Many stakeholders and communities benefit from being able to control their information through the acknowledgement of privacy zones.
Cohesion Debate (Schoeman, 1984)	Do the harms we are trying to avoid have anything in common or is privacy an umbrella term for a loosely related set of problems?	Schoeman (1984); Alderman and Kennedy (1995); Scolove (2006); Prosser (1960)	Not utilized in this chapter as the debate focuses on identifiable harms as tying together a notion of privacy.	A focus on harms is not necessary for a discussion about privacy.
Substitution Debate	Is there a right to privacy? Can we piece together other norms, rights, or values to substitute for privacy?	Schoeman (1984); Reiman (1995); Thompson (1975)	Highlights points of vulnerability or opportunities for creating value within a model of privacy zones.	The control view of privacy allows for acknowledging privacy claims which do not rise to the level of intellectual property claims.

with their stakeholders through blogs and press announcements including an apology.

While I argue privacy is distinct, useful (James, 1931), and not sufficiently substitutable, such examinations remain important by highlighting points of vulnerability or opportunities for creating value within our model of privacy zones. Attempts to build up the notion of privacy from different individual rights do a disservice to both our notion of privacy and those substitutable rights. The use of privacy zones and the control view of privacy makes these substitutable 'rights' illuminating potential points of vulnerability, such as the continuing control of information or full disclosure when gathering information; a discussion with stakeholders as to the unreasonableness of their privacy claims may be short lived.

As noted above, these debates shift in light of the arguments of this chapter. Some debates are transcended, such as the cohesion debate and

the rights debate, due to the approach and scope of this chapter to examine privacy as a common concern or source of social friction (Coase, 1960) to be resolved between interested parties. This argument assumes that parties can resolve differences through private ordering rather than through regulations. However, legal scholarship would have something to say about the underpinnings required to negotiate the privacy norms within a particular information space. Such social contract minimums such as informed consent and a right of exit could be considered moral minimums to negotiating privacy.

Yet other debates have been leveraged in this chapter to identify the overall value of privacy to many parties (the justification debate) or to highlight points of vulnerability and opportunity with the privacy model offered here (the substitution debate). All debates, research, and scholarship on privacy necessitate a definition of privacy and are therefore impacted by the treatment in this chapter.

Implications to Practice: From Lady Godiva and to Beacon

One strategy being pursued by organizations is to limit the quantity and length of time information is held within a privacy zone. Some search engines are limiting their exposure to privacy violations by deleting search results and user behavior within months; Google has taken a different tack by retaining data for two years (Lohr, 2007). Both companies and stakeholders benefit from balanced information collection and retention polices. Companies who retain information are vulnerable to violations of privacy through the misuse of information or third-party surveillance (Zeller, 2006). Yet, Facebook's current policy of indiscriminately collecting information through Beacon in addition to recent concerns as to the ability of members to delete their information from Facebook upon quitting (Rampell, 2008) combine to create privacy vulnerabilities to the organization and their members. Holding on to personal information within a negotiated zone of privacy places a burden on Facebook to maintain the security of that data. A definition of privacy—such as the access view—which suggests that individuals either provide necessary information for business transactions and relationships or retain their privacy does nothing to identify how business and stakeholders can develop a strategy to do both. By controlling information in privacy zones, individuals are able to both share and protect their information. Business can take steps in thinking about and managing these issues within an organization. How should organizations partner with individuals to share information with respect? How can technologies be designed to accommodate privacy? How can organizations put control in the hands of individuals to preserve their notion of privacy?

CONCLUSION

Understanding privacy helps us to develop technologies, norms, social contracts, and value systems to support a sustainable exchange of information. The underlying concept of privacy has not changed for centuries, but our approach to acknowledging privacy in our transactions, exchanges, and relationships must be revisited as our technological environment – what we can *do* with information – has evolved.

There is much to be said for all parties in business wanting to understand privacy. Consumers want to exchange information, simplify transactions, have books recommended, and save credit card information while retaining their privacy and the 'right to be left alone' (Warren and Brandeis, 1890). Furthermore, corporations want to limit their vulnerability to information leaks, hackers, or civil subpoenas and minimize the cost of preserving, storing, securing information while still lowering transaction costs and customizing services for customers and stakeholders. A definition of privacy which relies upon the control of information in privacy zones supports the possibility of all stakeholders to achieve their goals.

REFERENCES

Acquisti, A. (2002). Privacy and security of personal information: Economic incentives and technological solutions. Presented at the *1st SIMS Workshop on Economics and Information Security.* Berkley, CA

Acquisti, A., & Grossklag, S. J. (2004). Privacy attitudes and privacy behavior: Losses, gains, and hyperbolic discounting. In Camp, J., & Lewis, R. (Eds.), *The Economics of Information Security.* Boston: Kluwer.

Alderman, E., & Kennedy, C. (1995). *The right to privacy.* New York: Knopf.

Allen, A. L. (1988). *Uneasy Access. Privacy for women in a free society*. Totowa, NJ: Rowman and Littlefield.

AT&T revises privacy policy. (2006, June 22). Retrieved September 11, 2009, from *The New York Times* web site: http://query.nytimes.com/gst/fullpage.html?res=9F00E0DB1F31F931A1 5755C0A9609C8B63

Beacon. (2009). Retrieved May 29, 2009, from Facebook Web Site: http://www.facebook.com/business/?beacon

Bennett, C. J. (1992). *Regulating privacy*. Ithica, NY: Cornell University Press.

Brin, D. (1998). *The transparent society*. Reading, MA: Perseus Books.

Coase, R. (1960). The problem of social cost. *The Journal of Law & Economics*, *3*(1), 1–44. doi:10.1086/466560

Dash, E. (2007, January 19). Data Breach could affect millions of TJX shoppers. Retrieved September 12, 2009, from *The New York Times* web site. http://www.nytimes.com/2007/01/19/business/19data.html

Davidson, H. R. (1969). The legend of lady godiva. *Folklore*, *80*(2), 107–122.

DeCrew, J. (1997). *In pursuit of privacy*. Ithica, NY: Cornell University Press.

Edition, M. (2008). *GE loses consumers' personal records*. Retrieved September 12, 2009, from http://www.npr.org/templates/story/story.php?storyId=18212217

Elgesem, D. (1999). The structure of rights in directive 95/46/EC. *Ethics and Information Technology*, *1*, 283–293. doi:10.1023/A:1010076422893

Facebook. (2009). Retreived May 28, 2009, from Facebook web site: http://www.facebook.com/about.php

Foucoult, M. (1977). Discipline and punish: The birth of the prison. In Rabinow, P. (Ed.), *The Foucoult reader*. New York: Pantheon Books.

Fried, C. (1984). Privacy. In Schoeman, F. D. (Ed.), *Philosophical dimensions of privacy: An anthology*. Cambridge, UK: Cambridge University Press.

Godiva, L. (1950, January). Lady Godiva. *Western Folklore*, 77–78.

Hafner, K. (2006). Researchers yearn to use AOL logs but they hesitate. Retrieved September 12, 2009, from *The New York Times* web site: http://www.nytimes.com/2006/08/23/technology/23search.html

Hartland, E. (1890). Peeping Tom and Lady Godiva. *Folklore*, *1*(2), 207–226.

Helft, M. (2007). *Google zooms in too close for some*. Retrieved September 12, 2009 from http://www.nytimes.com/2007/06/01/technology/01private.html

Hermalin, B. E., & Katz, M. L. (2006). Privacy, property rights, and efficiency: The economics of privacy as secrecy. *Quantitative Marketing and Economics*, *4*(3), 209–239. doi:10.1007/s11129-005-9004-7

Holson, L. M. (2007a). *Verizon letter on privacy stirs debate*. Retrieved September 12, 2009, from http://www.nytimes.com/2007/10/16/business/16phone.html.

Holson, L. M. (2007b). *Privacy lost: These phones can find you*. Retrieved September 12, 2009, web site: http://www.nytimes.com/2007/10/23/technology/23mobile.html.

James, W. (1931). *Pragmatism: A new name for some old ways of thinking*. New York: Longmans, Green, and Co.

Jesdanun, A. (2007). Facebook retreat shows ad-targeting risk. *The Associated Press* Retrieved September 12, 2009, from http://www.washingtonpost.com/wp-dyn/content/article/2007/11/30/AR2007113001668_pf.html

Jiang, X. (2002). *Safeguard privacy in ubiquitous computing with decentralized information spaces: Bridging the technical and the social.* Presented at the 4th International Conference on Ubiquitous Computing (UBICOMP 2002), Gotenborg, Sweden.

Jiang, X., Hong, J. L., & Landay, J. A. (2002). Approximate information flows: Socially based modeling of privacy in Ubiquitous Computing. Presented at the 4th International Conference on Ubiquitous Computing (UBICOMP 2002), Gotenborg, Sweden

Jiang, X., & Landay, J. A. (2002). Modeling privacy control in context-aware systems. *Pervasive Computing, IEEE, 1*(3), 59–63. doi:10.1109/MPRV.2002.1037723

Johnson, D. J. (2001). *Computer ethics.* Upper Saddle River, NJ: Prentice Hall.

Lessig, L. (1998). *The architecture of privacy.* Retrieved September 12, 2009, from http://cyber.law.harvard.edu/works/lessig/architecture_priv.pdf

Lessig, L. (1999). *Code and other laws of cyberspace.* New York: Basic Books.

Liedtke, M. (2007). Facebook lets users block marketing tool. *The Associated Press.* Retrieved September 12, 2009, from SFGate.com

Lohr, S. (2007). *As its stock tops $600, Google cases growing risks.* Retrieved September 12, 2009, from http://www.nytimes.com/2007/10/13/technology/13google.html

Mitchell, D. (2007). *Online ads vs. privacy.* Retrieved September 12, 2009, from http://www.nytimes.com/2007/05/12/technology/12online.html

Moor, J. (1997, September). Towards a theory of privacy in the information age. *Computers & Society, 27*(3), 27–32. doi:10.1145/270858.270866

Nakashima, E. (2007). *Feeling betrayed, Facebook users force site to hone their privacy.* Retrieved September 12, 2009, from http://www.washingtonpost.com/wp-dyn/content/article/2007/11/29/AR2007112902503.html

Perez, J. C. (2007a). *Facebook admits ad service tracks logged-off users.* Retrieved September 12, 2009, from http://www.pcworld.com/article/140225/facebook_admits_ad_service_tracks_loggedoff_users.html

Perez, J. C. (2007b). *Facebook Tweaks beacon again; CEO apologizes.* Retrieved September 12, 2009, http://www.pcworld.com/article/140322/facebook_tweaks_beacon_again_ceo_apologizes.html

Perez, J. C. (2007c). *Facebook doesn't budge on Beacon's broad user tracking.* Retrieved September 12, 2009, from http://www.pcworld.com/article/140385/facebook_doesnt_budge_on_beacons_broad_user_tracking.html

Perez, J. C., & Gohring, N. (2007). *Facebook partners quiet on beacon fallout.* Retrieved September 12, 2009, from http://www.pcworld.com/businesscenter/article/140450/facebook_partners_quiet_on_beacon_fallout.htm

Posner, R. (1981). The economics of privacy. *The American Economic Review, 71*(2), 405–409.

Prosser, W. D. (1960). Privacy. *California Law Review, 48,* 383–396. doi:10.2307/3478805

Rachels, J. (1975). Why is privacy important? *Philosophy & Public Affairs, 4*(4), 323–333.

Rampell, C. (2008). *What Facebook knows that you don't.* Retrieved September 12, 2009, from http://www.washingtonpost.com/wp-dyn/content/article/2008/02/22/AR2008022202630.html

Reiman, J. H. (1975). Privacy, intimacy, and personhood. *Philosophy & Public Affairs, 6*(1), 26–44.

Reiman, J. H. (1995). Driving to the panopticon. *Santa Clara Computer and High-Technology Law Journal, 11*(1), 27–44.

Rosen, J. (2001). *The unwanted gaze: The destruction of privacy in America.* New York: Vintage Books.

Samarajiva, R. (1997). Interactivity as though privacy mattered. In Agre, P. E., & Rogenberg, M. (Eds.), *Technology and privacy: The new landscape.* Cambridge, MA: MIT Press.

Samuelson, P. (2000). Privacy as intellectual property? *Stanford Law Review, 52*(5), 1125–1173. doi:10.2307/1229511

Schoeman, F. (1984). Privacy: Philosophical dimensions of the literature. In Schoeman, F. D. (Ed.), *Philosophical dimensions of privacy: An anthology.* Cambridge, UK: Cambridge University Press. doi:10.1017/CBO9780511625138.002

Scolove, D. J. (2006). A taxonomy of privacy. *University of Pennsylvania Law Review, 154*(3), 477.

Scolove, D. J. (2007). 'I've got nothing to hide,' and other misunderstandings of privacy. *The San Diego Law Review, 44.*

Simmel, G., & Wolff, K. H. (1950). *The sociology of georg simmel.* Glencoe, IL: Free Press.

Singleton, S. (1998). Privacy as censorship. *Policy Analysis, 295.*

Smith, R. E. *Privacy: How to protect what's left of it.* Garden City, NJ: Doubleday Books.

Thomson, J. J. (1975)... *Philosophy & Public Affairs, 4*(4), 295–322.

Today, U. S. A. (2007). *Fired Wal-Mart worker reveals covert operations.* Retrieved September 12, 2009, from http://www.usatoday.com/money/industries/retail/2007-04-04-walmart-spying_N.htm

Travani, H. T., & Moor, J. H. (2001, March). Privacy protection, control of information, and privacy –enhancing technologies. *Computers & Society, 31*(1), 6–11. doi:10.1145/572277.572278

Warren, S. D., & Brandeis, L. D. (1890). The right to privacy. *Harvard Law Review, 4*(5), 193. doi:10.2307/1321160

Wasserstrom, R. A. (1978). Privacy: some arguments and assumptions. In Schoeman, F. D. (Ed.), *Philosophical dimensions of privacy: An anthology.* Cambridge, UK: Cambridge University Press.

Westin, A. (1967). *Privacy and freedom.* New York: Atheneum.

Westin, A. F. (1966). Science, privacy, and freedom: Issues And proposals for the 1970s. Part I-the current impact of surveillance on privacy. *Columbia Law Review, 66*(6). doi:10.2307/1120997

Zeller, T. (2006). *Your life as an open book.* Retrieved September 12, 2009, from http://www.nytimes.com/2006/08/12/technology/12privacy.html

ENDNOTES

[1] Compiled from Davidson (1969); Sidney (1890); "Lady Godiva" (1950).

[2] The latter example is from the movie <u>Old School</u> where a father, in search of the ability to speak freely with friends, would ask his son to merely cover his ears by saying "ear muffs" rather than remove the child from the room.

[3] In an attempt to incorporate individuals' control in any discussion of privacy—including access versions of privacy—scholars differentiate scope versus the definition of privacy with the latter including some attempts at control (Reiman, 1995), dismiss

consent as being implied in all interactions, relationships, and transactions (Singleton, 1998), or combine the two versions to create the control/restricted access version of privacy (Moor, 1997).

[4] An opt-out strategy defaults to the user being included in the program unless they specifically 'opt-out.' In other words, Beacon broadcasts would be sent to a user's identified friends unless the user opted out of the service. This is in contrast to the recommended opt-in strategy where users are not assumed to give consent and are asked to proactively opt-in.

Chapter 11
Against Strong Copyright in E-Business

D.E. Wittkower
Coastal Carolina University, USA

ABSTRACT

As digital media give increasing power to users—power to reproduce, share, remix, and otherwise make use of content—businesses based on content provision are forced to either turn to technological and legal means of disempowering users, or to change their business models. By looking at Lockean and Kantian theories as applied to intellectual property rights, we see that business is not justified in disempowering users in this way, and that these theories obligate e-business to find new business models. Utilitarian considerations support disempowering users in this way in some circumstances and for the time being, but also show that there is a general obligation to move to new business models. On these moral bases, as well as on practical bases, e-business ought to refrain from using the legally permitted strong copyright protections, and should instead find ways of doing business which support, value, and respect the technical capabilities that users have gained.

INTRODUCTION

One of the most ethically contentious areas of e-business is the assertion and enforcement of intellectual property rights by businesses based upon content provision. Businesses engaged in content delivery tend to view themselves as sellers of goods rather than service providers, and this difference in perspective has significant social, practical and ethical implications. By debunking the idea that strong copyright over content in a digital context is morally supported on Lockean or Kantian considerations, and by shifting the burden of proof on economic and utilitarian considerations against those who employ strong forms of copyright, the chapter demonstrates that there is generally a moral obligation to refrain from use of standard (or, *maximalist*) copyright protection in e-business. Thus, businesses based upon content provision should, for both moral and

DOI: 10.4018/978-1-61520-615-5.ch011

practical reasons, change their business model from the product to the service economy.

While it is well established in U.S. law that sweat-of-the-brow does not generate goods subject to copyrighting, there is still a strong intuition that a moral, if not a legal right over expressive works is generated through labour. By an investigation of the Lockean presumption against the right of exclusion, we can see that a viewpoint true to Locke's moral emphasis on the preservation of freedoms held within the state of nature would not support such a right given the current structure of digital media. Indeed, Locke himself, even though he wrote in a far different communications context, supported only limited intellectual property rights, which he justified on a utilitarian, not a labour-desert basis.

The Kantian basis of intellectual property rights is approached next. The Kantian view centers on author's rights, and is far removed from the current legal regime in place in the U.S. and spread internationally through the World Intellectual Property Organization, but again speaks to a strong intuition that a moral, if not a legal right to exclusion is generated in expressive works. Here, by looking at Kant's *Metaphysics of Morals*, as well as his essay "On the Wrongfulness of Unauthorized Publication," we see that the moral basis of the right of exclusion is founded upon the communicative relationship between an author and her public, and offers no support to the use of strong copyright law in our current communications context. Instead, a Kantian perspective would today support open-content models utilizing public domain dedication or GPL/Creative Commons licensing.

Finally, a utilitarian perspective—the explicit basis of intellectual property rights in the Anglo-American legal tradition—is considered. The moral basis being found in consequences rather than in natural or moral rights, the great success of the open-source movement is the primary consideration. By looking at recent history, it can be seen that the practical necessity of a right of exclusion over expressive works within our current communications context is suspect at best. The primary utilitarian considerations holding weight today have to do with the large role played by copyright-based industries in our economy, and the disutility that would be caused by undermining the basis of these industries. This consideration, while important, is counterbalanced by considerations of the loss of public rights required by the use of strong copyright.

Several examples of currently marginal but emerging business models are then presented, including shareware, subscription, patronage, and value-added delivery models. These examples illustrate how a conceptual shift in content-delivery-based e-business from a product to a service model is able to satisfy practical utilitarian and economic considerations while remaining true to Lockean and Kantian moral considerations. In conclusion, it is argued that e-business has an obligation, on the grounds previously explained, to refrain from the use of legally available strong copyright protection.

BACKGROUND

The current debate over copyright is a tangled mess. Concerns with individual and corporate rights are intertwined with concerns about economic and social effects, and within each of these kinds of concerns there are a great number of stakeholders whose interests are not readily ranked. The rights in question include those of various parties who may or may not be the same person: the author, the creator, the copyright holder, the purchaser, the consumer, the licensee, and the user. Within larger societal concerns, the field is no less crowded: we are concerned with the freedom of the market, the freedom of the culture, and the freedom of expression; and also with the benefit of the public, of the entertainment industry, of the national economy, and of the entrepreneurial impulse. We are rightly committed to all of these concerns, and the majority of differences in opinion between those who argue the strength of copyright protection should be diminished and those who defend full use and continuation of current copyright protections arise from little else than different views on the relative importance of these concerns. Ought we to view the rights of the author as determinative of the rights of the user, or the licensee? Ought we to view a largely corporate-owned culture as an acceptable price to pay for the health of our economy? To what extent does encouraging innovation through profit motive justify governmental protection of industry practices which disable free use of purchased goods, or which infringe upon the privacy of consumers and users?

In order to decide these issues, we tend to fall back to established positions. This takes place both practically and intellectually. Practically, we often respond to these shifts in architecture—to use Lawrence Lessig's term (2006)—by attempting to fit applications of technologies into the ruts already worn by previous architectures, maintaining the economic relations, rights, and privileges required by previous technical regimes. The movement towards Digital Rights Management (henceforth DRM) systems is straightforwardly an attempt to make the new media act like the old in order to keep stable the power relations between content producers, business, and the market. Intellectually, we also fall back to established positions; in the realm of copyright in particular, we fall back upon Lockean, Kantian, and utilitarian foundations of intellectual property rights.

The constitutionally granted power of Congress "[t]o promote the Progress of Science and useful Arts, by securing for limited Times to Authors and Inventors the exclusive Right to their respective Writings and Discoveries" (Federal Convention of 1787) has usually been interpreted in a strictly utilitarian manner. For example, we see Justice O'Connor's statement that "the 1976 revisions to the Copyright Act leave no doubt that originality, not "sweat of the brow," is the touchstone

of copyright protection in directories and other fact-based works" (*Feist Publications v. Rural Telephone Service*, 1991). This interpretation reinforces the implication, based upon the optional nature of Congress' granting of intellectual property protection, that the law recognizes these author's and inventor's rights as artificial rather than natural, as stated before by Chief Justice Hughes: "copyright is the creature of the federal statute passed in the exercise of the power vested in the Congress. As this Court has repeatedly said, the Congress did not sanction an existing right, but created a new one" (*Fox Film Corp. v. Doyal*, 1932). Nevertheless, it seems that Lockean intuitions may underlie recent legislative action, particularly the automatic extension of copyright over works already created provided in the Sonny Bono Copyright Term Extension Act of 1998 (henceforth CTEA), and the extreme length of the term created thereby. Justice Breyer's dissent in *Eldred v. Ashcroft* (2003), while it did not claim that the CTEA emerged from Lockean intuitions, provided many strong arguments that go towards establishing such a claim, such as that the CTEA would extend copyright terms to such an extent that these rights would serve the interest of providing income for authors and their descendents at the expense of public commerce in ideas and expressions, thus promoting to an unprecedented extent intellectual property owners' right of exclusion—the right to prevent others from utilizing intellectual property (henceforth IP) for profit, non-profit, or personal purposes—over the social and utilitarian benefits of releasing those rights.

These various foundations of intellectual property rights (henceforth IPR) are rational ones, and even if the motivation to provide IPR protection falls outside of the utilitarian intent that seems implied within the constitution, it is certainly possible that the legislation may still be constitutional even as the motivation behind it is not. These various foundations have, indeed, been part of the public dialog about IPR for some time, and our interest in non-utilitarian justifications of

IPR did not emerge from obscurity with the rise of digital media. And yet, the shifting architecture of the methods of articulation of IP provides radical new possibilities to authors, users, and owners of IP, and as we return to our Lockean, Kantian, and utilitarian intuitions, it is worthwhile to reconsider basic forms of the theories which underlie those intuitions in this new and changing social/technological milieu. What we find is remarkable.

In the following reconsideration of these theories, I attempt to provide evidence that, when we consider copyright protection over digital objects, not only can these traditional sources of justifications of IPR not be presupposed, but instead that, when digital objects are considered properly, the burden of proof ought to fall upon the copyright maximalists rather than the copyright minimalists. In this discussion, since I am trying to address what e-business should do today, the copyright minimalist and maximalist positions will not be approached as positions on the law—e.g. whether copyright terms should be extended to life of the author plus seventy years, as they are under the CTEA, or reduced to the earlier term of fourteen years, as in the Copyright Act of 1790—but instead only as positions on how to use the legal structures currently in place. In this context, the *copyright maximalist* position can be defined as the position which holds that it is appropriate to use the strongest copyright protection allowed under current law; a simple "all rights reserved" assertion of copyright. The copyright minimalist position, which I advocate here, can be defined as a position which holds that only a weaker protection can be morally justified, such as those provided by Creative Commons licenses or the GNU Public License. These less extensive uses of copyright law allow users to retain various rights, such as the right to share or reproduce content with attribution, the right to remix for non-commercial use, or the right to access and use source code so long as the resulting software is shared alike.

The case argued here is a limited one in a few important ways:

First, I do not hold that any of these ethical theories are right or wrong, or that any is more important than any other. I discuss these because they are the theories that most strongly influence public opinions and beliefs, regardless of how well- or ill-founded they are on a moral, philosophical, or constitutional basis. It is for this reason that I consider these perspectives rather than others, which may arguably be more correct, useful, or productive.

Second, I am addressing digital media, and digital media alone. Issues about analog IP in an increasingly digital world are beyond the scope of this inquiry.

Third, I address only copyright, and not other forms of IPR. Digital technologies have transformed IPR of all kinds in important ways, but only in the realm of copyright have they brought about widespread public skepticism regarding the justice of providing such rights. Thus, it is here that our intuitions seem to have the greatest—or at least most apparent—mismatch with current law.

Fourth, I will not discuss international issues in IPR legislation, but will concentrate on U.S. copyright law. IPR varies in significant ways in different nations, and enforcement varies even more, but U.S. copyright-based industries have been particularly aggressive in trying to export U.S. laws to other nations, both through industry groups such as the Business Software Alliance and the Recording Industry Association of America, and through the World Intellectual Property Organization. For this reason, U.S. law is of particular global relevance. Nevertheless, in order to maximize the relevance of these arguments to all parties in global e-business, I will not concentrate on legal structures, but instead the moral arguments that may or may not serve as a foundation to a variety of legal structures.

Fifth, and finally, I will not deal extensively with contemporary reformulations, rehabilitations, or reconstructions of the canonical ethical theories discussed. As valuable and powerful as such contemporary theories are, my intention here is not to prove that these theories are unable to support strong copyright protection, but rather to show that the basic intuitions which emerge from each of these traditional justifications of IPR are tied in important ways to pre-digital media. My goal is simply to shift burden of proof by raising doubts from within canonical versions of these theories, and for this, we need not show that such proof cannot be given.

THE ETHICS OF COPYRIGHT OVER DIGITAL GOODS

We will consider Lockean, Kantian, and utilitarian justifications of copyright in the light of digital modes of articulation. In order to avoid biased terminology, I will often refer to intellectual goods rather than intellectual property. Locke and Kant both offer rights-based justifications of intellectual property laws. When considering both Locke and Kant, we will see that their theories no longer imply a moral right that requires legal protection, and this will be sufficient to show that there is therefore no justification to use these legal protections, since they limit the rights and freedoms of others. Utilitarianism, however, even if it does support strong copyright, does not do so by asserting a moral right, but only because the granting of such an invented right of exclusion over intellectual goods is for the best for society as a whole. Here, it will be shown that this "copyright bargain" is questionable, but, due to the consequentialist reasoning involved, it can only be asserted that the burden of proof should be on those who would use strong copyright protection, not that such use will always be unjustifiable.

The Lockean Perspective

It is often suggested that the writings of John Locke—specifically the second of his *Two Treatises on Government*—may provide a justification for strong copyright laws, and public intuitions

supporting IPR are often based upon Locke's labour-desert theory of property rights. A critical investigation of the Lockean theory, taking digital media into account, reveals that it no longer supports the idea of a natural right which corresponds to copyright protection, and that it instead affirms the natural right to copy and to produce derivative works.

The famous and most oft-quoted passage, from *Chapter V: Of Property* (Locke, 2005), is as follows:

Though the earth, and all inferior creatures, be common to all men, yet every man has a property in his own person: this no body has any right to but himself. The labour of his body, and the work of his hands, we may say, are properly his. Whatsoever then he removes out of the state that nature hath provided, and left it in, he hath mixed his labour with, and joined to it something that is his own, and thereby makes it his property. It being by him removed from the common state nature hath placed it in, it hath by this labour something annexed to it, that excludes the common right of other men: for this labour being the unquestionable property of the labourer, no man but he can have a right to what that is once joined to, at least where there is enough, and as good, left in common for others. (Ch. V, §27)

"Mixing" of one's labour with the goods of the world, given to mankind in common by God, implies a right to the fruits of that labour, for, as Locke argues later, "if we will rightly estimate things as they come to our use, and cast up the several expenses about them, what in them is purely owing to nature, and what to labour, we shall find, that in most of them ninety-nine hundredths are wholly to be put on the account of labour" (2005, Ch. V, §40). If we follow Locke in holding that the value in the product springs from the labour mixed therein, and that the labour therein is by nature property of the labourer, the labourer then has a natural right to the value of the product with which he has mixed his labour.

This theory is a fine one when speaking of acorns and apples; when your labour provides you a good which has a use to you, whether that good emerges merely from collection of the bounty of nature or an industrial use of natural resources, you certainly have a moral claim over such a good. The idea that this theory can be applied to intellectual property, wherein what is collected and worked is not seed and soil but rather ideas, sounds, words, or facts of nature, is not an irrational one. It might additionally be argued that, the labour of the creation of such a good being of such a greater proportion than the labour involved in the subsequent utilization of it by parties other than the author, the author should retain a proportional right over such derivative works. This is certainly the viewpoint which has guided our extension of the Constitutional allowance of the ability of Congress to provide protection over "Writings and Discoveries" to the arguably extra-constitutional (cf. e.g. S. Vaidhyanathan, 2003) ability of Congress to legislate protection over works derivative of these writings and discoveries. There is, indeed, support from Locke (2005) on this point, for he argued that

[h]e that had as good left for his improvement, as was already taken up, needed not complain, ought not to meddle with what was already improved by another's labour: if he did, it is plain he desired the benefit of another's pains, which he had no right to, and not the ground which God had given him in common with others to labour on, and whereof there was as good left, as that already possessed, and more than he knew what to do with, or his industry could reach to. (Ch. V, §34)

Thus, if I should have words, ideas, sounds or facts of nature available to me which are as good as those already taken up by others, I have no right to meddle with those already improved by others. The Beatles do not have dominion over the sounds of which their music is made, and should I wish to make music of my own I have as much access as they to those sounds, and therefore have no right to trespass upon the particular ways in which they have been already cultivated.

We may first note a serious disanalogy that presents a problem for this application of Lockean property theory: digital objects, and songs in general, are non-rivalrous goods. When I take your acorns or apples, your cultivated land or, indeed, your book, my benefit is a rival of yours; I gain only through your loss. When I make a copy of your digital file or when I sing your song, you are not the less for it. Thus, when I copy software you have written, for example, I do not trespass upon your right to the fruits of your labour—I have in no way prevented you from reaping the benefits of your cultivation of mathematical facts. On the other hand, when you prevent me from sharing your music, making use of your source code, or remixing your movie, you prevent me from reworking those materials and being recognized for my addition of my labor.

It may be objected that I have lessened the profits that you might have realized through the sale of copies. This argument, however, is *ex post facto* and has no place here, for it already assumes that you have a right to control such copies, and this supposed right is precisely what is in question. Nevertheless, copies and derivative works are certainly a means of benefiting by the labour of another, and it may yet be that we owe to the author some share in the benefit we have therefrom taken.

If we should, for example, remix a Beatles song, we mix an amount of our own labour with the sounds, words, and ideas with which they have already mixed their labour. But these sounds, words and ideas are not found within virgin untouched nature. Do the Beatles then owe a portion of their benefit to, for example, the descendants of those who contributed to the invention of the modern guitar? To assert so would be ludicrous. The appropriation of the guitar has become akin to the appropriation of a fact of nature, for, just as Locke said regarding untamed nature, such appropriation being non-exclusive, it keeps nobody else from profiting from her own appropriation of the guitar. As Locke (2005) worded it:

Nor was this appropriation of any parcel of land, by improving it, any prejudice to any other man, since there was still enough, and as good left; and more than the yet unprovided could use. So that, in effect, there was never the less left for others because of his enclosure for himself: for he that leaves as much as another can make use of, does as good as take nothing at all. No body could think himself injured by the drinking of another man, though he took a good draught, who had a whole river of the same water left him to quench his thirst: and the case of land and water, where there is enough of both, is perfectly the same. (Ch.V, §33)

The Beatles do not infringe upon the natural rights of the inventors of the guitars they used, for they have not taken anything from them, or deprived them of anything whatsoever. They, further, have taken nothing from the commons by taking the instrument for their own use, for they in so doing have kept nobody else from doing likewise. The same can be said of a later artist using the music of the Beatles. DJ Dangermouse, for example, has taken up the work of those before him, mixing the Beatles' *White Album* and Jay-Z's *Black Album* in his *Grey Album*, but he has done so in a non-rivalrous, non-exclusive way, leaving, in effect, as much commonly available after as before his appropriation.[1]

It may be that we wrong the authors in some other way by such appropriation, but we do not tread upon their property rights in a Lockean perspective. For Locke, what is questionable is exclusion. This is necessary with regard to goods whose possession is rivalrous and exclusive, such as land, and it is this necessity that motivated

Locke's attempt to justify holding property. As Locke (2005) stated:

it is very clear, that God, as king David says, Psal. cxv. 16. has given the earth to the children of men; given it to mankind in common. But this being supposed, it seems to some a very great difficulty, how any one should ever come to have a property in any thing. (Ch. V, §1)

If you have mixed your labour with elements of the earth held in common, you have, he concludes, a natural right to possess that cultivated good. If that good is, however, of a kind which does not *require* exclusion in order for you to reap profits from it, the superaddition of exclusion can find no justification in Locke, for Locke seeks only to justify the necessary evil of exclusion, not to justify exclusion when wholly unnecessary. Thus, you have a natural right to the song that you write: I ought not to deprive you of the lyrics you have written down. However, you have no right to tell me not to sing it.

Tom Bell (2004) has put the point more generally:

[C]opyright and patent protection contradicts Locke's justification of property. By invoking state power, a copyright or patent owner can impose prior restraint, fines, imprisonment, and confiscation on those engaged in peaceful expression and the quiet enjoyment of the tangible property. Because it thus gags our voices, ties our hands, and demolishes our presses, the law of copyrights and patents violates the very rights that Locke defended. (p. 4)

If you should write a computer program, you have a right to object if I destroy your copy. In just the same way, if I should copy your program, I have taken nothing from you, and you have no right to keep me from free and full use of my copy of this program. We talk about property rights as a "bundle" which may or may not include certain particular rights, such as the right to exclude others from one's property, to use it, to profit from it, to rent it, or to sell it. Locke provides a strong defense of some rights in some ways, but his theory offers no support for the extension of the right of exclusion over copies of digital objects.

The Kantian Perspective

It may be argued that a Kantian justification for the extension of copyright over digital objects may be given on the basis that if we consider the maxim "I intend to copy this digital object" we shall find that it is not universalizable, for if everybody copies digital goods, nobody will be able to afford to produce further originals to copy. Furthermore, it may be argued that doing so does not respect the creator of the digital object as an end to herself, but treats her merely as a means. An additional Kantian justification of the extension of copyright over digital objects, one more in line with the continental European legislative tradition of author's rights, might be given in the argument that such objects are often or always expressive in nature, and the protection of the data which constitutes an authorial expression is required if the integrity of that expression is to be adequately protected. These arguments will be addressed in this order.

First, we ask, following Nissenbaum (1995), whether the maxim to copy my neighbor's software is universalizable. It seems to many people that such behavior is equivalent to theft,[2] but this ignores the non-rivalrous nature of digital objects.

Kantian moral theory claims that unethical actions are those which fail to accord with what Kant calls the "categorical imperative": That the maxim (or, the rule that one implicitly follows in the action) should be capable of being willed as a universal law of human behavior. So, approximately, we can ask whether it is possible for all other actors to take the same action we are considering. If it is not possible, or if a world in which everyone acted in such a manner would be undesirable, then we can tell that we are making an exception of ourselves in this action. We can consider this action beneficial only because most others refrain from doing as we are doing. Hence, Kant claims that when we cannot will the maxim of our action as a universal law, we are using others for our benefit.

Consider theft of some useful analog object from my neighbor. When I take valuable property, I am thereby given a good, and my neighbor loses an identical good. In one version of the famous Kantian argument I steal from my neighbor and thereby gain a loaf of bread, which is fine and good, but we discover that this is unfortunately not universalizable, for if everybody stole bread then those in possession of bread would stop leaving it out in the open where people can get at it, or perhaps bakers would stop making it altogether.

When I copy my neighbor's digital object, I am thereby given a benefit, and my neighbor is none the less for it, for digital objects are similar to an idea in that, in Thomas Jefferson's (1813) words, "he who receives an idea from me, receives instruction himself without lessening mine." For this reason already, we see that the categorical imperative does not provide as strong an argument against copying of digital goods as it does against the seizure of analog goods. If everybody copied their neighbors' digital goods, this *in itself* should in no way cause any reticence on anybody's part, unless it can additionally be reasonably expected that such an imagined law of nature would preclude the creation of further such goods, or would so greatly diminish their number or quality that we could not will to live in such a world. To make this more clear, consider Christine Korsgaard's (1996) restatement of wherein the wrongness of an immoral action lies according to Kant: "your action would become ineffectual for the achievement of your purpose if everyone (tried to) use it for that purpose" (p. 78).

Perhaps, although my neighbor is not directly harmed by my action, I am nevertheless using others because I am neglecting to support authors of digital objects, and the corporations that employ them, in order to allow for the continued creation of such products. The question then is whether, if everybody copied digital objects, new digital objects would continue to be created. The answer here is quite clear: there would be no shortage of digital goods. It is true that the profitability of large-scale digital good production would be greatly diminished, but this is not the only source of digital goods. The software released as freeware, under GNU general public license, under open-source BSD license, or otherwise copylefted, is sufficient to replace a wide variety of proprietary programs, and the availability of the source code of proprietary programs would provide ample resources to increase the number of such non-proprietary applications, and quite likely at an higher rate of increase than that provided through production under fully proprietary licensing. Similar licensing is being applied to other digital goods, such as music, video, and written works, and many authors feel that these licenses are preferable to standard copyrighting, which tends, in many important ways already discussed, to discourage innovation and creation. Thus, we see again that the categorical imperative, in its first formulation, is not at odds with the maxim to copy my neighbor's digital objects.

It may be objected that the maxim should read instead "I intend to copy *proprietary* digital objects whenever convenient," for the above confuses the copying of freeware or copylefted goods, which is equivalent to accepting a gift or continuing a conversation, with the piracy of commercially produced content, which is arguably equivalent to theft.

Admittedly, I have a duty to respect the law, as I cannot consistently will that everybody should break the law whenever they find it convenient to do so. This does itself establish a Kantian argument against so-called piracy, but not by means of any particular aspect of the action itself. In further support of this position, we may note that the creators of these digital products had a reasonable expectation of profit from its resale, and to deprive them of that profit which the law guarantees them would treat the programmers and corporate managers and CEOs as a mere means, for they no doubt have life plans which are dependent upon the income which our unlawful behavior would deprive them of. We are in a system wherein they are playing by the established rules, and for that reason, we have a duty to hold up our end of the bargain.

This is of course a conservative Kantianism, for it would be possible to argue that we never consented to this system in any meaningful sense, or perhaps that one ought not respect legislation which requires behavior which is immoral, an allegation which some have argued. Most notable is Richard Stallman's argument that copyright in fact requires us to act immorally and erodes basic social goods, an argument which he brings together by the slogan "cooperation is more important than copyright" (2004). A more modest line is taken by Helen Nissenbaum (1995), who argues that at least some situations exist in which it is moral to ignore copyright. The most comprehensive argument along these lines of which I am aware is Michael Perelman's *Steal this Idea* (2002), which makes a book-length case against intellectual property rights in general.

Even if we stay true to the conservative Kantianism, and respect laws in place simply because they are the laws in place, this is an *ex post facto* justification, for it tells us nothing about whether the law supports or inhibits moral behavior. If the copying of proprietary digital goods cannot be shown to be wrong outside of the fact of its illegality, we should conclude that it is wrong in the way that sitting at a whites-only lunch counter may have once been wrong: to transgress here is to neglect a duty to be lawful but is otherwise acceptable behavior, and thus the law itself should be considered questionable. If the transgression of this law violates no duty other than that of lawfulness, and if this law does not continue to fulfill the obligation in the service of which it was created, then we will conclude that copying proprietary digital objects is indeed immoral, but we must *also* conclude that *making use of* this law through asserting strong copyright is itself immoral, for it unnecessarily restricts the freedom of others.

When we look at the goals stated in the enactment of copyright law in the Anglo-American tradition, the justifications are (1) to ensure that the author of a useful expression is benefited, or at least not ruined, (2) to encourage the creation of such works, and thus (3) to benefit the public in general.[3] If we ignore, for a moment, the existence of digital media, we easily see a Kantian justification of such legislation. If we consider the creation of a useful and successful complex expression, say a manual of some kind, the production of this expression presumably requires a fair amount of labour while its reproduction requires relatively little. In this case, if we universalize the maxim of freely reprinting useful expressions, we see that this practice would not be sustainable, for nobody would find it worth their while to create such expressions. Thus, copyright law does have a clear and significant moral basis, especially if we also consider that to discourage such expressions would constitute a purposeful impediment to the realization of human potentialities, as in accord with Kant's third example in the second section of the *Groundwork*, 4:423 (1996a).

When we again take digital media into consideration we see that the law no longer has this moral basis, for the free reprinting/reproduction of digital goods can now be willed as a universal law. With regard to the obligation to assist the creator of such useful expressions, (1) above, we may easily note that the expenditures required in the creation of such complex expressions as software are no longer prohibitive, as evidenced by the fact that significant and increasing numbers of such creators choose not to receive recompense for their labour even when given the legal means to do so, and, further, that such recompense, in the cases of software and music at the least, is not limited to retail sales in its origin, but can profit-

ably be shifted into the service economy, as will be discussed further below. This itself is enough to establish that objectives (2) and (3) no longer require these laws, but it is worth noting that there is also significant evidence that copyright over digital objects is not only unnecessary but actually detrimental, discouraging both innovation and the public benefit which is to be gained from such innovation (e.g. Nadel, 2004; Perelman 2002).

We find more concrete discussion of the issue in Kant's article "On the Wrongfulness of Unauthorized publication of Books" and the related arguments in *The Metaphysics of Morals*.[4] Kant argues that writing does not constitute a plain object, but is instead also a form of speech, and thus an expression of the will of the author. Then there is the fact that the author chooses a certain publisher, a certain provider of the *"mute instrument for delivering the author's speech to the public"* (1996c, p. 30), which can then carry out the will of the author in the name of the author. This means that another publisher which would publish without authorization would express the will of the author in his speech, yet would do so against the will of the author, for the author *cannot* authorize more than one publisher to carry his expression to the public.

It is this last point that is least obvious and most crucial. Specifically, the argument is that the author cannot authorize more than one publisher as the instrument of their speech for "it would not be possible for an author to make a contract with one publisher with the reservation that he might allow someone besides to publish his work" (1996c, p. 31), because the two would "carry on the author's affair with one and the same entire public, [and] the work of one of them would have to make that of the other unprofitable and injurious to each of them" (1996c, p. 31). So, the author as a matter of *fact* cannot *actually* authorize two publishers to publish their speech, for no publisher would agree to such an arrangement. It is for this reason that another publisher cannot assume the permission of the author, for that permission cannot be granted, and thus they would bring the speech of the author to the public against the possible will of the author.

What is remarkable about this argument is that the claim of the authorized publisher against the unauthorized publisher is guaranteed by the impossibility of the consent of the author; an impossibility, but not a logical impossibility. We see, further, that with regard to digital objects this is no longer an impossibility, and thus the majority of the argument simply does not apply to digital objects.

Now we ask: Is it possible for an author of digital objects to authorize multiple publishers? Yes, it clearly is, for publishers need not be in competition with one another, though they speak to the same public, for they may not charge for the product at all. This conflict only arises with regard to proprietary digital goods.

Furthermore, Kant (1996c) argues that "if someone so alters another's book (abridges it, adds to it, or revises it) that it would even be a wrong to pass it off any longer in the name of the author of the original, then the revision in the editor's own name is not unauthorized publication and therefore not impermissible" (p. 35)[5] It seems this Kantian source does not support the kind of closed-source proprietary license currently prevalent, but instead supports only a license no more restrictive than the GNU GPL (General Public License), for if an adaptation is potentially sufficiently differentiated to be appropriately considered an independent work, then to prohibit the use of the first expression in the creation of an *ex hypothesi* different expression would clearly be outside of the realm of the author's rights.

So it seems that Kant may be used to support the judgment that it would be wrong to copy proprietary digital objects, for this could potentially be the act of a distributor acting in the name of the author, but against the will of the author. But it is far less clear that Kant would support making digital objects proprietary to begin with, since this is no longer necessary to express the communica-

tion of an author to her public. Even if Kant does not disallow making such goods proprietary, his writing and reasoning clearly does not support closed-source licensing, and even directly ridicules the very idea of contributory infringement.[6] So, on the basis of these arguments, it seems that Kant would require us to respect the chosen distributor of an expression unless the author authorizes multiple or unlimited distributors, but that authors and rightsholders are not justified in preventing the public from access to and use of code, or from the creation of derivative works.

The Utilitarian Perspective

Utilitarianism claims that the moral action is the action which produces the greatest net benefit for all parties involved. The utilitarian view, being based on the consequences for happiness or suffering, validates talk of "rights" only as a means to an end. Even fundamental rights, such as freedom of speech, are viewed as having an instrumental value rather than an absolute justification. Still, on a utilitarian view, some things, such as freedom of speech, are viewed as being so essential to happiness that they should not be abridged except in the most extreme cases. Hence, the utilitarian will claim that we should have a robust "right" to free speech because without it our society would be dysfunctional, ruled by prevailing opinion no longer subject to criticism, and we would feel unable to develop and express ourselves.

A utilitarian view of property rights in general is similar. Without the ability to own property, and the right to exclude others from use or enjoyment of it, we would lack many assets which seem to be basic to human happiness, ranging from the ability to enjoy the fruits of our labour to the inability to be secure in our plans for the future. If there is a utilitarian basis to rights over intellectual goods, there must be similar widespread and basic benefits for us, either individually or as a society. As we have already discussed, though, digital goods are non-rivalrous, so this right cannot be based on our individual ability to keep possession of our creation, for copies and derivative works do not remove our copies from our possession. Instead, a justification must instead come from a larger social benefit produced through the provision of this legal protection.

The most basic utilitarian intuition behind copyright protection is that the profit motive mobilized by the granting of a temporary monopoly provides a greater diversity and amount of intellectual goods *to the public*. Closely allied to this is the claim that such goods will be of higher quality, for producers of such goods will be better able to make investments of money and time, due to the possibility of remuneration granted through the artificial monopoly created by copyright, amounting to a market-based system of patronage.[7] This is what is referred to as the "copyright bargain": The right of exclusion is granted, for a limited time, not because there is any inherent benefit in this temporary monopoly, but because the temporary public loss of the right to use that copyrighted material is counterbalanced by a public benefit when that material falls out of copyright protection. This, of course, is only a good bargain when the goods produced remain of greater balance *after* the copyright term expires than the immediate value of the (presumably fewer and worse quality) goods which might have been produced in the absence of this legal protection.

The "copyright bargain," it seems, may have hitherto been a good deal for the public on the whole, however, with the widespread availability of digital technologies, it is possible for independent members of the public to produce goods of similar utility to those produced by the IP industries, and the production of these goods is not dependent upon granting a right of exclusion, due both to increasingly effective alternate methods of mobilization of profit motive and to the increasing relevance of non-profit-based motives as production costs fall. The utilities given to the public by the "copyright bargain" may now be realized in the absence of copyright to a far greater extent, and

thus the concrete benefit of copyright protection has been lessened. Concurrently, the social cost of effective copyright enforcement has risen—i.e. the "copyright bargain" is ever more onerous to the public in terms of opportunity costs, loss of freedoms, and imposition of externalities. Finally, it is reasonable to expect technological progress to continue to lessen the benefit of copyright protection relative to the absence of such protection, and to continue to increase the social costs of such protection.

The most basic and intuitive version of a utilitarian argument in favor of copyright goes approximately as follows. An exclusive right granted through copyright allowed for works to be created that would otherwise not have been. The expense required by producing and distributing copies of creative works was such that collective action was necessary; this was accomplished through granting the right of exclusion to authors, such that it could then be granted to corporations that were able to bear the costs of manufacture and sale which the author could not bear herself. Additionally, this has allowed for greater expenditures of both time and money on the part of the author, as publishing houses and their equivalents are able to use profits from other authors and prior works to speculatively support the creation of new works through book advances, record deals, upfront payment of actors, regular employment of software engineers, and so forth. Through this profit-based production, corporate support allows for the creation of works far more costly than would otherwise be produced. The concrete benefit which the public gains through these works form the basis of the copyright bargain as usually understood. The superadded value to these cultural products relative to those which would be produced in the absence of the collective advantage provided by corporate actions—be this benefit in terms of artistic or entertainment value, or simply in terms of the volume of works created—is meant to counterbalance the loss of rights to the public in their use of these works, which loss is mitigated by the temporary nature of this loss of rights. This bargain was, of course, a better deal for the public in the past, both due to the more limited term of copyright that was in place and to the far greater necessity of involving profit-motivated concentrations of capital.

The harms caused to the public by the "copyright bargain" rise sharply with widespread increased technological capabilities. As the means of production of industries based on intellectual goods have come increasingly within the possession of the public, industrial and private employment of ideas have become increasingly indistinguishable. It is impossible to tell whether I have sent somebody a music file as a form of private communication, as I might tell somebody about a book in recommending its purchase, or whether I have sent the music file as a form of industrial manufacture, as if I had *manufactured* a book in order that its purchase should be unnecessary. The priority of the industrial interpretation has been codified in law under the No Electronic Theft (NET) Act and, thus, the public is subjected to the strict requirements with which corporations were originally burdened in order to benefit the public. These basic forms of sociality are at risk, for as our lives become increasingly involved with digital technology, my inability to lend my neighbor a copy of e.g. some software, becomes an increasingly significant intrusion of business concerns into interpersonal and community relationships (Nissenbaum, 1995). The shrinking scope of fair use is, of course, also a significant concern, as it becomes increasingly difficult through DRM for the end-user to do simple space- and time-shifting, especially as DRM is given legal protection through the anti-circumvention provisions (§1201) of the Digital Millennium Copyright Act.

In order to enforce the artificial monopoly granted to the copyright holder, increasingly significant intrusions must be made on the public. The public is limited in use, sharing, and enjoyment of cultural artifacts, and even the astronomical

fines imposed seem insufficient to deter copyright infringement. From this, we should not simply conclude that music- and file-sharers are incapable of properly weighing risk and reward, but we should consider instead that sharing elements of culture with others may represent a human good as basic as freedom of speech, and that the file-sharer simply would prefer to put herself at risk of bankruptcy than to submit to a legal system which disallows her from this basic form of sociality.

These attempts at enforcement, furthermore, are not only onerous but also ineffective; perhaps doomed. In a global communications context, files can simply be hosted in localities where IPR are either weak or weakly enforced. The only options for a truly effective enforcement involve either (1) convincing the public that they ought not share, download, or remix IP, (2) spreading strong copyright law and strong copyright enforcement throughout all wired nations, or (3) monitoring individual internet connections to search for infringing packets. The first option seems both undesirable and unlikely: as noted above, sharing culture may be a human impulse basic to our social existence, and to convince the public otherwise may be difficult, and may be accomplished only at the expense of freedom of expression and a feeling of community. The second option seems an insurmountable task, unless strong moral arguments can be given why all nations should adopt strong copyright protection, and, as I have argued, sufficiently strong moral arguments may be lacking. The third option requires a thoroughgoing invasion of public rights that is hard to imagine could possibly represent the greatest good for the greatest number. The feeling of surveillance is deeply unsettling, and injurious to our privacy and security—no amount of big-budget movies and new pop singles seems to outweigh the disutility of knowing that our communications are being monitored.

Although the industrial dependence on strong copyright protection carries these disutilities, any lessening of the use of copyright protection may carry with it significant upheaval in IP-based industries, and these disutilities on the other side of the equation must be considered as well. We cannot ignore that the industrial production of culture employs a great many Americans, and thus supports many families and provides a driving force in our national economy. Intellectual products provide emotionally and economically rewarding work to a great many—not just studio and label heads, actors, musicians, programmers, and writers, but also caterers, personal assistants, studio musicians, limousine drivers, costume designers, best boys, lawyers, and lobbyists—and are a major export commodity, most notably in the movie, music, and software industries. As industry lobbyist Jack Valenti once hyperbolically stated this point, "By leading all other manufacturing sectors in their contribution to the American marketplace, the copyright industries are this nation's most treasured assets... Protection of our intellectual property from all forms of theft, in particularly [sic] online thievery and optical disc piracy, must take precedent [sic] if the United States is to continue to lead the world's economy" (Motion Picture Association of America, 2002).

It is, however, one thing to argue that utilitarian considerations support copyright protection *simpliciter*, and quite another to argue that utilitarian considerations support the status quo due to the disutility of social upheaval that may follow from an otherwise beneficial change in policy—especially if a similarly disutile social upheaval is required by the maintenance of the status quo, due to the shifting architecture discussed above. Indeed, on a strictly economic basis the structure of intellectual property lends itself to public production and ownership, even considered independently from the social and legal costs necessitated by enforcement given public availability of the relevant means of production. Intellectual property based products, as they have a high fixed cost and a negligible marginal cost, are of a type of good whose production costs are

best socialized rather than left to the private sector. As Michael Perelman (1991) stated,

[w]ithin [Kenneth] Arrow's logic, computer software is an ideal public good. Once produced, software code costs virtually nothing to duplicate. One can even read Arrow's analysis of the economics of information as an economic justification for the piracy of computer software; that is, software piracy, generously interpreted, approximates the price structure that pure neo-classical economics implicitly recommends, assuming that software vendors are marketing nothing more than the information embodied in the program. (p. 194)

This is not the place to ask about whether it is best to create intellectual goods through market forces, however. Our question here is whether utilitarian considerations justify using the strong copyright laws currently provided. If the utilitarian defense of our current IP legal regime is simply that we are dependent upon it, rather than that it is and will continue to be for the greatest good, then we are perhaps justified in continuing to use copyright, but only if we do so in a way that minimizes and our dependence upon it, and minimizes the burdens placed upon end users and consumers. If this is right, in other words, we are economically dependent upon a system which will cause increasing social harms, and the proper solution is to decrease our dependency until we can afford to no longer cause these harms.

There is an ancient and probably apocryphal story about Hero of Alexandria. Having invented a steam engine—an *aeolipile*, he called it—he went to his king, excited by its great productive potential. The king saw this as a problem, rather than a solution. "What," he asked, "would we do with all the slaves?" In a situation where we are economically dependent upon a legal regime, and technological changes make that regime unjustifiable, we should not use the disutility of social upheaval as a reason to avoid change. The situation here is of course less morally dire, but structurally similar. To continue using strong copyright is to continue our dependence upon an increasingly harmful legal regime, and, therefore, while it may be justifiable in some cases and temporarily, the burden of proof in any particular case should be *against* those who would argue that strong copyright is for the best, especially when enforced through DRM.

FUTURE RESEARCH DIRECTIONS

Shareware licensing represents a viable business model that is currently underutilized. Those illegally downloading music or movies often state that if they find something they truly enjoy, they will buy the CD or DVD in order to support the artists. This is nothing but a spontaneous public employment of the shareware model over currently proprietary goods. Musicians such as John Mayer and Jonathan Coulton have used online distribution networks in order to gain notoriety, and established musicians gain from the so-called "piracy" of their back catalogue, as Janis Ian has noted (2005). Many artists already produce most of their profits from live performances, and the diminution or elimination of copyright laws would aid this by reducing the artificial scarcity in reproductions of music, thereby freeing expendable income to be spent on entertainment to be given over directly to artists. Some artists such as Radiohead and Trent Reznor have decided to make their music freely downloadable, and have shown that it is possible to regard the musician as the provider of a service, supported through ticket sales, donations, and upselling, rather than as the provider of a product (the album or track) for sale. There is no reason that these methods of utilizing an open-source shareware model, already successful in some areas of music and software development, could not be adapted to other kinds of intellectual production. As public radio demonstrates, there is no need to ensure that each and every user pays for a service in order

to be able to fund the provision of that service to all who wish it.

A sponsorship model of funding is another underutilized strategy, and one which provides a patch to one of the major problems of the shareware model: the risk inherent in speculative creation of a good requiring a significant financial outlay. An example of sponsorship funding can be found in Maria Schneider's Grammy-winning album "Concert in the Garden." The album was produced through artistShare.net, which offers artists the ability to offer fans the opportunity to sponsor the production of works, similar to the way in which institutions or individuals might commission a work, but on something closer to a grass-roots model. As Maria Schneider put it, "This project was funded with the help of my fans and distributed entirely through my own website. I feel very proud of taking that first step and incredibly grateful that it has proven to work so well!" (AllAboutJazz.com, 2004). The artistShare model could equally well be applied to other areas of intellectual production, allowing businesses currently dependent upon sales for income to find an alternate and sustainable source of revenue.

The shift from a sales- to a service-based business model is already emerging in various ways in the marketplace. In software, Red Hat offers the clearest example of the potential for growth in this business model, and Richard Stallman (2004) has argued for this as a more general economic model. Red Hat has enjoyed great success selling subscriptions, support, training, and customization of open-source and free software. As open-source and uncommodified software becomes more prevalent, we can expect demand for software servicing to increase perhaps precipitously, not only because of the obvious challenges of implementation or the obvious advantages of customization, but because of liability issues. One of the great benefits which software service providers are able to offer businesses is the ability to take a share of legal liability in cases of software-based damages, such as compromised client personal information or data loss. In this way, software service firms are able to act as informational insurance brokers, offering a greater degree of financial security, in addition to the improved data security which customization and a direct service relationships offer, as compared to centralized, mass produced software sellers.

MIT's OpenCourseWare provides a similar example of rethinking intellectual property as something to be serviced rather than sold. MIT OpenCourseWare

- Is a publication of MIT course materials
- Does not require any registration
- Is not a degree-granting or certificate-granting activity
- Does not provide access to MIT faculty (2007)

When MIT first started this open-source approach to their course materials, which makes available online all manner of information from syllabi to lecture notes to tests, there was reportedly some confusion about why MIT would give away the very goods it was in the business of selling. As Hal Abelson (2004) explained at an academic open-source conference, MIT decided the information covered in a course was not a competitive good with enrollment within that course—MIT, in other words, does not view itself as a seller of information, but rather as a supplier of services, which include not only access to information, but also presentation of that information within a particular environment of other students, accessible and responsive professors, labs, discussion groups, and various other educational resources.

CONCLUSION

Now that the means of industrial production of intellectual property are firmly in public hands, any reasonable assessment of the future of monopolistic and exclusive employment of intel-

lectual goods will include increased regulation and encroachment upon the everyday lives and personal activities and projects of the public at large. The public is already denied many benefits of intellectual goods. As technological advances continue, these denials will have to become more extreme as means of circumvention become more powerful. Furthermore, the benefits which these denials remove from the public will become greater as the productive abilities which would otherwise be in public hands increase in power and as their application continues to expand in breadth through the increasing range of uses of digital technology. In the absence of any moral right of authors requiring the right of exclusion—a moral right which the Lockean and Kantian perspectives no longer supply—this increasing disutility is unjustifiable.

Given recent technological advances, the copyright bargain removes significant benefits from the many in order to benefit the few, and furthermore does so by inflicting harms upon the many by restricting basic rights, by breaking down social bonds, and by preventing the public from freely realizing their creative impulses. Given the likelihood that such advances will continue, we can expect both the benefits lost and the harms gained to steadily and continually increase. Whatever benefits the copyright bargain still provides may be nearby and certain, but they are impure, and the harms brought about through the copyright bargain are nearby, certain, fecund, and wide-ranging.

As e-business models emerge, and as businesses increasingly become e-businesses, entrepreneurs, managers, and others involved in business have a responsibility to find revenue models which are not dependent upon asserting unjustified rights and removing rights and freedoms from others. There are many such models, as discussed in the foregoing, but the business environment is currently in flux, and these models are unstable. There are surely other such models yet to be developed. There are, however, some principles that businesses should follow.

Given that consumers are now producers and distributors; given that anyone can create, edit, publish, and remix intellectual goods; and given that the technical capabilities of the public are likely only to further increase, businesses that wish to survive as content providers must recognize that they can no longer be gatekeepers, and that they need to ask what they have to offer the consumer, rather than asking what they deserve from the consumer. Members of the public must be treated as partners and clients, not consumers. This is the primary principle I would suggest for how to use copyright in an ethical and sustainably profitable way: *think service, not product.* Digital media means consumers do not need content producers anymore: they can produce their own content if you make yours too difficult to use, restrictive, or otherwise unattractive. Don't ask how to get your user/audience to pay to access content; ask what you can do to make them value your business and the service you provide in that content delivery.

How can you do this?

1. *Don't tell your customers how to enjoy your product.*
2. *Open discussions, don't close them.*
3. *Remember that the user is supporting you, even if she might not be paying to do so*

Don't lock users into a particular device or software program if you don't have to. Don't make them feel like you're trying to take their information, privacy, or ability to choose. It's unjustifiable, and the public is increasingly unwilling to put up with it, even if it is legal. Both morally and practically, it's a bad way to do business. Give them the freedoms they need to enjoy your service on their own terms. Give them the opportunity to support you because they like your service and your content rather than trying to force them to support you if they want access or use. This is a time of great opportunity for e-businesses based in content provision, but only if business can respond in a positive and proactive way to technological changes; can recognize and respect the power that users now have; and chooses to treat the user base as partners, whether or not they have paid for the privilege of popularizing your content and bringing you publicity.

REFERENCES

Abelson, H. (2004, November). *Universities, the Internet, and the Intellectual Commons.* Paper presented at the Conference on the Intellectual Commons, Orono, ME.

AllAboutJazz.com. (2004). *Maria Schneider - 4 Grammy nods and NOT ONE RETAIL SALE!* Retrieved from http://www.allaboutjazz.com/php/news.php?id=4785

A&M Records v. Napster, No.00-16401, D.C. No. CV-99-05183-MHP (2001).

Assemblée nationale française. (n.d.). *Law on the Intellectual Property Code* (Legislative Part) (No. 92-597 of July 1, 1992, as last amended by Law No. 97-283 of March 27, 1997). Retrieved from http://www.wipo.int/clea/en/details.jsp?id=1610

Bell, T. (2002). Indelicate Imbalancing in Copyright and Patent Law, In A. Thierer and C.W. Crews Jr. (Eds.), *Copy Fights: the Future of Intellectual Property in the Information Age* (pp. 1-17). Washington, DC: Cato Institute.

Deutscher Bundestag. *Law on Copyright and Neighboring Rights (Copyright Law) of September 9, 1965, as last amended by the Law of July 22, 1997.* Retrieved from http://www.wipo.int/clea/en/details.jsp?id=1032

Eldred v. Ashcroft. 1 U.S. 537 (2003).

Federal Convention of 1787. (1787). *The Constitution of the United States of America.* Retrieved from http://www.law.cornell.edu/constitution/constitution.overview.html

Feist Publications, Inc. v. Rural Telephone Service Co. 499 U.S. 340 (1991).

Fox Film Corp. v. Doyal, 286 U.S. 123 (1932).

Free Software Foundation. (2002). *Selling Free Software.* Retrieved May 6, 2009, from http://www.gnu.org/philosophy/selling.html

Hull, G. (2009). Clearing the Rubbish: Locke, the Waste Proviso, and the Moral Justification of Intellectual Property. *Public Affairs Quarterly, 23*(1), 67–93. Retrieved from http://papers.ssrn.com/sol3/papers.cfm?abstract_id=1082597.

Ian, J. (2005). *The Internet Debacle—An Alternate View.* Retrieved May 6, 2009, from http://www.janisian.com/article-internet_debacle.html

Jefferson, T. (1813). *Thomas Jefferson To Isaac McPherson, Monticello, August 13th, 1813.* Retrieved from http://www.temple.edu/lawschool/dpost/mcphersonletter.html

Kant, I. (1996a). Groundwork of the Metaphysics of Morals. In Gregor, M. (Ed.), *Practical Philosophy (Cambridge Edition of the Works of Immanuel Kant)* (pp. 37–108). Cambridge, UK: Cambridge University Press.

Kant, I. (1996b). The Metaphysics of Morals. In Gregor, M. (Ed.), *Practical Philosophy (Cambridge Edition of the Works of Immanuel Kant)* (pp. 353–604). Cambridge, UK: Cambridge University Press.

Kant, I. (1996c). On the Wrongfulness of Unauthorized Publication of Books. In Gregor, M. (Ed.), *Practical Philosophy (Cambridge Edition of the Works of Immanuel Kant)* (pp. 23–36). Cambridge, UK: Cambridge University Press.

Korsgaard, C. (1996). *Creating the Kingdom of Ends.* Cambridge, UK: Cambridge University Press.

Landes, W., & Posner, R. (2003). *The Economic Structure of Intellectual Property Law.* Cambridge, UK: Belknap.

Lessig, L. (2006). *Code 2.0.* New York: Basic Books. Retrieved from http://codev2.cc/download+remix/

Locke, J. (2005). *Two Treatises of Government.* Project Gutenberg. Retrieved from http://www.gutenberg.org/ebooks/7370

MIT OpenCourseWare. (2007). *MIT OpenCourse-Ware.* Retrieved September 4, 2007, from http://ocw.mit.edu/index.html Archived at http://web.archive.org/web/20070904051417/http://ocw.mit.edu/index.html

Motion Picture Association of America. (2002). *Study shows copyright industries as largest contributor to the U.S. Economy.* Retrieved December 3, 2005, from http://mpaa.org/copyright/2002_04_22.htm Archived at http://web.archive.org/web/20051203235151/http://www.mpaa.org/copyright/2002_04_22.htm

Motion Picture Association of America. (2005). *Internet Piracy.* Retrieved May 6, 2009, from http://www.mpaa.org/piracy_internet.asp

Nadel, M. (2004). How Current Copyright Law Discourages Creative Output: The Overlooked Impact of Marketing. *Berkeley Technology Law Journal, 19*, 785-856. Available at SSRN: http://ssrn.com/abstract=489762

Nissenbaum, H. (1995). Should I Copy My Neighbor's Software. In Johnson, D., & Nissenbaum, H. (Eds.), *Computers, Ethics, and Social Values* (pp. 200–212). Upper Saddle River, NJ: Prentice Hall.

Parliament of England. (1710). *The Statute of Anne.* Retrieved from http://www.copyrighthistory.com/anne.html

Perelman, M. (1991). *Information, Social Relations and the Economics of High Technology.* New York: St. Martin's.

Perelman, M. (2002). *Steal this Idea: Intellectual Property Rights and the Corporate Confiscation of Creativity.* New York: Palgrave.

Stallman, R. (2004). Why Software Should Not Have Owners. *The GNU Project and the Free Software Foundation*. Retrieved from http://www.gnu.org/philosophy/why-free.html

U.S. Department of Justice. (1998). *The No Electronic Theft ("NET") Act: Relevant portions of 17 U.S.C. and 18 U.S.C. as amended (redlined)*. Retrieved from http://www.usdoj.gov/criminal/cybercrime/17-18red.htm

Vaidhyanathan, S. (2003). *Copyrights and Copywrongs: The Rise of Intellectual Property and How It Threatens Creativity*. New York: New York University Press.

ENDOTES

[1] The argument I have presented in this section is based on the "enough and as good" portion of the Lockean proviso. For a valuable alternate approach, based upon the spoilage proviso, see Gordon Hull's "Clearing the Rubbish: Locke, the Waste Proviso, and the Moral Justification of Intellectual Property" (2009).

[2] E.g. "Piracy is theft, and pirates are thieves, plain and simple. Downloading a movie off of the Internet is the same as taking a DVD off a store shelf without paying for it" (Motion Picture Association of America, 2005). This false identity has in important ways been codified in American law under the No Electronic Theft Act of 1997 (henceforth, the NET Act). This act, besides implicitly equating copyright infringement and theft through its very name, strikes a provision in 17 U.S.C. § 506 which formerly required commercial advantage or private financial gain as a criterion for a criminal infringement offense. Under the law as modified by the NET Act, criminal infringement may be established by the willful reproduction or distribution of copyrighted material of a total retail value exceed $1000 within any 180 day period or by the receipt of commercial advantage or private financial gain. Further, the NET Act added a creative definition to 17 U.S.C. §101, stating that "the term 'financial gain' includes the receipt, or expectation of receipt, of anything of value, including the receipt of other copyrighted works" (U.S. Department of Justice, 1998). This definition states that the receipt of copyrighted works is an illegal financial gain, thereby implying that piracy is a form of theft.

[3] As stated in the Statute of Anne: "Whereas printers, book sellers, and other persons have of late frequently taken the liberty of printing... or causing to be printed... books and other writings without the consent of the authors or proprietors of such books and writings, to the very great detriment, and too often to the ruin of them and their families; for preventing therefore such practices for the future, and for the encouragement of learned men to compose and write useful books; may it please your Majesty..." (Parliament of England, 1710) Cf. also Constitution of the United States, Article 1, §8, clause 8 (Federal Convention of 1787).

[4] *The Metaphysics of Morals* does not add anything of significance to these issues to his earlier comments in "On the Wrongfulness [etc.]." For those who wish to compare the two, or who do not have easy access to the less common essay, the argument addressed in following is presented in briefer form in *The Metaphysics of Morals*, §31, II., 6:289-91 (1996b, pgs. 437-8).

[5] This idea is also present in both French and German law (Assemblée nationale française, 1997; Deutscher Bundestag, 1997).

[6] In a surprising footnote, Kant asks rhetorically "Would a publisher really venture to bind everyone buying the book he publishes to the condition that the buyer would be pros-

ecuted for misappropriating another's goods entrusted to him if the copy sold were used for unauthorized publication, whether intentionally or even by negligence?" (1996c, p. 29fn). This now may constitute "contributory infringement," wherein one may be found guilty of copyright infringement simply by offering assistance to the practice of copyright infringement in situations in which the abettor knew or ought to have known of the infringing activities. This was a primary charge in the Napster case (*A&M Records v. Napster*, 2001), and has since been used in threats directed at individuals, universities, and internet service providers.

[7] A more complete list of possible utilitarian justifications might look like this: copyright protection, in exchange for an acceptable loss of social freedoms, (1) provides a greater diversity and amount of intellectual goods to the public, (2) allows for creation of goods requiring a huge initial investment, which would not otherwise be produced, (3) allows for profit-motivated responsiveness to the desires of the public, (4) diminishes search costs on the part of users/consumers through effective marketing and distribution, (5) effectively and efficiently preserves and distributes older and less popular goods of cultural value, (6) provides rewarding labour to many Americans, (7) represents a valuable export good for the U.S. market, (8) discourages wasteful rent-seeking expenditures. Here, I only am able to address the most central intuitions about a utilitarian justification of IPR: that the copyright bargain is necessary to adequately encourage socially and economically foundational production. Landes and Posner (2003) discuss all of these possible utilitarian justifications, including those that are less obvious or intuitive, however they note that "while we discuss a number of issues relating to intellectual property rights in computer software and to the impact of the Internet on intellectual property law, readers who believe that these are *the* central issues of that law today will be disappointed with our coverage" (p. 7).

Chapter 12
A Case for Consumer Virtual Property

Matt Hettche
Christopher Newport University, USA

ABSTRACT

While the Internet is generally regarded as a tool of consumer empowerment, recent innovations in e-marketing signal a disparity in the quality of knowledge that the e-buyer and e-seller each bring to the exchange process. Armed with sophisticated consumer tracking programs and advanced data mining techniques, the e-seller's competitive advantage for anticipating consumer preference is quickly outpacing the e-buyer's ability to negotiate fair terms for an equal trade. This chapter considers the possible threat that aggressive forms of electronic surveillance pose for a market economy in e-commerce and offers a framework for how marketing practitioners can protect consumer autonomy online. Using John Locke's classic social contract theory as a model, I argue that information created by an end-user's online activity is a form of 'virtual property' that in turn establishes a consumer's right to privacy online.

INTRODUCTION

As a modern business practice marketing requires, if not depends on, a certain degree of consumer autonomy. If a consumer's ability to participate in the exchange process is significantly undermined, there is an important sense in which marketing ceases to be a relevant business function. A consumer's freedom to choose is a basic requirement for marketplace competition. If a buyer's decision to enter

into a sales agreement is manipulated or coerced, the conditions for a fair exchange are jeopardized. Practitioners of marketing, therefore, have a special interest in both preserving and advancing consumer autonomy. (Lippke, 1989; Heath, 2005)

In the traditional marketplace, one way for preserving consumer autonomy has been the personal relationships that develop between buyer and seller in the exchange process. Physical geography and the necessities of life dictate standards for ongoing sustainable commerce. The integrity of the marketplace, in this respect is grounded upon the

DOI: 10.4018/978-1-61520-615-5.ch012

assumption that trading partners will (or at least in theory, could) meet again in the future. On the traditional model, consumer satisfaction and merchant success are interdependent variables situated within a larger social context.

In our postindustrial capitalist society, however, the means for protecting consumer autonomy has shifted away from personal relationship buying and selling. Advancements in technology and communication allow buyers and sellers to interact in ways that transcend traditional social constructs. The level of production and vast scale of distribution supported by the modern global corporate system place an increased emphasis on brand image and a firm's ability to provide consistent and predictable customer service throughout multiple channels of distribution.

The central problem considered in this chapter is whether the type of transactions that take place in e-commerce can support the principles of a market economy, if certain forms of electronic surveillance become the norm. The worry, put succinctly, is that if the e-seller's ability to predict an e-buyer's purchase disposition greatly exceeds the e-buyer ability to evaluate marketplace conditions, the chances for fair trade diminish considerably. Having detailed knowledge about a consumer's interests and lifestyle choices, for example, can easily lead to instances of manipulation and exploitation. In a worst-case scenario, I will argue, aggressive e-marketing techniques represent a form of predatory capitalism that significantly undermines consumer autonomy (in general) and broadens the susceptibility of certain at-risk consumers (in particular). The solution this chapter outlines involves using a notion of 'consumer virtual property' to demarcate the ethical boundaries for fair information transfer within the e-business environment. Using John Locke's classic social contract theory as a model, I will explore how his concept of the 'state of nature' and his labor theory of private property has unique application for protecting consumer privacy online.

THE MORAL AND SOCIAL CONDITIONS OF E-COMMERCE

One truly amazing aspect of e-commerce is the incredible speed with which buyers and sellers are able to connect with one another. In *The World is Flat*, Thomas Friedman suggests the way information is created, delivered, and exchanged online has both a "leveling" and "democratizing" effect for the global economy. (Friedman, 2004) The web's leveling effect allows unprecedented opportunities for individuals to communicate across vast distances. For the first time in history, the average person can author digital content, reach a diverse audience, and sustain a meaningful dialogue with like-minded individuals. Traditional media outlets, such as newspapers and network television, are increasingly in direct competition with novices who self publish on the web. The web's democratizing effect, in turn, is manifest in the way information is accessed and created. Online content is quite literally produced "by the people, for the people" in a way that (good or bad) breaks with longstanding printing and publishing traditions.

From a marketing perspective, the Internet unquestionably provides increased opportunities for buyers and sellers to interact. E-commerce is an evolving process that facilitates buying and selling in an online digital medium. Products and services are either traded directly (i.e., electronically) or indirectly by establishing the conditions for a future exchange in the real 'brick-and-mortar' world. In ways that even possibly surpass traditional commerce, e-commerce supports the principles of free market enterprise by facilitating transparency, low transaction costs, rational engagement between trading partners, and open competition. In contemporary e-commerce, buyers and sellers throughout the world have the opportunity to enter freely into trade agreements, negotiate price, and establish mutually beneficial conditions of sale. (Brodie, Winklhofer, Coviello, & Johnston, 2007)

At present, the typical relationship formed between the buyer and seller in e-business is socially indeterminate. Who a person is and how that person's social identity is set within a broader social context is rarely disclosed in an e-commerce sales transaction. Independent of any profiling measures, an e-consumer's personal preferences and means of financial support are not easily discerned. The populations converging online to do business are simply too large and diverse to know with any great certainty who a person is and how a person is situated within a larger social community. The e-consumer's social indeterminacy is further affected by the very fluidity of an end-user's identity when online. The purpose and goals of our online activity often range considerably: a particular online session can either be well-planned with a particular objective or it can be dictated by an arbitrary playful whim. We might seek product information, or entertainment, or personal connection (or all three) in any one given online session. Simply put, our online personalities quickly change, morph, and bifurcate, all depending on our ability and motivation to manage multiple interests.

Another notable aspect of e-commerce is the very loose, but still detectable, forms of regulation that underlie the exchange process. To a large degree, the majority of business transactions that take place online are private 'single-price-taker' trades where no actively observing third party can mediate potential disputes. Compensation and due process for failed sales contracts are at least possible, however, if the trading partners are motivated to seek an independent arbiter. For example, quite often the consumer can appeal to his/her credit granting institution to block or stop payment for a sales transaction that has gone bad. Sellers, in turn, can report to third party credit agencies about any buyer malfeasance, such as insufficient payment for product/services rendered. In some cases, if participants are trading within a sufficiently transparent context, government and law enforcement can intervene to resolve conflicts.

These sorts of interventions are most common when a product or service is deemed illegal or socially inappropriate. In recent years, for example, child pornography, the distribution of controlled substances, and instances of cyber-bullying (i.e., intimidating/threatening communication directed at a private innocent citizen) have warranted the involvement of government authorities.

To be sure, part of the success of e-commerce in modern society is tied to the highly evolved direct-to-consumer business infrastructure that has grown up in the last thirty years. Two instruments in particular that have facilitated the buying and selling of products online are (1) the modern consumer credit industry and (2) the ground-to-air 'overnight' shipping industry. Both predate the arrival of the World Wide Web in mass culture and each has subsequently grown as a result of the web's rapid adoption. The pace and scale of e-commerce are a direct correlate of consumers obtaining credit and having a reliable means for delivery.

Marketing techniques that rely on aggressive forms of surveillance and profiling, however, threaten the competitive equilibrium that exists between buyer and seller. For our purposes, aggressive e-marketing is defined as the set of tools and techniques that shift the balance of power in the exchange process away from the buyer to the seller by utilizing forms of technology which are neither transparent nor accessible to the buyer. If the information collected by an e-seller about an e-buyer is disproportionate to the information available to the e-buyer about the market environment, the transparency of the sales agreement (or what is being agreed to) is undermined. Arguably, the more detailed the information is about a consumer's private preference, the more opportunity there is for exploiting those preferences for profit. One classic example of how aggressive forms of information surveillance can lead to unfortunate and ethically suspect results involves Maryland's public health commission. (Quinn, 2009; Robischom, 1997) In 1993, the State of Maryland, in an attempt to cut costs related to its health/ medical programs, decided to utilize a digital medical records format for some of its residents. One of the administrators of the program, who was incidentally also a bank executive, became aware through the database that some of his loan customers were undergoing treatment for cancer. Using information he gained from the database, he called in the loans of these customers with cancer out of fear they would not be able to repay their loans because of mounting medical costs and/or death. Undoubtedly, both the amount and detailed nature of digital information, including the incredible speed and accuracy with which it can be recalled, poses potential security risks to those exchanging information in a digital medium.

Arguably, it is therefore useful to identify a worst-case and best-case scenario for how electronic surveillance and data mining can possibly affect consumer autonomy. Anticipating how the tools of database marketers can be applied both positively and negatively helps establish a frame of reference for constructing normative guidelines for responsible marketing practices.

- *Worst-case scenario:* A web company builds a complete longitudinal profile of a particular end-user/consumer. An electronic history of an individual's lifestyle and personal choices are recorded in detail, including information about regular purchases, tastes in literature, film, music, religion, politics, and sexual preference. An end-user's search behavior of financial investments and health/medical concerns are tracked and help provide a metric for assessing the consumer's tolerance for risk. By further monitoring electronic bank statements, credit card transactions, and personal email, a profile of the consumer's health and well-being is constructed. No longer simply a member of a targeted group or demographic, the individual consumer is the subject of direct surveillance.

Information collected is used to predict when an individual consumer is most vulnerable to an exploitive sales agreement.

- *Best case scenario*: Firm-side surveillance of online consumer activity reveals basic trends of consumer wants and needs. Tracking consumer preferences helps identify opportunities for expanding existing markets and developing new products and services. By preserving end-user anonymity and clustering private preferences into statistical norms, consumer behavior is extrapolated and generalized across several different market sectors. The exchange of information between the e-buyer and e-seller is open, by consent, and support the principles of free market enterprise. Marketing, as an essential business function, strives to maximize consumer well-being for the sustainability and efficiency of modern commerce.

While these two scenarios represent two sides of a continuum, it is perhaps remarkable how close each is to becoming a reality. Certainly, aspects of both are already in existence. In truth, from a technology standpoint, there is little difference in what a company or firm would need to pursue an aggressive verses an ethically responsible program of consumer data mining. Companies like 'Acxiom, ' of Little Rock, Arkansas and 'Cisco Sytems, Inc.,' compile huge databases of consumer information from census data, tax records, insurance claims, credit card receipts, and customer loyalty cards. (Goodman, Rushkoff, & Dretzin, 2003)

Serving a clientele of local, state, and federal governments, politicians, and commercial businesses, these companies can produce very specialized marketing lists that are segmented according to very precise parameters. (Jacobs & Stone, 2008)

The fundamental challenge for marketing practitioners, therefore, is how to promote a course of action for modern business that will more likely favor a best case rather than worse-case scenario? What, if anything, can influence the standards within e-business to promote the responsible use of consumer information? The answer I offer in this chapter, involves a notion of 'consumer virtual property,' that is modeled in large part from John Locke's social contract theory.

LOCKE'S STATE OF NATURE AND E-COMMERCE

A standard way to introduce Locke's concept of the state of nature, and one I follow here, is to contrast Locke's concept with that of his predecessor, Thomas Hobbes. Hobbes's view is a mainstay of the social contract tradition and is typically interpreted as a form of egoism: the self-serving behavior of human beings, when combined with the natural scarcity of resources available in nature, leads human beings into a state of war/ state of nature. In his *Leviathan* (1651), Hobbes describes pre-political society as one dominated by fear, insecurity, and the brute exercise of might. Distribution of goods in Hobbes's state of nature is dictated by the strength and ingenuity of each individual hoping to secure marginal material gain. Hobbes believes that as available resources decline, inevitable competition will occur between those seeking benefit and advantage. For Hobbes, human beings are mechanistically determined to act in ways that will satisfy their interests, needs, and desires. The scant availability of resources and the inevitable conflicts that ensue for possession and control define a set of social conditions where life for all is "solitary, poor, nasty, brutish, and short." (Hobbes, [1651] 1990, p. 89) According to Hobbes, only if humans act rationally and resign their natural rights to an ultimate authority can they extract themselves from such a deplorable natural state.

Locke's view of the state of nature, in contrast, assumes a much more optimistic view about human

motivation and the social environment. At one point in the *Second Treatise*, for example, Locke even refers to the state of nature as "a state of peace, goodwill, mutual assistance, and preservation." (Locke, [1690] 2008, p. 280) Human beings, aided with rational faculties and natural compassion for others, pursue goods to sustain their life, interests and general convenience. While competition and conflicts are bound to occur, Locke believes that it is not simply human nature and environmental scarcity that spawn war and motivates humans to seek political compromise. Rather, for Locke, political society and the desire for government emerges from a practical need to increase convenience and efficiency in life. Locke believes, in particular, it is because inhabitants within the state of nature desire impartial umpires to judge and enforce contracts that a political solution comes into being.

Following A. John Simmon's commentary, there are three unique features of Locke's concept of the state of nature worth noting. (Simmons, 1989) The first is that the state of nature, although a pre-political environment where civil law has little or no reach, is not devoid of morality altogether. In Locke's state of nature, human beings possess natural rights and a "rule of reason" that establish a basic set of duties and obligations. According to Locke, each human has the right to pursue his own "life, health, liberty and possessions" and all are "free" (in a fundamental moral sense) to do so. (Locke, [1690] 2008, p. 271) Lockean natural rights also stipulate that human beings have a duty not to waste materials and resources that would otherwise be available to the common public and humans should only consume nature's bounty to the extent that there is "enough and as good left for others." These natural obligations are sometimes referred to as the 'no waste condition,' and 'sufficiency proviso,' respectfully. (Locke, [1690] 2008, pp. 287-9; Tavani, 2005, p. 88)

A second key feature of Locke's concept of the state of nature is how the concept (as descriptive of a particular context, situation, or condition) individuates members of a group rather than describes a group or society as a whole. In this sense, Locke's concept of the state of nature is individualistic and anchored to a person's unique identity through time. (Simmons, 1989) For Locke, reason is a human's most important asset for survival. In contrast to Hobbes, where human beings are seen as egoistic and mechanistically determined, Locke's view of human beings assumes a sufficient degree of free will and personal efficacy. Locke is firmly committed to the Enlightenment ideal that people can and should be in control of their own lives. For Locke, people can think through things for themselves and act in a responsible manner to guide their own destiny. Guidance from a pre-established authority, such as the church or state, is simply not required for pursuing a happy and meaningful life.

A third feature of Locke's concept of the state of nature important to mention is its relational aspect (or the extent to which the state of nature stands as a relation that holds between an individual and others). Descriptive of a particular type of interaction, the state of nature is a condition in which two or more individuals find themselves. As a concept, it extends into the past as well as present moment of human interaction. A person can be within the state of nature with regard to one relationship, and yet in a different context, perhaps with the very same person, remain outside the state of nature. For Locke, the state of nature is less of a social or anthropological period of development for human beings as it is a construct that defines how individuals (living now) are related to one another. In the *Second Treatise*, Locke identifies children, princes of sovereign nations, and madmen as typical participants of the state of nature. (Locke, [1690] 2008) In this respect, the key for civil society to come about is the consent of interacting parties to oblige the impartial authority of a legitimate government. For Locke, more precisely, it is ultimately the mutual consent of the involved parties that extracts them from their natural condition.

Locke's view of the state of nature is largely informed by his theological commitments. For Locke, nature as a whole is considered a gift that descends from God, the creator. Humans are free to exploit the resources of the planet, including animals, for the preservation, enjoyment, and convenience of their own life. On Locke's view, humans are granted dominion over nature insofar as humans have the liberty to consume and appropriate the materials of nature for their own interests and projects. Restricted only by the 'no waste principle' and the 'sufficiency proviso,' humans are free to do what they want, when they want, with the available bounty of nature.

Applying Locke's 17th century concept of the state of nature to the digital environment is initially plausible, given the widespread availability of information online and the relative ease end-users can access it. The way digital information is created, copied, transferred, and stored, essentially guarantees no spoilage or violation of a Lockean 'no waste principle.' Using web page publication as a model, the underlying structure of the Internet where information is uploaded, downloaded, accessed, and copied establishes a digital commons that supports both benign consumption and civil exchange. Apart from the very real problem of e-waste (i.e., the production, consumption, and disposal of computer and high tech communication devices), the mere participation of interacting online remains in principle a harmless activity that empowers each end-user to seek advantages in life for his/her own convenience. The Internet conceived as a digital commons, therefore, remains open for end-users to access how often they want, whenever they want, for whatever purpose they want, as long as an end-user's access does not conflict with the natural rights of others.

Relationships typical of e-business also bear a striking relationship with those occurring in a Lockean state of nature. Similar to a Lockean state of nature, e-business assumes a basic level of human rights, presupposing equality, fairness, and the value of possession in the marketplace.

Exchanges that occur online are predicated upon normative expectations that mirror those in any healthy capitalist system of exchange. Although perhaps not always realized, there are moral expectations that the e-buyer and the e-seller each bring to the exchange process.

E-commerce, as a contemporary activity, also assumes a certain degree of individualism, again similar to a Lockean state of nature. The ability of individuals to seek out and maintain relationships online requires a basic level of rational competence in order to successfully enter and execute contracts. In fact, the 'single-price-taker' model, central to the functioning of a capitalist economy, is reinforced by the one-to-one interaction typical of an e-commerce exchange. Although there are occasions when groups of individuals share one online session from a single internet portal, the vast majority of those interacting online are engaged through a single-user (personal computer) internet terminal.

Relationships within e-business, much like those in a Lockean state of nature, also exhibit varying degrees of independence from outside regulation. The typical e-commerce transaction requires a certain amount of good faith on part of participating parties to both motivate and complete an exchange. And although some of the relationships in e-commerce are independent of any regulating body or authority, the relative peace and chances for success depend on the individual context that the sales transaction takes place. For example, activities that transpire between consumer to consumer trading, such as bartering on social networking sites, often function at the limits of *caveat emptor*. In this respect, the relative degree to which the state of nature, as a relation between individuals, expresses itself remains a function of how isolated the exchange process is from public view. To be sure, some forms of e-commerce, where incentives for the buyer and seller are mutually present and transparent, have a better chance of success than when the motives of trading partners are not as obvious.

THE E-CONSUMER AND A LABOR THEORY OF PROPERTY

The modern e-consumer does not easily conform to the image of an exploited *bourgeois* of industrial age capitalism. The abilities and opportunities of the e-consumer to seek out information and find new trading partners demonstrate a level of empowerment that arguably far exceeds the disadvantaged worker/consumer of Marx's milieu. Today's e-consumer patiently initiates commercial exchanges that transpire over vast geographic distances that often require days, if not weeks, to complete. The e-consumer's confidence to manage and benefit from relationships formed online also signals a level of independence that would seem to support the conditions of a sustainable market economy. The typical e-consumer not only exemplifies a tolerance for risk, by routinely releasing personal information within an electronic medium, but also demonstrates a willingness to experiment with new products and forms of distribution.

The situation unfolding for modern e-consumers, however, presents a challenge to the level of empowerment and independence they currently seem to enjoy. For not only do the techniques of surveillance, data mining, and behavioral profiling threaten continued consumer autonomy, but there is growing evidence that the very businesses participating in a high tech marketplace are engaging in anti-competitive behavior. In a recent article, Joseph Sirgy and Chenting Su argue that consumer sovereignty is an increasingly untenable concept given how buyers and sellers interact in a high tech marketplace (Sirgy and Su, 2000). Sirgy and Su maintain that the considerable resource advantages of large firms to research and develop new products create enormous incentives for companies to seek cooperation in the production process. The rapid pace and sophistication that new products are introduced into the marketplace, for example, undermine a traditional *quid pro quo* model of producer competition. They maintain that the development of high tech products descends from "different experts and a history of trial and error, product development decisions, and technological improvements.... [And] [n]o one person has all the expertise to fully develop a technological innovation from concept generation to the commercialization stage." (Sirgy and Su, 2000, p. 7)

Consumer sovereignty is further threatened by the appearance of highly similar competing products, saturating the marketplace, and thereby thwarting the consumer's decision making ability. Labeled by Barry Schwartz as the "paradox of choice," consumers in the high tech marketplace are confronted with a myriad of product and brand choices that are difficult to discriminate. (Schwartz, 2004). According to Sirgy and Su, consumers lack both the motivation and ability to maximize their utility in the marketplace, and often exhibit behavior of "information avoidance" (i.e., ignore available information because of its complexity or confusing nature) or engage in "artificial brand/product comparison" (i.e., compare two objects/brands with an arbitrary or inconsistent criteria). (Sirgy and Su, 2000, p. 7)

So while e-commerce begins as a promising extension of free market exchange, it quickly becomes threatened by the very nature of modern production. The dual threat of non-competitive production and aggressive marketing techniques not only disadvantages the modern consumer, in general, but broadens the susceptibility of certain at-risk consumers. Groups such as low-literate consumers, mentally ill consumers, and ageing consumers will have an increased difficult time coping with marketplace decisions as the reach and growth of e-commerce expands. So in addition to creating ethical guidelines for marketing, directed at what George Brenkert calls the "specially vulnerable" consumer, there needs to be measures for protecting e-consumer privacy so that the lure of predatory business practice is kept in check. (Brenkert, 2009).

Locke's labor theory of private property can be useful for outlining how consumers can retain and defend a set of natural rights within the online exchange process. In Locke's *Second Treatise*, the philosophical problem that motivates his labor theory of property involves the dilemma of how anything can be owned privately when, in the state of nature, everything is held in common and all human beings are equal. (Locke [1690] 2008,) Put simply, if everybody has an equal right to nature's bounty, how can anybody legitimately claim one piece of it for his/her own? Locke's answer involves a series of inductive inferences, starting with the idea that (1) each person retains ownership of his/her body; and (2) each person is responsible for his/her labor. Using a metaphor of "mixing," Locke maintains that a person's effort and ingenuity, when combined with materials gained from nature, creates private property. As long as the laboring individual does not violate the 'no waste condition' and 'sufficiency proviso,' Locke believes the individual is morally entitled to the fruits of his/her own labor. When presenting his account, Locke appeals to the objects of apples, acorns, and fish to illustrate his point. His central idea is that because an individual can successfully and safely obtain these objects of nature by virtue of his/her own labor, the resulting reward is private possession.

THE STRUCTURE OF VIRTUAL PROPERTY

There is something certainly initially plausible about using property to ground rights in a system of capitalism. Within the classic liberal tradition of economics, for example, property and contractual obligation are considered bedrock for the function and success of an economy. For the political philosopher John Hospers property is considered the best hedge humans have for securing an uncertain future. (Hospers, [1974] 2005) Property rights not only establish a basis for noninterference from authoritarian rule but they allow for standardization and consistency when property is transferred from one owner to the next. In his philosophical analysis of property, John Christman identifies two forms of property that typically operate in free markets. The first is what he calls property "as a set of autonomy interests," and second is property "as a set of income interests." (Chrisman, 1994, p. 7) Whereas autonomy interests are those features of human existence that foster a person's ability to plan and execute projects in life, income interests designate and direct the distribution of wealth in society. Both sets of interests have their place within the proper functioning of a capitalist economy; however, according to Christman, there is an important sense in which autonomy interests have a *prima facie* moral priority over income interests when a conflict of rights occur. Christman's insight, and one I apply here to virtual property, is that some rights of possession are basic to the preservation and conduct of one's life, whereas others (*viz.* income interests) are merely a product of the economic system itself.

Virtual property comes into existence through a variety of means. As Internet and computer technology evolve, virtual property will undoubted itself change. At present, however, there are three common end-user activities that epitomize virtual property, as such. The first is clickstream data. During a given online session, an electronic history of web browsing activity is created by recording consecutive discrete digital inputs, such as clicks of a mouse and strokes of a keyboard. Tracking an end-user's clickstream data is made possible by accessing an end-user's "cookies" and/or *java script* that is stored locally on the end-user's machine and then transmitted via a "web beacon" to a remote server or external platform. (Cisco Systems, 2009) More recently, the emergence of 'Adware' and 'Spyware' programs have made it possible for third parties to monitor an end-user's online activity (often operating without either the end-user's knowledge or consent). (Dobosz, Green, and Sisler, 2006; Palmer, 2004) The creation of clickstream data represents a form of labor insofar as the intentional activity of visiting one cyber location, and moving on to the next,

presupposes an end-user's conscious activity and actual pursuit of interests.

A second form of virtual property brought about through an end-user's Internet activity is the input of syntactically complete information [SCI hereafter]. SCI requires a unique combination of digital inputs to bring about the next source of information, cyber event, or online activity. Input of SCI includes correctly typing a website's URL address in the address pane of a web browser, satisfying uniqueness requirements or "signature strips" when creating an online profile or account, and successfully initiating and then redirecting the results of an optimized search engine. SCI is a form of labor in the sense that it presupposes and relies upon a pre-existing knowledge base of the end-user. While also a form of clickstream data, SCI requires detailed prior knowledge of the web's infrastructure and communication protocol.

A third form of virtual property is the input of personal identity information [PII hereafter]. PII involves the set of details that can be used to pick out a single person from a larger population. It might include an end-user's geographical location, age, education level, occupation, gender, sexual preference, Social Security number, e-mail address, and income. Familiar to anyone who has had to fill out an online application for employment, credit, or housing, PII is the set of facts that individuates a person as a unique member of society. PII not only requires the end-user/consumer to manage, remember, and apply select forms of information at different times, but it requires a basic level of civil engagement to sustain and keep it active. In this sense, PII differs slightly the other two forms of cyber labor/activity in that it requires real world employment and/or underlying support. PII includes all the sorts of things in a person's life that makes a cyber identity possible, such as the economic stability to maintain a computer, Internet connection, and place of residence, *etc.*

According to Chrisman, ownership as a concept functions as a triadic relation, connecting three things: (1) the owner; (2) the thing owned; and (3) everybody else (i.e., society). (Christman, 1994,) If somebody owns something, the parameters of that claim involves not only connecting the thing owned with the person who owns, but it also obligates others in society to respect and acknowledge the unique relational features established by the process and maintenance of a thing's possession. Of course, ownership does not entail unlimited or unrestricted use of an object by its owner. Using a classic example, simply because a person has ownership of a gun does not entitle her to use the gun in any way that she wishes, such as in shooting an innocent person. Gun ownership involves a range of claims and entitlements. Its use for personal security and recreation are a prerogative of the owner, all things being equal.

Similarly, virtual property takes on a triadic structure as well, connecting the end-user with her online digital footprints, in a way that involves special acknowledgement and noninterference from others in society. The information generated by an end-user's online activity should remain the private possession of the end-user as long as its creation does not actively or directly undermine the autonomy interests of others. From a Lockean standpoint, an end-user's virtual property demarcates a range of interests and activities that accompany the pursuit of a meaningful and productive life. The projects and plans of a person's life often unfold in an ongoing process of self-discovery. Having the freedom and privacy to access information related to our interests, therefore, is fundamental in the construction and pursuit of our goals in life. Framing an understanding of what the world is like, what others have done before us, and how our interests are situated within a broader context, provides depth and context for self-understanding. Insofar as an end-user's online activity is an instance of information gathering, fantasy, or scientific inquiry, ownership of the resulting virtual property is morally justified by appealing to an end-user's/consumers autonomy interests. Succinctly put, others have an obliga-

tion to respect the special relationship that an end-user has with her own clickstream data, SCI, and PII. Appropriation of this information without an end-user's knowledge and consent too easily thwarts an end-user's autonomy interests at the cost of advancing another's income interests. In practical terms, the plausible way to promote a best-case scenario for marketing in e-commerce is to acknowledge and protect consumer virtual property.

Objections to 'Consumer Virtual Property'

Whether virtual property can serve as a viable normative concept for the protection of e-consumer autonomy depends in part on how well it can stand up to some basic philosophical objections. Three of the more pressing worries confronting the account offered here involve the role property, the concept of work, and the ontological reality of virtual property (itself).

The first objection stipulates that using a notion of property to support consumer autonomy is problematic because the very criteria by which virtual property is established for the consumer applies equally (if not more so) to the business firms who engage in electronic surveillance and behavioral profiling. That is to say, if the criteria of labor and the Lockean obligations of 'no waste' and 'sufficiency proviso' apply, who is not to say that the information gained from aggressive forms in marketing is not (itself) a form of virtual property? On a relative comparison, for example, the number of people and the amount of man hours to assemble programs of electronic surveillance far exceeds the "labor" of an individual end-user's clickstream data. The intellectual objects and virtual property produced by a firm's research into consumer behavior requires an extensive amount of planning, investment, research and development. Single end-user labor production, in contrast, is often the result of unplanned, impulsive and non-goal directed behavior. If consumer virtual property is a plausible concept, then acknowledgment of the marketer's property should likewise be granted. Given the wide disparity of worth between an e-consumer's and e-marketer's property, as merely established by the degree and type of labor involved in the production of each, it appears that consumer virtual property is simply too weak to establish a consumer's right to privacy.

A second objection against consumer virtual property involves questioning the type of labor that our account attributes to a typical end-user. The central insight motivating this objection is that not just any form of human activity counts as work. There are many human behaviors that involve intention, effort, and goal-directedness but do not fit with a common understanding of what work is and what it involves. Among Lockean scholars, this worry sometimes emerges in discussion of Locke's 'mixing metaphor.' The idea here is that simply adding one's effort and energy to a natural object does not warrant full rights of possession. Work and labor, as a human activity, involves a certain amount of insight and commitment to future security. Within the e-business environment, therefore, simply because an end-user can go online and create a series of digital footprints does not (in itself) qualify as a form of work that can support full possession rights. Simply put, there is nothing unique or special about an end-user's general online activity that qualifies it as a form of labor.

A third philosophical objection that can be raised against consumer virtual property involves questioning its ontological reality and its subsequent applicability for establishing consumer rights. The worry here is that since virtual property, as it is instantiated in the physical world, is simply a collection of ones and zeros stored on a computer's memory or hard drive, there is nothing actually taken or stolen when it is the subject of electronic surveillance. As digital information, virtual property can be harmlessly replicated and then transmitted with relative ease. In truth, the objection continues, virtual property does not involve a meaningful sense of transfer (i.e., as an object that leaves one person's possession and enters another's). With virtual property it is difficult to discern the exact boundaries of an owner's possession and control. Similar worries are often raised in discussions of intellectual property. (Zemer, 2006; Spinello 2006; and Tavani, 2005) Since intellectual property involves the creation of non-tangible abstract ideas, images or sounds, questions emerge of it reality and locus of control. Skeptics of using a Lockean labor theory of property for intellectual property, for example, often cite Locke's exclusive use of physical objects in his account of original natural property. They claim that Locke and his labor theory of property refer only to a very special and restricted case of property, far removed from contemporary discussions of electronic media. In a similar fashion, therefore, our attempt to ground a consumer's right to online privacy in digital virtual property appears similarly displaced.

Noteworthy of all three objections presented above is how they remain independent of any one given ethical outlook or perspective. It is less the coherence of Locke's account of original natural property that is called into question as it is our precise application of Locke's account to e-business. All three objections also remain silent or neutral about the actual existence of a market economy in e-commerce. Each objection, for example, assumes no deep skepticism about the value of property and the sustainability of modern capitalism. The objections outlined above, therefore, aim to expose problems and inconsistencies internal to our account of consumer virtual property.

Replies to Objections

The general defense I offer here for consumer virtual property is fashioned from the idea that work/labor is a fundamental value producing human activity. In our postindustrial capitalist society, boundaries traditionally separating the home, market, and workplace are becoming increasingly obscured. Computer technology and the Internet have created a plethora of opportunities for individuals to merge the different aspects of their life into one streamlined production schedule. The dominant role of computer technology in modern life is changing the way humans interact with each other, as well as the way they plan for the future. As the different aspects of our social and work-related lives intertwine, it understandable that there will be competing desires for privacy, security, efficiency, entertainment, convenience, and information. The underlying intuition of consumer virtual property, therefore, is that as our complex social identities evolve in modern society, it is our work and creative energy that will help establish the criteria for protecting our autonomy interests in an online environment.

To reply to the objection that questions the relative worth of consumer virtual property when compared to e-marketer's virtual property, I think it is important to recall that labor/work is not the only means to create value in the marketplace. That is to say, there are certainly other ways that property is both generated and assessed. For Locke, it is true, labor does establish a form of private possession, but it is important not to forget the context of his discussion. Recall, Locke's focus in the *Second Treatise* is with the question of original natural property: how can private possession come about in the context were all are equal and everything is held in common. The e-business environment, although resembling aspects of Locke's state of nature, does not completely reduce to an egalitarian property free environment. So to say that more labor (or more involved forms of labor) go into the production of an e-marketer's virtual property when compared to the production of consumer virtual property, misses the point about alternative value production.

One concrete example for how consumer virtual property involves alternative forms of value is to consider how marketers are already willing to pay (or change something of value) for an end-user's private information. Google, for example, provides e-mail service with virtually unlimited storage space for anyone agreeing to its terms of service. As a Gmail user, the end-user foregoes a right to privacy when sending or receiving e-mail messages. In exchange for server space and a web based e-mail application, Gmail has exclusive access to market messages and products to its mail users. Marketing messages in Gmail appear within the border or frame of an end-user's e-mail webpage windowpane. Moreover, there are other e-marketing companies that also provide services for trade to gain access to end-user's web surfing behavior. In some airports and train stations, for example, end-users obtain free high speed Internet access in exchange for revealing their clickstream data.

To answer the second objection above, concerning whether end-user activity really counts as work and therefore is able to establish the moral basis for private possession, we need only consider again the evolving character of our online personalities. The increasingly dominant role that computer technology plays in modern life forces us to re-examine the boundaries of work, play, private interest, and intellectual curiosity. Part of the problem is that old standards for evaluating work simply do not apply in the high tech digital environment. To arbitrarily dismiss some forms of Internet activity as work, but not others, ignores the sheer complexity of maintaining autonomy interests in an online environment.

To answer the third objection that questions the ontological reality of consumer virtual property it is perhaps helpful to contrast virtual property, on one hand, with intellectual property, on the other. Unlike intellectual property where the essence and extension of an idea is difficult to track in space and time, consumer virtual property has (as its anchor) the physical 'brick-and-mortar' life of the end-user. Recall, the form of virtual property known as PII is predicated upon the real life management and civic engagement of the computer end-user. PII is the set of facts that individuates a person as a unique member of society; it requires an end-user to manage, remember, and apply select forms of information at different times. In fact, if the information created by an end-user did not vertically report his or her true interests and preferences, the data created would not be something valued by e-marketers. That is to say, only because there is a real concrete existence in the physical world, managing information and generating capital, does the digital profile of an end-user become an object of interest for the e-marketer.

One common thread of all three objections presented above is the attempt to marginalize online consumer behavior as something unworthy of serious privacy protection. The basic idea here is that concerns for privacy are too often exaggerated, and as it turns out, more closely guarded in theory than in practice. Truth be told, most end-users report a high level of concern for online privacy despite engaging in behaviors both online and off-line that involve a liberal release of their personal information into the public space. According to a study by The Pew Internet and American Life Project, "American Internet users overwhelmingly want the presumption of privacy when they go online....[However]... a great many Internet users do not know the basics of how their online activities are observed and they do not use available tools to protect themselves." (Fox, 2000, p. 2) Consumer online behavior, therefore, represents a form of cognitive dissonance. Consumers think, believe, and self-report one thing about privacy, but do something completely different in their day-to-day life. Moreover, the bottom-line challenge from staunch supporters of electronic surveillance and behavioral profiling is that you simply cannot make people buy stuff that they don't want or don't need, regardless of how much information you collect about a consumer's personal preference. Even if a marketer acquires complete knowledge of a consumer's behavior, it doesn't undermine the powerful force of a consumer's choice to pick one thing over another in the marketplace.

While there is certainly something encouraging about such appeals to personal responsibility and consumer individualism, the issue of information transparency (as it relates to the ethics of consumer surveillance) is really only one threat to the sustainability of a market economy in e-commerce. As we have noted, there are other pressures in the marketplace, including noncompetitive behavior among producers, which increase the chances for predatory business practices to occur. It is, therefore, the combination of emerging problems acting at the same time that significantly undermines a free market in society. The mortgage and housing crisis in the United States in 2009, for example, was a product of not just one destabilizing business practice. Several aspects of the house buying and selling process were affected simultaneously, culminating in marketplace collapse, requiring massive and unprecedented government intervention. The success of e-commerce, therefore, depends on the ability of business managers to protect against several threats at one time. Consumer virtual property is simply one tool marketing practitioners have for promoting ethical and sustainable business transactions online.

FUTURE RESEARCH DIRECTIONS

The concern for consumer autonomy in marketing ethics has received increased interest in recent years. Two separate streams of consumer research where the notion of consumer virtual property is likely to be relevant are (1) consumer well-being and quality of life studies; and (2) participatory action research [PAR].

Research programs promoting consumer well-being and quality of life identify the consumer as a stakeholder in modern business. (Sirgy and Lee, 2008; Gibbs, 2004) A perennial theme in this normative marketing ethics approach is to identify policies and metrics that can positively affect consumer interests. Identifying methods for establishing virtual property rights, therefore, might be one natural extension of advancing consumer well-being and quality of life in the online e-business environment.

PAR aims to empower the individual consumer through the marketing research process. (Ozanne and Saatcioglu, 2008) Central to this approach is the idea that both the researcher and consumer should mutually benefit from the marketing process. The consumer-respondent, for example, is not merely a statistic, measured once, and used solely for the means of obtaining information. In PAR a concerted effort is made to return to the information gathering environment and discern if the consumer-respondent is any better off after the marketing research project than before. In this light, future projects for advancing consumer virtual property might include PAR opportunities to learn what end-users already know about online privacy and then construct educational forums at a later date that enable the same end-users to learn more about the online digital environment.

CONCLUSION

Individual choices and lifestyle preferences revealed by an end-user's online activity can potentially transform the way products are created, introduced, and promoted in the marketplace. Online advertising and consumer-side research of products and services reveal expanded opportunities for the flow and transparency of information. In a best case scenario, e-marketing and ethical forms of consumer surveillance can strengthen free market enterprise in the digital environment. What I have attempted to argue for in this chapter is that consumer virtual property is a useful tool for avoiding a worst-case scenario where predatory business practice exploits various 'at risk' consumers. Very simply, the labor and creative energy of an end-user's online activity establishes a form of private property that e-sellers and e-marketers are morally obligated to respect.

REFERENCES

Brenkert, G. (2009). Marketing and the Vulnerable. In Beauchamp, T., Bowie, N., & Arnold, D. (Eds.), *Ethical Theory and Business* (8th ed., pp. 297–306). Upper Saddle River, NJ: Pearson/ Prentice Hall.

Brodie, R., Winklhofer, H., Coviello, N., & Johnston, W. (2007). Is e-marketing coming of age? An examination of the penetration of e-marketing and firm performance. *Journal of Interactive Marketing*, *21*(1), 2–21. doi:10.1002/dir.20071

Christman, J. (1994). *The Myth of Property: Toward an Egalitarian Theory of Ownership*. New York: Oxford University Press.

Cisco Systems, Inc. (2009). *Online Privacy Statement*. Retrieved June 13, 2009, from http://www. cisco.com/web/siteassets/legal/privacy.html

Dobosz, B., Green, K., & Sisler, G. (2006). Behavioral Marketing: Security and Privacy Issues. *Journal of Information Privacy & Security*, *2*(4), 45–59.

Fox, S. (2000). Trust and privacy online: Why Americans want to rewrite the rules. *The Pew Internet and American Life Project*. Retrieved June 13, 2009, from http://www.pewinternet.org

Friedman, T. (2005). *The World is Flat: a brief history of the globalized world in the twenty–first century*. London: Allen Lane.

Gibbs, P. (2004). Marketing and the Notion of Well-Being. *Business Ethics. European Review (Chichester, England)*, *13*(1), 5–13. doi:10.1111/ j.1467-8608.2004.00344.x

Goodman, B., & Rushkoff, D. (Writers), Goodman, B., & Dretzin, R. (Directors). (2003). The Persuaders [Television series episode]. In D. Fanning (Producer), PBS Frontline. Boston: WGBH.

Heath, J. (2005). Liberal Autonomy and Consumer Sovereignty. In, John Christman (ed.), Autonomy and the Challenges to Liberalism: New Essays (204-225). Cambridge, UK.Cambridge University Press, Hobbes, T. [1651] (1991). Leviathan. R. Tuck (Ed.) New York: Cambridge University Press.

Hospers, J. [1974] (2005). What Libertarianism Is. In James Sterba (Ed.), Morality in Practice. Belmont, CA: Thomson-Wadsworth.

Jacobs, R., & Stone, B. (2008). *Successful Direct Marketing Methods: Interactive, Database, and Customer Marketing for the Multichannel Communications Age*. New York: McGraw-Hill.

Lippke, R. L. (1989). Advertising and the Social Conditions of Autonomy. *Business & Professional Ethics Journal*, *8*(4), 35–58.

Locke, J. [1690] (2008). Cambridge Texts in the History of Political Thought: Two Treatises of Government. Edited, introduction, and notes by P. Laslett (Student Edition). New York: Cambridge University Press.

Ozanne, J. L., & Saatcioglu, B. (2008). Participatory Action Research. *The Journal of Consumer Research*, *35*(3), 423–439. doi:10.1086/586911

Palmer, D. (2005). Pop-Ups, Cookies, and Spam: Toward a Deeper Analysis of the Ethics Significance of Internet Marketing Practices. *Journal of Business Ethics 58*, 271-280.Quinn, M. (2005). *Ethics for the Information Age*. Boston, MA: Pearson Addison Wesley Inc.Robischom, N. (1997). Rx for Medical Privacy. *Netly News*, Sept 3.Schwartz, B. (2004). *The Paradox of Choice: why more is less*. New York: Ecco. Simmons, J. A. (1989). Locke's State of Nature. *Political Theory*, *17*(3), 449–470.

Sirgy, J., & Lee, D. (2008). Well-being Marketing: An Ethical Business Philosophy for Consumer Goods Firms. *Journal of Business Ethics*, *77*(4), 377–403. doi:10.1007/s10551-007-9363-y

Sirgy, J., & Su, C. (2000). The Ethics of Consumer Sovereignty in the Age of High Tech. *Journal of Business Ethics*, *28*, 1–14. doi:10.1023/A:1006285701103

Spinello, R. A. (2006). *Cyberethics: Morality and Law in Cyberspace* (3rd ed.). Boston, MA: Jones and Bartlett Publishers.

Tavani, H. T. (2005). Locke, Intellectual Property Rights, and the Information Commons. *Ethics and Information Technology*, *7*, 87–97. doi:10.1007/s10676-005-4584-1

Zemer, L. (2006). The Making of a New Copyright Lockean. *Harvard Journal of Law & Public Policy*, *29*(3), 891–947.

Section 5
Ethical Issues in Public Policy and Communication

Chapter 13
Fairness and the Internet Sales Tax:
A Contractarian Perspective

James Brian Coleman
Central Michigan University, USA

ABSTRACT

One of the most noticeable features of online business transactions in the United States is the absence of a sales tax on interstate purchases. Consumers are not expected to pay taxes on Internet purchases across state lines, and businesses are not expected to collect taxes for such purchases. The absence of an interstate Internet sales tax (shortened hereafter to "Internet tax," except where noted) has been both praised as an incentive to promote business, and condemned as the cause of serious revenue loss by municipalities throughout the nation. In this chapter, I shall present an analysis of current proposals about instituting a sales tax on Internet purchases. Both sides of the debate argue for their position on grounds of fairness to the businesses who, were an internet tax to be levied, would be expected to collect and remit it to the state and local governments. So, I will focus on the concept of fairness in this discussion. I will consider the claims about the fairness or unfairness of the Internet tax from the standpoint of contractarianism: a group of philosophical theories that focus on questions concerning the just distribution of social resources. I will make use specifically of the views of John Rawls and David Gauthier to analyze the positions on the Internet tax. I shall conclude by arguing in favor of the Internet tax on contractarian grounds.

BACKGROUND OF THE DEBATE

Since the US Supreme Court's 1992 Quill v. North Dakota decision, there is a moratorium on collecting taxes on e-business transactions across state borders. In that case, the North Dakota Tax Commissioner attempted to collect a use tax (a variety of the sales tax, generally filed voluntarily along with state tax returns) for sales in North Dakota from the Quill Corporation, a mail order based office-supply retailer based in a different state. The trial court ruled in favor of Quill, after which

DOI: 10.4018/978-1-61520-615-5.ch013

the decision was appealed to the state's Supreme Court. The North Dakota Supreme Court decided in the state's favor. The case was then appealed to the U.S. Supreme Court. The U.S. Supreme Court reversed the State Supreme Court's decision, deciding that, as long as the company has no brick-and-mortar presence in the state where the item is purchased, Quill owed North Dakota no duty to collect and remit a tax on purchases made by citizens in that state. This decision followed a test proposed in a previous U.S. Supreme Court decision, Complete Auto Transit v. Brady. According to that test, a business must have "nexus," that is, a physical presence of some sort, in the state where the sales tax is collected. Lacking such a nexus, there is no constitutional basis for the sales tax. The effect of the Quill decision, then, is that demanding that interstate companies collect and remit taxes places an unconstitutional burden on interstate commerce. The Court noted, however, that Congress could weigh in on the legality of a use tax in interstate commerce. A given state, that is, may legislate certain conditions under which taxes may be collected from interstate companies: Justice John Paul Stephens writes, "No matter how we evaluate the burdens that use taxes impose on interstate commerce, Congress remains free to disagree with our conclusions" (Quill v. Heitkamp, 1992). Lacking such legislation, however, the Constitution's Commerce Clause forbids North Dakota from collecting a use tax from Quill Corporation.

The implication of the Quill decision for e-business is that the majority of Internet sales take place across state lines, like the transactions involved in that case. For this reason, the Court's decision is considered to hold for Internet transactions as well as traditional mail-order purchases. What generates controversy about the Quill decision is the Court's allowance that it is within Congress's authority to determine whether to institute an Internet tax. Congress may, in other words, decide that an Internet tax must be collected.

Opponents of the Internet tax argue that keeping e-business tax-free promotes online retail, and that to levy an Internet sales tax would negatively affect e-businesses and thus the overall economy (Neil, 1996). Opponents argue further that the sheer complexity of instituting an Internet sales tax would be overwhelming, given the nearly 7,500 tax zones across the U.S. The Internet retailer Amazon.com claims, in a 2008 lawsuit against New York State's proposed sales tax bill on e-business, that no clear rule for an Internet tax can be formulated that does not violate the Quill v. North Dakota ruling (Hansell, 2008). Supporters of the Internet tax argue that the growth of tax-free online retail as a portion of overall retail sales deprives state and local government of much-needed revenue (McQuivey and DeMaullin, 2000). The percentage of the overall retail sector made up by e-commerce was approximately 6% in 2008 (Vollman, 2008). Most analysts expect the percentage to increase with time. Subtracting the tax revenue from this percentage of the transactions leaves states and local areas with a substantial shortfall, to the detriment of police departments, parks, and other municipal services. Further, although there are considerable complications involved with instituting a sales tax, given the various tax zones, there is nevertheless good reason to suppose that these difficulties can be resolved.

Opponents of the Internet Tax

Both opponents and supporters of the Internet tax tie their positions into questions about the fairness of an Internet tax. Opponents of the sales tax believe that instituting an e-business sales tax would be unfair to the individual businesses. In a discussion characteristic of the anti-tax position, Aaron Lucas of the libertarian think-tank the Cato Institute writes,

The... most important reason to limit state taxing authority over remote transactions, however, is

fairness. When a local business collects sales taxes, there is a clear link among taxes paid, services provided and legislative representation. Local firms benefit from police and fire protection, roads, waste collection and other services, so it's proper that they help cover those costs. Remote sellers don't enjoy any of those services, and shipping companies already pay taxes to cover their use of public goods. To force a wholly out-of-state business to collect taxes would be "taxation without representation," pure and simple (Lucas, 2000).

An Internet sales tax, on this view, goes against any of the traditional reasons for levying a tax. Legal scholar Deborah Schenk (Schenk, 1996) writes, "The government levies the vast majority of taxes in order to provide revenue to operate. But the tax system also can be utilized to compensate the government for using certain of its facilities ('user fees'), to provide incentives to engage in certain activities, or to penalize those who take certain actions" (page 611). In Lucas' view, taxation makes no sense in relation to Internet transactions, since they benefit from none of the costs that taxation is intended to cover. Hence, it is "taxation without representation." I shall refer to this as the "no representation" view. Suppose that a book seller, Sarah, makes a sale to a customer, Jane, in a different state. According to the "no representation" view, Sarah benefits in no way from the taxes that would be collected from Jane. The upkeep of the infrastructure of Sarah's state is unaffected by the taxes she returns to Jane's state. This is unfair to Sarah, since the taxes she returns to Jane's state do not contribute anything to Sarah. The argument of the "no representation" view, then, is that the sole condition under which taxation could be fair is the provision of services to the taxed individual (understood here to mean both the customer and the business). But an Internet sales tax would provide no such service to the taxed out-of-state individual. Hence, such a tax is unfair, and thus unjustifiable.

The "no representation" view initially appears to assume, mistakenly, that the businesses are themselves the taxpayers. In fact, the businesses in effect act as tax collectors on the purchases of their consumers. The businesses then remit them to the state or municipality in which the purchases were made. Since this is the case, one might object, there is no unfairness involved here, for it is the consumer rather than the business benefiting from the services provided by the taxes.

But the "no representation" view does not depend on this claim. For, according to that view, the imposition of tax-collection on Sarah's business is unfair simply because Sarah receives no compensating benefit for that service. Even if there is no monetary loss, the "no representation" view maintains, the demand that the business act as a tax collector is unfair. Further, purveyors of this view may add, there is a monetary loss, since demanding sales taxes potentially reduces purchases from tax-shy customers. Regardless of the effect of these qualifications, they undoubtedly weaken the initial "no taxation without representation" claim. For, the imposition of a service-provision seems relatively benign compared to the collection of taxes for no benefit to the taxed party. Nevertheless, it is a burden that becomes more considerable when one considers the sheer number of tax-zones in which a business may sell their goods. Without standardized tax rates, a business faces the potential task of determining and remitting taxes to thousands of tax-collecting zones. This is certainly a potentially unfair burden. Hence, I shall hereafter regard the "no representation" view as being premised on the notion that, even if there were no monetary loss, the effort of collecting and remitting taxes to multiple tax-zones is itself a potentially unfair burden. While noting that the view might be more properly called the "no tax collection without due consideration for burdens imposed," I shall continue to use Aaron Lucas' "no representation" terminology in this discussion.

Supporters of the Internet Tax

Supporters of the Internet sales tax focus on the unfairness to the tax collecting municipalities of not having an Internet sales tax. If e-business is projected to grow as a percentage of the total retail sector in the near future, the supporters worry, this could impose increasing fiscal demands from reduced tax revenues for state and local regions, as well as local businesses. For example, in a recent Minnesota State Senate Taxes Committee meeting, Kati Gallagher of the Midwest Bookseller's Association testified that,

Untaxed Internet shopping is causing locally owned Main Street businesses to close in both small towns and large cities.... Internet competition is often cited as the cause. Business, sales, and income taxes from these closed businesses are then not paid to their communities and states. Public services, education, road repair, etc. go unfunded. Unemployment rises.... It's a vicious circle. I think we cannot yet imagine the long-range effects of allowing Internet retailers with nexus in Minnesota to bleed our local businesses dry through the unfair competitive advantage of not having to charge tax (Grogan, 2009).

In short, the absence of a sales tax on interstate e-business transactions might have positive effects on Internet commerce, but only at the expense of brick-and-mortar locations. The customers of brick-and-mortar locations make use of the resources that taxation makes possible—the local infrastructure—where e-businesses require only a computer and an Internet connection. When the e-businesses transactions go untaxed, the burden of paying for the infrastructure falls to the brick-and-mortar stores that compete with e-business for customers. This, says Gallagher, is an unfair burden. To return to the example used in the "no representation" view: when Sarah profits from a sale to Jane in a different state, Sarah's state sees no contribution of this sale to its infrastructure. But

without such contributions, Sarah's (and Jane's) state's budget becomes progressively smaller. The brick and mortar retailers in Sarah's state remain to pay in to the increasingly poorer state revenue; and as fewer brick and mortar retailers remain, the revenue decreases even more. The reason why this situation seems likely to continue, according to the position Gallagher presents, is that as a result of the lack of an Internet tax, the e-business retailers can easily undercut their brick and mortar competitors prices. The lack of such a tax results in an unfair competitive edge, which results in the general weakening of the state's revenue base.

I shall refer to this position as the "unfair competition" view, since a side effect of the tax policy is to put brick and mortar stores at an unfair competitive disadvantage with respect to Internet retailers. The argument of the "unfair competition" view is that, to the extent that a policy provides an advantage to one party and a disadvantage to another, with no relation to the situation, wealth or income of either, it is unfair. But having no Internet tax is unfair in just this sense, since businesses with a nexus in one state are put at a distinct disadvantage by the lack of an Internet sales tax. Where the local bookseller collects and supplied taxes on all of its brick-and-mortar sales, the Internet bookseller collects no taxes on the majority of its sales. The revenue would have been generated for the area by brick-and-mortar purchases diminishes significantly, leaving the remaining retail stores to bear an unequal amount of the burden of generating sales tax revenue for the region. Hence, there should be a tax on Internet sales transactions.

According to the "no representation" view, then, it is unfair to tax businesses that maintain no nexus within the state, since the proposal that they should compute and pay in to the revenue of several thousand tax zones is one that cannot be demanded of in-state brick and mortar locations. The "unfair competition" view, on the other hand, points out the unfair competitive advantage of a business with a nexus in a given state earning profits, but not being required to collect and submit taxes on a substantial amount of its sales, in contrast to brick and mortar stores. Where the "no representation" view cites the unfairness of being burdened with the task of collecting and submitting sales taxes to multiple tax-zones, the "unfair competition" view cites the unfairness of Internet businesses profiting from the losses of brick-and-mortar stores in other states. Furthermore, in contrast to retailers in brick-and-mortar locations, Internet retailers contribute to the gradual weakening of the tax base in their own states. For as customers increase their purchases from e-businesses, the competing brick-and-mortar locations lose business. As they lose business, the tax revenue generated by the purchases disappears. But then everyone in the state or municipality, including those operating the e-business, loses the advantages provided by that revenue. Hence, on the "unfair competition" view, the Internet retailer loses the same benefits as the brick-and-mortar retailers. Furthermore, the e-business retailer fails to provide a service that all brick-and-mortar locations do: collecting and returning revenue to the state or municipality. The Internet retailer merely takes advantage of the services provided by such revenues, but does not take up a fair share of the burden of submitting them. The untaxed e-business is, then, a financial free rider according to the "unfair competition" view: someone who takes advantage of goods or services for which he or she does not pay. In effect, the untaxed e-business practitioner is, understood in these terms, a thief of public services. From the perspective of the "no representation" view, the opposite situation prevails: to demand tax-collection would in effect rob them of time and resources.

If one attempts to maintain both views, then, the e-business practitioner is either a thief when not taxed or the victim of theft when taxed. Is this "damned if you do, damned if you don't" result unavoidable? It would be unavoidable, of course, only if both views were correct. In the following

section, I shall discuss which view, if either, is correct by considering in greater detail the philosophical implications of the concept of fairness.

CONTRACTARIAN VIEWS OF FAIRNESS

As I observed in the previous section, the competing claims of the "no representation" and "unfair competition" views are premised on the concept of fairness. According to the "no representation" view, an Internet tax is unfair to the business collecting the taxes, since the burden of doing so is one that intrastate (as opposed to interstate) transactions do not share. The "unfair competition" view, however, regards the absence of an Internet tax as an imposition on the brick-and-mortar business, which then bears the sole responsibility of contributing sales tax revenue to the local and state governments. Since both views cannot be correct, there must be some way of determining which is best equipped to support its claims about the fairness of an Internet tax. This section, then, will address the concept of fairness from the philosophical perspective of contractarian philosophy, and then continue to address the fairness of the Internet tax from that perspective.

Before considering directly the fairness of the Internet tax, some general comments about the concept of fairness are in order. The concept of fairness is complex. Related concepts include justice, equality, and rights. In one sense, to say that an outcome or situation is fair is to say that it is just: no injustice (moral or legal) has been committed against a person or group (Kickul, 2001). In another sense, equal distribution is sufficient for a fair outcome: provided that no person or group has been unreasonably disadvantaged by an outcome or situation, it is fair (Saunders, 2008). Another, more minimal sense would be that fairness is achieved when no person's or group's rights have been violated (Barclay & Markel, 2009). The concept becomes still more complex when considering the various overlaps between the concepts of equality, justice, and rights (Guest, 2005)

I shall focus primarily on fairness as it relates to the concept of justice. Traditional philosophical discussions of the relation between the concepts of fairness and justice tend to focus on distributive justice in particular (Sterba, 1980). Distributive justice is the justice of distributing goods and services among recipients. A taxation policy is distributively just, for instance, when no particular person or group's resources or properties are imposed on disproportionately to their wealth, income, etc. Among recent discussions of distributive justice, one philosophical tradition tends to dominate: contractarianism. There are two opposing varieties of contractarianism particularly relevant to the current issue: John Rawls' liberalism, and David Gauthier's mutual advantage theory.

Contractarianism is the view that the principles constitutive of a just and reasonable society are best understood as those that hypothetically rational agents would agree to, in precisely the same way they would agree to a well-designed contract. Different types of traditional contractarianism conceive of the terms of the contract differently. Hobbes' contractarianism regarded the function of the contracted principles as a prevention of a universal and perpetual condition of war (Hobbes, 1996). Locke conceived of the contract as the basis on which the people consent to be governed (Locke, 1980). Rousseau conceived of the contract as the expression of a "general will" of the governed people (Rousseau, 1987). Each of these traditional views is characterized by the notion that the best way of attaining these outcomes is by conceiving of the rules of a just political body as the results of a contract between hypothetically rational agents (Boucher, 1994). More recent forms of contractarianism like Rawls's and Gauthier's extend the conception of a rational contractor as the source of reasonable

principles of justice and morality (Kukathas & and Pettit, 1990; Kymlicka, 2002).

Rawls' Liberal Contractarianism

According to Rawls' liberal contractarianism, the principles constitutive of justice are determined by reference to a hypothetical scenario in which the individuals who determine those principles deliberate about them. In Rawls' view, the best way to ensure that people who deliberate about the principles of justice do so with fairness as a criterion is to imagine oneself uninfluenced by one's actual social circumstances. In order to determine what policies are just, Rawls says, the deliberators consider what sort of rules they would decide to have enacted if they knew nothing about the position that they would occupy in the society in which those policies were enacted. Rawls refers to this scenario as "the original position": "The original position is, one might say, the appropriate initial status quo, and thus the fundamental agreements reached in it are fair" (Rawls, 1971, p. 12). A crucial part of the "original position" is the impartiality of the deliberators. They may be among the most or the least privileged; in the original position, they do not know which. Rawls writes,

I assume that the parties are situated behind a veil of ignorance. They do not know how the various alternatives will affect their own particular case and they're obliged to evaluate principles solely on the basis of general considerations. First of all no-one knows his place in society; nor does he know his fortune in the distribution of natural assets and abilities, his intelligence and strength, and the like. Nor again, does anyone know his conception of the good, the particulars of his rational plan of life, or even the special features of his psychology such as his aversion to risk or liability to optimism or pessimism. More than this, I assume that the parties do not know the particular circumstances of their own society... It is taken for granted, however, that they know general facts about human society. They understand political affairs and the principles of economic theory; they know the basis of social organization and the laws of human psychology. Indeed, the parties are presumed to know whatever general facts affect the choice of the principles of justice (1971, pp 136-137).

The "veil of ignorance" Rawls describes is a heuristic device for evaluating the reasonableness of the principles of justice. The main aim of the original position is to arrive at principles that can be considered fair: that is, principles that, while allowing for some inequalities, would place no-one in a position so disadvantaged that no-one in the original position would consent to occupy it. Fairness, on Rawls' view, is a function of consent through impartial deliberation about principles.

To return to the conflict between the two views of taxation: According to the "no representation" view, an internet sales tax is unfair because it provides no benefits to the businesses for their services in collecting and remitting those taxes. According to the "unfair competition" view, the Internet tax would be fair because it would make internet businesses pay for the same benefits as brick-and-mortar stores within their states. What would Rawls' contractarianism say about these views? The following seems likely. The first step would be to determine whether the principles enunciated in the respective defenses of the views are justifiable from the original position. The main principle of the "no representation" view it is unfair to demand of a business that it collects and remits taxes to areas in which it is not located, and for which it receives no benefits for the service. This principle is clearly unjustifiable from the original position. For if one were to consider impartially the results of a taxation policy intended to maintain the municipal services throughout the nation's tax zones, few would likely be willing to find themselves after deliberation living in areas short-changed by the current policy. The

e-business retailers like the brick-and-mortar retailers wish to have roads to travel on, police to prevent crime, and parks to walk in. A policy that substantially reduces funds for all of those things is unlikely to receive approval from behind the veil of ignorance.

The "unfair competition" view fairs better than the "no represenetation" view. The principle of that view is that no tax policy that provides an initial advantage to one party but not to another comparable party is fair. Contrapositively, a tax policy is fair to the extent that it extends the same treatment to all comparable parties. When considered impartially, the likely outcome would be that the "unfair competition's" principle would likely pass the veil of ignorance test: to the extent that any taxes are desirable, they should be distributed in a way that does not make everyone worse-off. But maintaining the current refusal to tax Internet sales could lead to a situation in which everyone could end up worse-off. So, Rawls' contractarian approach indicates an overall approval of the "unfair competition" view's position.

Two objections may be raised to the Rawlsian response to the "no representation" view. First, one of the main concerns of the "no representation" view was that requiring an e-business to contribute to the revenue of other states imposes an unfair burden: namely, saddling the business owner with the responsibility of sorting out taxes for several thousand potential tax-zones. Such a demand might appear excessively demanding. What would a Rawlsian analysis say to this problem? The following is likely. While it might be a burden to have to submit taxes to multiple locations, it is not necessarily an unfair burden. First, the e-business owner lives in a state that could potentially benefit from such a policy—that is, by collecting taxes from the Internet retailers in other states. Further, the e-business person benefits from making interstate sales in a way brick and mortar equivalents do not benefit: by a substantially increased customer base. One might argue, along Rawlsian lines, that this benefit is offset in a fair manner by the responsibility to calculate taxes from different tax-zones. Second, the complexity of the task could be simplified by other means. One option would be to purchase software that sorts out the tax-codes of the various tax zones in which one's products are purchased. Another, more ambitious but desirable option, is to work towards a simplification of the United States system of taxation. While this is clearly the more difficult of the two options, it would also be the more generally beneficial of the two. In short, the "no representation" view's concern about the task of remitting taxes to various tax-zones is not by itself sufficient to win approval from deliberators in the original position.

Gauthier's Mutual Advantage Contractarianism

There is second, more substantial objection to this analysis of the "no representation" view. Rawls' original position, one could argue, obscures the importance of different and legitimate individual preferences. Some people may wish for a different social outcome than others: some people may prefer, for instance, a greater degree of risk in the resulting society than do others. The "no representation" view might be defended from the Rawlsian analysis by observing that one reasonable outcome could be that keeping internet transactions sales-tax free could be beneficial would increase overall wealth. Even if the potential for economic losers is arguably decreased in the Rawlsian picture, so is the potential for economic winners. If the overall wealth of a community is augmented by no-sales-tax policies, the "no representation" proponent might argue, this could offset the possible negative effects of reduced municipal tax revenue. In short, the "no representation" view could be seen as an expression of the desire for a strongly competitive marketplace, where people can compromise and bargain for their individual preferences. David Gauthier's "mutual advantage" contractarianism offers a response to Rawls on this issue.

According to Gauthier, the best means of securing reasonable principles of justice and morality more generally is to determine what rules are most likely to result in the maximization of the aggregate of individual interests. According to Gauthier, some individuals desire a greater degree of risk in economic dealings than others. Since this is a legitimate desire, the best way to decide on fair principles of morality among the risk-takers as well as the risk-averse is compromise through rational bargaining. The fundamental principles of morality in Gauthier's view come in the form of rational agreement. In this respect, Gauthier's views are a departure from Rawls', since according to Rawls, the principles of justice are not decided through bargaining, but through impartial deliberation. The compromises reached through bargaining, on Gauthier's view, result in constraints on voluntary action. One person's desire to engage in unregulated commercial activities might conflict with another person's desire to maintain revenues for the community through taxation. The two parties have conflicting desires. But they could reach agreement through rational discussion and compromise. Gauthier explains,

...each person can see the benefit, to herself, or participating with her fellows in practices requiring each to refrain from the direct endeavor to maximize her own utility, when such mutual restraint is mutually advantageous. No one, of course, can have reason to accept any unilateral constraint on her maximizing behavior; each benefits from, and only from, the constraint accepted by her fellows. But if one benefits more from a constraint on others than one loses by being constrained oneself, one may have reason to accept a practice requiring everyone, including oneself, to exhibit such a constraint. We may represent such a practice as capable of gaining unanimous agreement among rational persons who were choosing the terms on which they would interact with each other. And this agreement is the basis of morality (2007, p. 620).

The background scenario against which Gauthier sketches his position is, of course, a Prisoner's Dilemma. The salient point Gauthier takes from Prisoner's Dilemma situations is that, while it is possible to maximize one's own interests at the expense of others, it is not mutually advantageous to do so. The most desirable outcome has to be attained in terms of social coordination of individual interests, as the acceptance of "second-best" outcomes, of constraints on voluntary action. The agreement to accept such self-imposed constraints is the basis, according to Gauthier, of morality itself.

How would Gauthier's proposal affect the debate between the two views of Internet taxation? Initially, it might seem that the "no representation" view would fare better on Gauthier's than on Rawls' view. For the "no representation" regards as unfair the imposition of an Internet tax on interstate commerce; further, proponents of this view could argue, lower taxes might promote increased wealth. On Gauthier's view, the propensity to take risks must be taken into account in the bargaining process, for this is a rational desire. For this reason, it seems that Gauthier's view would be friendlier to the "no representation" view than Rawls'.

And yet this result is not clear-cut. For an essential part of engaging in bargaining is the acceptance of some or all parties of mutual constraints for mutual advantage. In the case of tax policy, there could be constraints one accepts not for the sake of a specific benefit to oneself, but in exchange for another benefit. The "no representation" view assumes that an Internet sales tax would be unfair to the Internet business, since that business won't gain from the benefits of that tax. But if one assumes that not all taxes one pays ought always to be strictly for one's own direct benefit, the results might be different.

Suppose that an online business, say bookseller A, is located in state X, but sells the majority of its stock to residents in state Y. Suppose there is another bookseller business, B, residing in state Y, which sells the majority of its stock to residents in

state X. Neither A nor B are taxed for online sales to residents of their non-residing state. Neither state X nor Y receives the tax money for their respective municipalities from these sales. According to the "no representation" view, this is as it should be, since neither A nor B resides in the locations of their customer-base, and it is unfair to demand that they be saddled with the responsibility to remit taxes to multiple tax-zones. For neither A nor B would see the benefit of the taxes they would give back to the state governments on the out-of-state sales. But this is mistaken, for reasons mentioned earlier. Even though neither A nor B themselves benefit directly from the taxes they collect and submit to other states, there are clear benefits to be gained from having that revenue collected in both states. If, as the "unfair competition" view maintains, the absence of an internet sales tax reduces the revenue for the services all state residents benefit from—police, parks and recreation, and so forth—then A and B do stand to benefit from the collection of a sales tax. In other words, both A and B benefit indirectly from an Internet sales tax. If one assumes that the objections about the trouble involved with collecting and submitting taxes can be overcome, there is little to an Internet tax that Gauthier's mutual advantage contractarianism would find objectionable.

Hence neither sort of contractarian theory tends to support the "no representation" view of Internet taxation. But how does the "unfair competition" view fare? It passes Rawls' original position test, as we have seen. Further examination shows that its basic premise passes Gauthier's "mutual advantage" test as well. The "unfair competition" view claims that given the purpose of a sales tax is to fund public works, police, and so forth, everyone making sales in the area or municipality (depending on the state laws) should carry their fair share. But this is consistent with Gauthier's theory; for if the sales tax is an agreed-upon constraint, then it is arbitrary to exempt online retailers and not brick and mortar ones.

Proposals for Future Research

As a subject of research in business ethics, taxation receives relatively little attention. This is particularly the case with the issue of Internet taxation. The proposal of this chapter is that contractarian philosophy can provide a useful means of delving into the difficulties and intricacies of the issues involving Internet taxation. Other forms of ethical analysis are, of course, possible. If one regards the fundamental issues at stake in Internet taxation as involving benefits and harms, some form of utilitarian analysis could be particularly useful (Brink, 2006). But if the question is fairness and just distribution, contractarianism is particularly is particularly suitable. If one pursues a contractarian approach, further research could investigate what form of contractarianism is best suited to address the issues. Differing conceptions of the overall goals of economic development, and of the overall good of society, may lead to one form of analysis to be favored over another. Libertarian conceptions of the good of society and individuals in it will tend in the direction of Gauthier's analysis (Gauthier, 1986). Gauthier emphasizes concerns central to libertarian strains of thought, in particular the focus on individual initiative, risk, and bargaining. Robert Nozick's classic work may be useful here as well (Nozick, 1974). Liberalism, on the other hand, tends to focus on the rights and liberties of particular people and groups, and tends to regard government as an important element in the furthering of these rights and liberties. If the overall goal of economic activity follows a more liberal direction, Rawls' analysis would be favored.

CONCLUSION

Those who pay taxes rarely see taxation in a favorable light, since it involves the forfeiture of income with relatively little assurance about the where that income will go. But the effects of taxation are often favorable: roads, public works, and so forth are their results. Taxation, then, is a compromise. The onus on the legislative bodies is to propose taxes that respect the compromises made by the subjects of their policies. Gauthier's mutual advantage test is relevant here. If the basis for morality consists in the self-imposed constraints on voluntary action, then one could argue that the Internet sales tax could be precisely a constraint of this sort. Even if the taxed business sees no direct benefit from the taxes, as the "no representation" view claims, that business receives services that either it or other Internet businesses in its municipality support with tax revenue. If the best-case business scenario is tax-free enterprise, the effects of such enterprise on municipal up-keep could be devastating. Hence, the Internet tax, if applied with care, could be mutually advantageous.

This discussion is not intended as an endorsement of either version of contractarianism discussed above. I have assumed, however, that the fairness at issue in the Internet tax debate is a matter of distributive justice. Contractarian theories provide a particularly clear perspective on matters of distributive justice; so I have relied on them for my analysis of the two views of taxation. The further question of the truth or overall adequacy of these theories is not a question I have intended to pursue here. I have merely intended to show that one of the views of taxation discussed above works better than the other on both theories. Between the "no representation" and the "unfair competition" view, the balance of reasons tends to favor the latter. If fairness is taken as the main consideration of the Internet sales tax, then the claims of the differing views ought to be examined on that basis. But the "no representation" view fails to show that an Internet tax really would be unfair. It seems neither more nor less fair than the standard forms of sales tax. But the core idea of the "unfair competition" view, that it is unfair and arbitrary to exclude one type of business from a sales tax but not another, similar business, is supported by both contractarian positions examined above.

REFERENCES

Barclay, L., & Markel, K. (2009). Ethical Fairness and Human Rights: The Treatment of Employees with Psychiatric Disabilities. *Journal of Business Ethics*, *85*(3), 333–345. doi:10.1007/s10551-008-9773-5

Boucher, D. (1994). Introduction. In Boucher, D. (Ed.), *The Social Contract from Hobbes to Rawls*. New York: Routledge. doi:10.4324/9780203392928

Brink, D. O. (2006). Some Forms and Limits of Consequentialism. In Copp, D. (Ed.), *The Oxford Handbook of Ethical Theory*. New York: Oxford University Press.

Corp, Q. v. Heitkamp, 504 U.S. 298 (1992). Retrieved from http://laws.findlaw.com/us/504/298.html

Gauthier, D. (1986). *Morals by Agreement*. New York: Oxford University Press.

Gauthier, D. (2007). Why Contractarianism? In Theory, E. (Ed.), *Russ Shafer-Landau* (pp. 620–630). Malden, MA: Blackwell Publishing.

Grogan, D. (2009, April 23). *Minnesota Internet Sales Tax Bills Rolled Into Omnibus Tax Bills* Bookselling This Week, American Booksellers Association. Retrieved from: http://news.bookweb.org/news/6761.html

Guest, S. (2005). Integrity, Equality, and Justice. *Revue Internationale de Philosophie*, *59*(3), 335–362.

Hansell, S. (2008, May 1). Amazon Sues New York State to Void Tax Rules. *The New York Times*. Retrieved from http://bits.blogs.nytimes.com/2008/05/01/amazon-sues-new-york-state-to-void-sales-tax-rules/

Hobbes, T. (1996). *Leviathan* (Gaskin, J. C. A., Ed.). New York: Oxford University Press.

Kickul, J. (2001). When Organizations Break Their Promises: Employee Reactions to Unfair Processes and Treatment. *Journal of Business Ethics*, *29*(4), 289–307. doi:10.1023/A:1010734616208

Kukathas, C., & Pettit, P. (1990). *Rawls: A Theory of Justice and its Critics*. Stanford, CA: Stanford University Press.

Kymlica, W. (2002). *Contemporary Political Philosophy: An Introduction* (2nd ed.). New York: Oxford University Press.

Locke, J. (1980). *Second Treatise of Government* (Macpherson, C. B., Ed.). Indianapolis, IN: Hackett Publishing Company.

Lucas, A. (2000, Feb. 2). *Safeguarding Internet Tax Fairness*. Center for Trade Policy Studies. CATO Institute. Retrieved from: http://www.freetrade.org/node/350

McQuivey, J., & DeMaulin, G. (2000). *States Lose Half a Billion in Taxes to Web Retail*. New York: Forrester Research, Inc.

Neil, P. (1996). Electronic Commerce: Taxation without Clarification. New York: Klynveld, Peat, Marwick, and Gordeler (KPMG).

Nozick, R. (1974). *Anarchy, State, Utopia*. New York: Basic Books.

Powell, D. C. (2000). Internet Taxation and U.S. Intergovernmental Relations: From Quill to the Present. *Publius*, *30*(1), 39–51. doi:10.2307/3331120

Rawls, J. (1971). *A Theory of Justice*. Cambridge, MA: Harvard University Press.

Rousseau, J.-J. (1987). *The Basic Political Writings*. Indianapolis, IN: Hackett Publishing Company.

Saunders, Ben. (2008). The Equality of Lotteries. *Philosophy: Journal of the Royal Institute of Philosophy*, *83*(325), 359-372.

Schenk, D. (1996). Taxation. In New York University School of Law (Ed.), Fundamentals of American Law. New York: Oxford University Press.

Chapter 14
Ethical Issues Arising from the Usage of Electronic Communications in the Workplace

Fernando A. A. Lagraña
Webster University Geneva, Switzerland & Grenoble École de Management, France

ABSTRACT

E-mail has become the most popular communication tool in the professional environment. Electronic communications, because of their specific nature, raise a number of ethical issues: e-mail communications are distance, asynchronous, text-based, and interactive computer-mediated communications and allow for storage, retrieval, broadcast and manipulation of messages. These specificities give rise to misunderstanding, misconduct in the absence of the interlocutors, information and mail overload, as well as privacy infringement and misuse of shared computing resources. Inexperience explains some users' unethical behavior. Other forms of unethical behavior find their roots in corporate culture, internal competition and management styles. E-businesses, as early adopters of information and communication technologies, are being particularly exposed to such behaviors, since they rely heavily on electronic communications. They should therefore assess their internal situation and develop and enforce e-mail policies accordingly.

INTRODUCTION

The digitization of information and communication technologies (ICTs), the world-wide extension of ICT-based networks, services and applications, and in particular of the Internet and of the World Wide Web, have paved the road and made possible the correlated development of e-business. The web and

the Internet have also changed the way we communicate and interrelate, both in our private sphere and in the office.

During the past decade, according to a survey conducted by Dimension Data, electronic mail (e-mail) has become the most popular communication tool in the professional environment, outpacing fixed and mobile telephony. 96% of the researched organizations declared that they offered access to e-mail in the workplace to their employees, 91%

DOI: 10.4018/978-1-61520-615-5.ch014

to a conventional fixed-line telephone line, and 86% to a professional mobile phone. On the user side, 99% of employees declared they were using e-mail professionally, against 80% for fixed-line telephony and 76% for mobile phone use (Dimension Data, 2007).

Because of the specific nature of e-businesses, as early adopters of ICTs as the underlying infrastructure and tools supporting their business models, the trend towards a heavier usage of computer-mediated communications (CMC) and in particular e-mail is particularly visible in the e-commerce and on-line industry.

There is little doubt that electronic communications, and in particular e-mail, have introduced a paradigm shift in management, organizational and working methods, as well as in business performance, as they have in the economy in general. While ICTs have dramatically improved business-to-business (B2B) or business-to-consumers (B2C) communications, they have also significantly impacted our day-to-day personal and professional lives. In particular, many organizations and their employees seem to have been overwhelmed by a number of issues arising from the usage of electronic communications in the professional environment.

The adoption of technologies is not morally neutral, and the emergence of the new electronic communication systems has come along with, or has favored, new attitudes and behaviors, giving rise to new ethical concerns. Previous research on ethical issues in e-business mainly addresses the relation of e-businesses to their external environment. Issues such as data mining and profiling, customer and business-critical information protection and privacy, intellectual property rights in a digital economy, or advertizing and spamming, to name but a few, are well covered in the existing literature (Danna & Gandy, 2002; Davenport & Harris 2007; Palmer, 2005; Roman, 2007; Stead & Gilbert, 2001). However, most of the issues explored relate to B2B or B2C relations, or to e-

businesses within their strategic, regulatory and legal environments.

In this chapter, we should take a slightly different perspective as we shall observe business entities *from the inside*. Considering that business ethics and ethical behavior find their roots within the internal corporate culture and practice, this chapter focuses on the ethical issues that arise from the usage of electronic communications in the workplace, and specifically upon the use of e-mail.

Ethical issues are not specific to e-business *per se*. However, e-businesses have specific characteristics, in particular because their model relies heavily on electronic communications, which make them more sensitive to some ethical issues, and in particular to some unethical behaviors. Focusing on interpersonal communication within organizations, this chapter will look at these issues from three different perspectives, e.g. from the managerial, employee, and interpersonal points of view.

The managerial perspective addresses issues such as employee monitoring, private usage of company resources, as well as the implementation of favorable working environments and of pertinent internal policies. The employee perspective encompasses issues such as privacy, autonomy, and mail overload. Finally, the interpersonal perspective covers issues such as the role of e-mail in professional communications, mutual respect (in a team or competitive environment), and misunderstanding in communications. A section of this chapter is also dedicated to the review, definition and description of the most common, although questionable on-line behaviors.

The various behaviors described are also put in context of theory. In particular, three specific avenues are explored: modernity and communication with the "absent other", the right of privacy, and information overload. In this chapter, we will consider e-mail within the theory of modernity (Giddens, 1990), and in particular how the specific characteristics of e-mail regarding space (distance) and time (asynchrony) develop a new form of communication between persons who are neither in the same location nor communicate at the same time. We will also address one of the most challenging managerial issues regarding e-mail, which is the balance between the right to privacy of the employees and the need to monitor the usage of employers' ICT resources. And we will also tackle the issue of mail and information overload and of its consequences on performance.

While exploring these theoretical considerations, the aim of this chapter is nevertheless to provide a practitioner perspective, since ultimately ethics should address behavior and not merely theory. Therefore, the last part of the chapter goes back to practice and provides some guidelines on how to address the various ethical issues arising from the usage of electronic communications in the workplace and on how to foster more ethical behavior, from a managerial, employee and interpersonal perspective.

BACKGROUND

Ethics in Practice: Electronic Communications

What do ethics, and business ethics, mean in terms of electronic communications? This section presents some general principles of ethics, as well as the specific elements regarding ICTs in general and electronic communications in the workplace in particular.

The purpose of ethics is to provide tools to help discern what people should do and how they should behave or, in a simplified – and possibly simplistic – way, what is good and what is bad behavior. As such, theoretical or fundamental ethics are the playing ground for philosophers and theologians and indeed, ethical theory finds its origins in various religious and philosophical views.

Ethics deals with behavior, e.g. with action rather than thoughts or feelings. This is specifically relevant in the business environment. Ethics is a practice rather than a science or an epistemology. Ethics is not a science since its ultimate purpose is not to solely to develop knowledge (it is about virtues or virtuous behavior). Ethics is not simply an epistemology, which would consider each virtue as a science (by example the science of justice) and aim only at developing knowledge about these virtues (Aristotle, 1994). And ethics deals with actual behavior: as one may feel like acting in a certain way, one may also analyze the ethical "quality" of such behavior, and then decide to act in a one way or another. Only the eventual action will be considered from an ethical perspective, e.g. the action is considered from an ethical perspective rather than the thoughts, feelings or opinions.

Action or behavior must be voluntary. An action taken under constraint is not subject to ethical evaluation. Furthermore, as Aristotle points out in his *Eudemian Ethics,* "what he [a man] does in and by ignorance, he does it involuntarily" (Chapter 9, 3). We will see later in this chapter that many e-mail users behave improperly because they do not understand the tool properly, and in such cases it is important to determine whether their action is voluntary or not. Even if their action is involuntary due to ignorance, however, we must also determine if their lack of knowledge is the result of their own negligence.

Ethics is about how we treat each other and how we treat common resources. Demonstrating mutual respect, empathy, trying to understand each other and be understood, ensuring a proper working environment is considered ethical behavior. Getting one's right part of shared resources (such as Internet bandwidth) rather than overusing such resources because they are provided for free, is also considered ethical behavior (Hardin, 1968; Huberman & Lukose, 1997).

Ethics varies with culture (including philosophy and religion). What is acceptable behavior in one culture might be unethical or illegal in another. Legal behavior is a behavior respecting the laws in force in a specific jurisdiction (typically a national or regional territory where the behavior actually occurs). Ethical behavior is a behavior respecting the moral rules shared by a specific community (culture, philosophy, religion, nation, etc.) In our collective societies, where laws are devised to protect the common good and the smooth functioning of the communities, what is illegal is generally considered as unethical (Aristotle, 2008). On the other hand, the opposite is not directly true. We should review some examples later in this chapter, such as employee monitoring by example, which is legal, but considered by many as being unethical (Miller and Weckert, 2000).

Ethics also needs to take into account what can (technically) possibly be done or not, since some behaviors depend upon technological development. It is therefore useful to understand the technical setting before addressing the correlated ethical issues (Langford, 1996). Ethics evolve with technology and in particular, since the development and usage of ICTs, new ethics issues have appeared.

What is E-Mail?

The Webster's New World College Dictionary defines electronic mail, usually abbreviated as e-mail, as a communication system used to send messages between computers or terminals connected *via* telecommunication networks, and in particular the Internet. E-mail(s) also refers to the messages sent through such communication systems. To e-mail refers also to the action of sending such messages.

E-Mail Structure

The Internet Engineering Task Force (IETF) is the main standardization body defining the technical characteristics of the Internet. IETF standards are called RFCs (Requests for Comment) since they are produced through an open consultation process to which any expert can contribute. In its standard-setting capacity, the IETF has defined the structure of e-mail messages in RFC 5322 (IETF, 2008).

E-mail messages consist of an envelope and contents, as described as follows.

E-Mail Envelope

The envelope contains the information needed to transmit and deliver the message.

On the *sender side*, the envelope includes the address of the recipient(s) (Destination Address Field), which can be nominative or in the form of a distribution list. The recipients can be either primary (direct) recipients (their address will appear in the "To:" field), or secondary recipients and receive a "carbon copy" of the message (an analogy to typewritten correspondence) and their address will be listed in the "Cc:" field. The sender may also decide to send the message without disclosing the addresses of some of the recipients: this is the purpose of the "Bcc:" field. Recipients whose addresses are listed in this field will receive a "blind carbon copy" i.e. they will receive the message without the other recipients of the message knowing it.

The envelope also includes the "Subject:" field, which allows the sender to inform the recipients about the subject (purpose) of the message (*Example:* "Subject: Trip to Biarritz"). Most e-mail user agents (the software programs used to manage e-mail) keep the subject field of incoming messages when the recipient either replies to them or forwards them to other addressees. An indication is added before the original subject in the "Subject" field, indicating that the message is a reply (*Example:* "Subject: *Re:* Trip to Biarritz") or is forwarded (*Example:* "Subject: *Fwd:* Trip to Biarritz").

Finally, the envelope also contains the e-mail address of the sender (Originator Address Field).

When the sender uses several e-mail accounts this specifically allows the sender to indicate which one of her addresses the recipients should use when replying to the message.

Once the message is sent, the e-mail agent will automatically add a date and time stamp to the envelope of the outgoing message. This element will appear in the envelope of the message on the *recipient side,* in addition to the ones included by the sender, and with the exception of the addresses of users receiving a blind carbon copy of the message.

E-Mail Body (Content)

The body of the message contains the message to be transmitted to the addressees. Originally, e-mail systems only permitted the transmission of text messages. Since these early days, e-mail technology has considerably evolved and now allows users to send any type of material – documents, pictures, videos, audio files, web pages, etc. in the form of attachments or embedded in the body of the message, according to their format and nature.

When the recipient of a message replies to it, or forwards it to other recipients, most e-mail agents offer the possibility to include the original message in the body of the new message to be created, in addition to the inclusion of the "Re:" and "Fwd:" elements in the subject field. Generally, e-mail agents offer the possibility to select the parameters identifying the original message (font, color, side line, indent, etc.) and differentiating it from the elements added at the later stage by the next redactor.

Why Such Details about the Nature of E-Mail?

Langford (1996) considers that "what is ethically appropriate must reflect what is technically possible". As for most electronic documents, both the envelope and the body can be electronically manipulated and altered. Practice shows that this opportunity is used (and abused) and that all of the elements of e-mail messages are subject to unethical behaviors. These behaviors can take the form of misquotes, modifications of dates, of recipient lists, and other types of alterations (forgery). The knowledge of the technical characteristics of e-mail systems hence provides the basis for the analysis of their usage.

E-Mail Specific Characteristics

Communications *via* e-mail are characterized by a number of specific properties, which differentiate them from more traditional forms of professional communications such as typewritten correspondence – office memoranda and the like, traditional post or telephone conversations (Garton & Wellman, 1993; Akrich, Méadel & Paravel, 2000) or emerging forms of near real-time communications such as instant messaging and short-message systems.

Comparing E-Mail to Traditional Communication Media

Asynchrony

E-mail communications are asynchronous. This means that there is no need for the sender and the recipient to be using their e-mail systems simultaneously, contrary to telephone communications (with the exception of answering machines and more recently voicemail). E-mail systems are based on store-and-retrieve technologies. When a message is sent and reaches its destination, it is stored in a mail server, where the recipient can retrieve it at any time through a mail agent. This characteristic makes e-mail resemble the traditional typewritten and postal correspondence.

E-mail asynchrony gives the false impression that e-mail users can read and manage their messages at the time they freely choose. While this might be true in theory, practice shows that asynchrony may be a cause of mail overload. Since most e-mail systems offer remote access through the Internet, asynchrony also contributes to blurring the barriers between home and office time.

Interactivity and Instantaneity

E-mail systems allow near real-time interactive communications, in particular within the immediate (company-wide) professional environment. Modern businesses, and in particular e-businesses, are organized around an underlying broadband high-speed telecommunication infrastructure which allows quasi-instantaneous transmission and reception of e-mail messages (at least text-only messages). This instantaneity of e-mail communications can be used for interactive, conversation-like communications, organized in "threads".

Instantaneity of e-mail makes e-mail communication similar to face-to-face communications, at least in terms of chronology. This explains why some users consider e-mail as short-lived, or transitory. However, once written and sent, e-mail messages are stored, archived and can be retrieved, re-used, as well as forwarded with practically no limit in time.

Broadcast and Narrowcast

The envelope of the e-mail message allows for several Destination Address Fields, giving the sender the possibility to define addressees at will. Messages can be sent to individual addresses *(narrowcast)*, or to distribution lists, mail exploders (generic addresses which forward received mails automatically), and newsgroups, etc. *(broadcast)*. This makes it possible to send or forward any message to a virtually unlimited number of recipients, as long as their e-mail addresses are available to the originator.

This raises the ethical concern of privacy, or more precisely the issue of private communication. When sending a message to a single recipient, a user may genuinely consider being engaged in a private conversation. However, it frequently happens that such "private" e-mail messages, sent "for your eyes only" are broadcast to a wider audience, voluntarily or not. Similar cases also occur when the "Reply to all" feature of e-mail messaging is used improperly.

Storage

Since e-mail systems are based on store-and-retrieve technology, storage of e-mail messages is an integral part of e-mail services, as it is ensured and guaranteed even before recipients open the messages they receive. Compared to traditional communications systems, such as typewritten correspondence and telephony, e-mail systems offer safe and easy-to-use storage facilities to their users. Further, because of the electronic nature of e-mail, indexing, sorting and retrieving e-mail messages is far more efficient, fast and reliable.

In the professional environment, the storage and archiving of e-mail servers are provided and controlled by the employer, generally through the corporate information system service. This raises the ethical issue of privacy infringement by individuals having privileged access to employee correspondence, either for managerial or technical reasons.

Text-Based

Despite the fact that new developments allow the inclusion in e-mail messages of any type documents, including audio, still and moving pictures (video) attachments, e-mail is and remains essentially a text-based medium.

This raises the issue of user articulation, and their ability to get their message across properly in a written form of communication, in order to avoid misunderstanding or emotion escalation. This is particularly true in multicultural environments, when backgrounds, languages or levels of education vary among users.

Why Elaborate on these Technical Characteristics?

In the mid 1990s, the specific characteristics of e-mail were considered, with good reason, a considerable technological advance and even a paradigm shift in computer-mediated communications, which was making McLuhan's metaphor of the *global village* a reality (McLuhan, 1962). A number of economies engaged in vast development programs of the then called information highways. E-mail was coined as the "killer" application, while traditional mail was labeled "snail" mail. With its apparent user-friendliness and its low cost, e-mail was considered to be *the solution* for modern communications.

However, as we shall analyze later in this chapter, the specific characteristics of e-mail, its differences, as well as sometimes its similarities with other communication media, are exposing e-mail users to a number of drawbacks and harmful experiences.

Still, before exploring these aspects, further development is needed regarding the specific challenges facing e-mail as an interpersonal communication tool.

E-Mail Communications: The Challenges

The Main Challenge of Communications: Misunderstanding

In 1872, the widely applauded French writer Gustave Flaubert – famous for his novel *Madame Bovary*, wrote in a letter to an unknown female friend of his: We are all in a desert. Nobody understands anybody *("Nous sommes tous dans un désert. Personne ne comprend personne.")*. However, the feeling of loneliness and of unavoidable misunderstanding is not limited to romantic writers and desperate artists looking for a friendly soul. Rather than being an exception, misunderstanding in communications, and in particular in telecommunications (etymologically "distance communications"), is the common rule (Karsenty, 2008).

Definitions and Causes
Humphrey-Jones (1986) defines misunderstanding as the failure of an attempt of communication which occurs because what the speaker wants to express differs from what the listener believes has been expressed (mentioned in House, Kasper

& Ross (2003)). For Weigand (1999), misunderstanding is a kind of understanding which deviates partially or totally from what the speaker wanted to communicate.

This is a real challenge in the professional environment, since communication occupies a central place at work (Karsenty & Lacoste, 2004). Organizations should therefore pay particular attention to misunderstandings and their consequences and should do their best to avoid them. Studies in workplace communication ergonomics have identified three main causes for misunderstanding (Falzon, 1989):

a. The absence of a properly operational communication channel,
b. The absence of a common vocabulary, and
c. An exclusive information context.

Firstly, misunderstanding may occur if the communication channel doesn't offer the proper level of quality of service and doesn't preserve the integrity of the message across the system boundary. In such instances, the *syntax* of the message is not preserved (Carlile, 2002). While this is obviously a challenge for e-mail communications, we should assume that e-businesses are supported by proper communication systems, where the information is carried over the networks and the communication channels without physical alteration.

Secondly, more challenging and frequent are instances when the recipient doesn't understand the *semantics* of the message, its intended meaning. According to Karsenty, this happens in the majority of communications (Karsenty, 2008). For example, there is general agreement on the fact that intercultural communication might generate misunderstanding (Bennett, 1998), in particular when cultural differences include the usage of different languages. Different communication styles, linked to different cultures, may also generate difficulties (House, 2003), as may differences in competence areas or in levels of competence (Enquist & Makrygiannis, 1998).

Thirdly, an exclusive information context may occur when the sender and the recipient are in different functional units, or at different levels of responsibility, or simply don't have access to the same sources of information. These instances may occur when some individuals decide to retain information, when managerial policy segregates the access to certain information according to the position in the company, or when the corporate cultural is such that there is a clear separation among different units or departments (production and marketing, administration and design, etc.) (Peyrelong & Accart, 2002; Sproull & Kiesler, 1986). Typical examples are communications where some participants don't have access to sensitive information that a company is not willing to disseminate.

In addition to these causes of misunderstanding, Karsenty (2008) has identified a number of situations which *facilitate* misunderstanding, including:

a. Distance communications, when the interlocutors don't share the same visual perception of the physical environment.
b. Asynchronous communications, which reduce the possibilities of direct interactions to control mutual understanding. These include all business communications using paper or electronic media.
c. Communications between individuals with different areas and/or levels of competence.
d. Man-machine communications.

As e-mail communications combine these four characteristics, the probability of misunderstanding when using this channel is high.

Mutual comprehension is the first step towards mutual respect. By offering a favorable ground to misunderstanding, e-mail communications may constitute a barrier to a proper dialogue, and hinder the establishment of a climate of confidence and mutual respect. Therefore, users of electronic communications must be particularly vigilant to ensure that their messages come properly across, and are not misunderstood or misinterpreted. Furthermore, e-mail recipients must ensure that they carefully read and properly understand the messages they receive. Rogers and Kinget (1966) consider that the best attitude to establish the proper communication climate, is the one of empathetic communication, where each interlocutor makes all possible efforts to ensure that he understands the other properly, and that the messages received are interpreted as positively as possible.

Three Specific Challenges of Computer-Mediated Communications

Among the various challenges facing organizations using computer-mediated communications, this chapter explores three specific avenues. The first section explores how e-mail illustrates the theory of modernity developed by Anthony Giddens (1990) and what lessons can be learned from this theory and applied to electronic communications. The second section explores a typical issue of e-mail, mail overload. The third section elaborates on an issue of high ethical interest, the right to privacy.

Modernity and Communication with the Absent Other

A characteristic of distance asynchronous communications such as e-mail is the absence, both in time and space, of the interlocutor. In other words, when one redacts an e-mail message, the recipient of the message is neither physically present (absence in space), nor reading the message while the redactor is drafting it (absence in time). Giddens (1990) considers this situation as the essence of modernity. Modernity is characterized by a separation of time and space, and by the further separation of *space* from *place*, where place should be understood in geographical and physical terms, while space should be approached

Table 1. Specific properties of e-mail compared to traditional communication media

Medium Properties	E-mail	Mail	Phone
Asynchrony	Yes	Yes	No[a]
Instantaneity	High	No	Highest
Interactivity	High	Low	Highest
Broadcast	Yes	Yes (Mailing)	Low (Teleconference)
Storage	Electronic	Manual	Recording
Manipulation	Natural	Possible	No
Text-based	Yes	Yes	No

[a] With the exception of answering machines and voicemail.

in functional and environmental terms. This separation is perfectly reflected in the nature of e-mail. E-mail communications "foster relations between 'absent others', locally distant from any given situation of face-to-face communication" (p. 18). Modern organizations need to adapt themselves to this new societal paradigm, and have to renew their forms of coordination and cooperation, not only because they operate in multiple sites and locations, but also because in single campuses, they manage an increasing number of employees who act as 'absent others' linked by electronic communications systems.

In modern societies, the separation of space and place and the communication with the absent other, rather than being an exception, becomes common behavior. A notable consequence is that face-to-face communications, particularly in the workplace, are replaced by computer-mediated communications, such as e-mail (Sarbaugh-Thompson & Feldman, 1998), and that the overall volume of all forms of communication decreases. Therefore, co-workers communicate less, and do so through communication channels that expose them to a higher level of misunderstanding.

Communicating with absent others by e-mail does not only increase the risk of misunderstanding. It also affects the quality of the interpersonal communication. E-mail communications are essentially text-based. This makes this medium

less "rich" than other forms of communication (Carnevale & Probst, 1997). A major drawback is the lack of non-verbal cues, which makes it very difficult to communicate emotions (Byron, 2008).

In face-to-face communications, the interlocutors use body language such as head nodding, eye contact, smile, or any types of gestures or corporal attitudes to provide elements of information in addition to the verbal message. This is also partially true in phone conversations, where speed, intonation, pauses can transmit emotions and meta-information. However, this is not possible in e-mail. This difficulty may lead to conflict escalation (Friedman & Currall, 2003) because of the instantaneity and interactivity of e-mail, technical characteristics which didn't exist at the time of postal or typewritten communication, which were much slower communication channels (See Table 1).

Sproull & Kiesler note that e-mail does also provide weak social context cues, and that people focus more on themselves than on others when drafting messages, thus making empathic communication less probable. Further, as we become less aware of our audience, we become secure in our anonymity and less bound by social norms (Lea & Spears, 1991). For example, employees have a tendency to be less formal and respectful to their supervisors when they communicate by e-mail, compared to face-to-face conversations.

These characteristics may affect mutual respect as well as the way coworkers – both employees and employers – treat each other, in the professional context, a common concern in business ethics research. Ethical behavior encompasses making efforts to be understood by the recipients of one's messages and considering positively the messages one receives. This also implies mutual respect, even when the interlocutor is an "absent other".

Mail Overload

In January 2008, a Canadian government ministry sent out a directive to its employees urging them to relax and not to use their Blackberry® at night or on weekends and holidays, and asking them to implement a Blackberry® "blackout" from 7 p.m. to 7 a.m. In the memorandum sent to all staff of the department, Deputy Minister Richard Fadden wrote: "Work/life quality is a priority for me and this organization because achieving it benefits us both as individuals and as a department." (Fadden, 2008)

In April 2007, venture capitalist Fred Wilson declared a 21st century kind of bankruptcy. He declared on his blog: "I am so far behind on e-mail that I am declaring bankruptcy. If you've sent me an e-mail... you might want to send it again. I am starting over."

These real-life examples illustrate some possible measures against a growing phenomenon: mail overload. The number of e-mail messages is one factor cited by many people as a cause of professional overload (Edmunds & Morris, 2000). E-mail becomes a potential stressor in itself, in addition to being a stress conduit (Taylor, Fieldman & Altman, 2008), because of the amount of effort that an employee must deploy to reduce superfluous information to a manageable level (Dawley & Anthony, 2003), and because overload goes generally is accompanied by a perception of urgency (Isaac, Campoy & Kalika, 2007). The more a user has to dig into an unmanageable number of e-mail messages, the higher the urgency, the more difficult it is to act in a reflexive, reasoned way,

the higher the stress. There is no virtuous cycle here, and Taylor, Fieldman & Altman show that an unbalanced use of e-mail may have an adverse impact upon well-being, stress and productivity.

As illustrated by the case of the Canadian government ministry, one response to mail overload is to deal with e-mail messages from home during the weekend, a way of doing called "weekend catch-up". This is facilitated by the ubiquity of ICTs. Nowadays, nothing is easier than connecting to the office environment from home and doing some work on-line. Since late 2006, PricewaterhouseCoopers employees logging on to work e-mail on Saturdays and Sundays have been greeted with a pop-up window that says: "It's the weekend. Help reduce weekend e-mail overload for both you and your colleagues by working off-line." The alert adds that workers who want to get a head start on writing messages should hold off sending them until Monday unless urgent. "By sending an e-mail," says Human Resources Communications Director Geneviève Girault, "you're encouraging... people to respond." (McGregor, 2008).

These various examples drawn from practice demonstrate a particular phenomenon: e-mail overload is generally not generated by *spam*, e.g. unsolicited e-mail messages from external sources. Its origins are to be found *inside* the organization. Most of these messages come from colleagues, supervisors, etc. In an analogy with the concept of "friendly fire", we should call this concept "friendly spam": friendly spam originators are coworkers rather than external correspondents.

Unethical behavior may come from a lack of expertise in the use of technology. In that sense, not providing appropriate e-mail training or not investing in one's ability to use e-mail properly can also be considered as unethical behavior, by not respecting the principle of precaution to avoid harmful consequences of one's poor performance.

The Right to Privacy

Ethical concerns about privacy are relatively recent in the history of ethical theory. This concept could probably not appear in the founding texts on ethics of the Greek philosophers such as Aristotle or the stoics, at the time when the philosophical *theoria* was cosmos-centric rather than human- or individual-centric (Ferry, 2006). It could however have been present during the 18th century in the work of the humanists and founding fathers of the modern philosophy such as Jean-Jacques Rousseau. Surprisingly, it wasn't, and this issue was first explored in the second part of the 19th century (Warren & Brandeis, 1890), when the right "to be let alone" was eventually recognized, discussed, and protected.

Since then the right to privacy has become a major moral right. Miller and Weckert (2000) consider that there is a fundamental moral obligation to respect the individual's right to privacy. Concerns regarding privacy protection and infringement have grown with the development of ICTs, and the fact that technology advances facilitate new unethical behaviors (Loch, Conger & Oz, 1998), by the ability to collect personal data and private information. Moor (1990) defines privacy as the fact that "an individual has privacy if and only if information related to this individual is protected from intrusion, observation and surveillance by others." (Moor, 1990, p.76)

On-line businesses are early adopters and articulate users of electronic communications. During the last two decades, the technology in use has gradually moved from telephony to on-line communication systems, and in particular e-mail, for internal and external communications, and thus for daily activities of employees. This has been accompanied by a change in the "philosophy of privacy". While telephone and postal systems favor the individual right to privacy through the secret of correspondence, company rights tend to prevail in e-mail systems, and limitations to individuals' privacy are common. Since ICTs provide new tools for employers to monitor employees' computer and e-mail usage, the ethical challenge is high. How can the willingness to monitor em-

ployees' performance and activity – through the monitoring of their computer and e-mail usage – be balanced with the individual's right to privacy?

From a managerial perspective, one should have a close look into the matter. Nussbaum and DuRivage (1986) note that computer monitoring increases worker stress, reduces job satisfaction, and ultimately goes against the management goal of increased productivity. Langford (1996) argues that the Internet was created and developed around a paradigm of freedom and that Internet users are resisting external control. Therefore, while managers have a legitimate right to monitor *work*, they shouldn't be allowed to monitor *workers*. Miller and Weckert (2000) conclude: "employing people does not confer the right to monitor their private conversations, whether those conversations be in person or via e-mail".

A Managerial Perspective: Internet as a Shared Resource

While on the one hand employees may raise concerns regarding their legitimate right to privacy, on the other hand employers may have parallel concerns regarding the usage of company resources for private purposes. According to a recent study (Monster, 2009), more than half of the European employees send personal e-mail or surf the Internet during working hours.

While it is almost impossible to avoid private usage of computer resources in the workplace, employees must be aware of the limitations set up by their employers. The most efficient way of doing so is to develop and enforce an appropriate policy (Womack, L., Braswell, D. E., & Harmon, W. K., 2004), clearly indicating the objectives of the employers in developing such policy, generally to protect company resources, data, and image.

Improper behaviors should be explicitly designated, such as the access, storage and dissemination of material that is inappropriate or illegal, by example:

a.   language, messages, or other material that are fraudulent, obscene, abusive, derogatory, or inflammatory,

b.   material that is pornographic or sexually explicit,

c.   material promoting sexual exploitation, discrimination, racism, hate, or violence,

d.   information concerning gambling, drugs or weapons,

e.   material which violates copyright or intellectual property rights.

A balance must be struck the company's interests and the rights of its employees. Again, ethical behavior is tempered behavior, with the right mix of rights and obligations.

COMMON UNETHICAL BEHAVIORS IN E-MAIL USAGE

In the previous section, we considered general challenges facing e-mail communications: abuse of shared resources, misunderstanding, mail overload, privacy infringement, or usage of company resources for private matters. In this section, we provide an analysis of the most common unethical behaviors found in e-mail practice, based on a review of literature, general and business press, as well as on the analysis of the author's own experience in the field of ICTs.

A variety of unethical behaviors identified by practice are presented, including a short definition, description and/or comment. Table 2 intends to categorize these behaviors in relation to the e-mail challenges described in the previous section, and introduces the issue of mutual respect. We then discuss whether these behaviors are unethical or not, if they are intentional or denote an inexperience in the usage of e-mail, and if they denote a personal attitude or a corporate culture. Finally, we propose several types of possible solutions.

Definitions & Descriptions

- **[Harmful] Address Disclosure:** *Harmful Address Disclosure* occurs when someone discloses someone else's e-mail address for harmful purpose, by example by subscribing the address owner to an on-line newsletter or to an advertisement site, with a view to have him/her receiving spam, and eventually making his/her mailbox unmanageable.

- **Ambiguity:** *Ambiguity* occurs when the sender of a message has intentionally drafted it ambiguously, with a view to create delays, to cause *confusion*, or to mislead the message recipient(s). Generally, a sender of such a message would later claim that the message was crystal clear.

- **[E-mail] Bankruptcy:** *Bankruptcy* occurs when someone is submerged underneath an endless pile of e-mail messages in the inbox and considers that the only way out is to declare e-mail bankruptcy, e.g. to delete all messages from the inbox and (preferably) to inform all message senders of what has occurred. Informing the senders of the deleted messages is considered a form of respect.

- **Broadcast:** *Broadcast* occurs when someone sends a message to too many addressees, including those who might have only a marginal interest in the content of the e-mail. This creates overload, demonstrates poor self-confidence and lack of respect towards the recipients. *Broadcast* is a cause of friendly spam.

- **"CC:":** The *CC:* (or BCC:) lists of recipients are dangerous tools. Some senders use these lists to exercise pressure on, or discredit the recipients through undue *broadcast* or indirect power flaunting (the most common application in that case being to copy the recipient's direct supervisor). In most cultures, this behavior is considered as a lack of maturity, professionalism and self-confidence. It is also considered to be offensive.

- **Click and Rush:** *Click and Rush* occurs when the author of message rushes (physically) to the office of the intended recipient immediately after having clicked on the "Send" button, or immediately calls the intended recipient, "to check if [the recipient] has received the message", possibly to discuss it. This behavior denotes a certain (possibly cultural) attitude towards time, and specifically the fact that the originator of the message doesn't consider e-mail as an asynchronous communication channel. This behavior creates disruptions in the recipient's agenda, overload, and might be perceived as a lack of respect.
- **Confusion:** *Confusion* occurs when the recipient of a message voluntarily pretends not to understand the content or substance of a received message, when he/she actually does. This behavior is in essence comparable to *denial*. It may generate overload and put time pressure on the originator of the message who would have to draft the message again or find another communication channel to get understood.
- **Debate:** *Debate* occurs when correspondents start an e-mail discussion and pursue it for too long. E-mail is a poor communication channel for debating, in particular when issues are complex, emotional (see *Flaming*), or when correspondents don't know each other well.
- **Denial (of receipt):** *Denial* occurs when the recipient of an e-mail message wrongly and intentionally claims that he or she has not received the message. The recipient may then claim by example that he/she couldn't complete an action or an instruction because he/she was not aware of the message.

- **[Message] Delay:** *Delay* occurs when the sender of a message intentionally delays delivering a message in order to deprive its intended recipient with time-critical information, with a view to have the recipient loose an opportunity, face, or credit. The sender generally claims afterwards that the recipient was indeed properly informed, that the delay might have occurred because of a server breakdown, or for any other reason out of his/her control.
- **Fast Forward:** *Fast Forward* occurs when the participant to a thread (a series of messages on the same topic), forwards the whole content of a message (including the historical elements of the discussion still visible at the bottom of the message), to a third party, e.g. a recipient who was not involved in the thread. This action can be intentional or by mistake, but may generate embarrassing situation, when someone should for example be informed of the result of a discussion, but not of its content.
- **Flaming:** *Flaming* is the act of posting or sending offensive (angry, insulting) messages. Such messages are called "flames". They generally appear in discussion forums. They may nevertheless appear when people use e-mail for *debates*.
- **Forgery:** *Forgery* occurs when someone voluntary modifies the meta-information of a message before forwarding it, by modifying the date, time, author, subject, routing, list of recipients of the message, with the purpose to abuse the recipient(s). This highly unethical behavior denotes generally a potential conflict within the organization.
- **Gossip:** We all know what *gossip* is about. While *gossip* can't probably be avoided around a coffee machine, it should never be transmitted in writing by e-mail. And in particular, *gossip* shouldn't be *broadcast*.

Table 2. Categorizing e-mail unethical behaviors according to their potential organizational impact

	Misusing Shared Common Resources	Misusing Mail for Emotional Discussions	Generating E-mail overload	Lacking Mutual Respect	Generating Time Pressure	Creating Mis-understanding
Address Disclosure			1			
Ambiguity			2	2	2	1
Bankruptcy			1 a)	2		
Broadcast	1		1	2		
CC:			2	1		
Click & Rush			2	1	1	
Confusion			2	1	2	
Debate			1		1	
Denial				1	2	
Delay				1	2	
Fast Forward				1		2
Flaming		1		1		
Forgery				1		
Gossip				1		
Grammar				2		1
Humor				1		2
Infection	1			2		
Indifference				1		
Interpretation				2		1
Masquerade				1		
Misidentification				1	2	
Misquote				1		
Omission				1		
Power Flaunting		2		1		
Reply (to All)	2		1	2		
Sarcasm		2		1		2
Sent (by Mistake)				1		
Silence				1		
Thank You Mails			1	2		
Weekend Catch-Up			1 b)			

a) Bankruptcy is a consequence, rather than a cause of e-mail overload.

b) Weekend Catch-Up is both a consequence and a cause of e-mail overload.

1: Primary (high) impact 2: Secondary (lower) impact

- **Grammar:** Should *grammar* be included in the list of ethical issues? If the purpose of e-mail is to communicate properly, we may agree. Poor *grammar* may create misunderstanding. In certain cultures, it might also be consider as a lack of respect to the recipient(s), in particular for those who consider that e-mail *is really* a written form of communication.

- **Humor:** *Humor* is a difficult issue. How would someone consider that humor is unethical? Pierre Desproges, a famous French humorist, once said: "One can laugh about anything, but not with everyone". One culture's humor might be offending to another culture. One individual's humor might be offensive to many others.

- **Infection:** *Infection* occurs when a message contains viruses. Checking regularly for viruses is not only good for one's computer; this is also a sign of respect for one's correspondents.

- **Indifference:** *Indifference* is a new form of *power flaunting*, which has appeared with the introduction of electronic communications. *Indifference* takes the form of "cold", emotionless messages, that employees receive providing them with carrier-critical information, including transfer(s) and termination. In traditional management such information would only have been transmitted in person.

- **Interpretation:** *Interpretation* of e-mail messages is probably the most important and the most difficult task in electronic communications and often leads to *misunderstanding*. A proper (ethical) behavior would be to interpret e-mail messages as positively as possible.

- **Masquerade:** *Masquerade* occurs when someone uses another's e-mail account, without permission, to send messages, for the purpose of deceiving the recipient(s) by pretending to be the owner of the account. In some countries, *masquerade* is illegal, as a form of identity theft.

- **Misidentification:** *Misidentification* occurs when someone intentionally provides a wrong address to a correspondent, to

avoid (or slow down) further communication with this correspondent.

- **Misquote:** *Misquote* occurs when the recipient of a message alters and uses part of it, without clearly indicating the modifications (deletions, replacements, adjunctions), with a view to transform the original message of the sender for whatever reason (by example to use the result for one's own credit or to discredit the sender of the original message). Modifications in messages should be clearly indicated by using ellipsis [...] or keywords such as <snip>, or short clarifications.

- **[Abusive] Omission:** *Omission* occurs when the sender of an e-mail message intentionally omits someone from the list of recipients. This frequently happens in competitive environments, when mission-critical information is shared among mission stakeholders, and someone intentionally excludes one (or several) of them.

- **Power Flaunting:** *Power flaunting* occurs when the sender of message reminds overtly the recipient(s) that he or she has organizational power that he/she might exercise (on all or some of them). In most organizational cultures, this behavior hurts, in several cultures, it is considered as offending and unethical.

- **Reply (to All):** *Reply to all* is probably used under time pressure as a way to act fast and easily. It is justified in some cases but unnecessary in most communications. Using it with parsimony is a sign of respect to (potential) recipients and avoids increasing inboxes (and servers) *overload*.

- **Sarcasm:** *Sarcasm* is usually obvious in face-to-face or co-presence conversation, because the interlocutors are able to provide emotional signals through verbal (voice tone) and body language. E-mail is a poor communication channel when non-verbal cues are needed to interpret the substantial message.

- **Sent (by Mistake):** This occurs when someone sends an embarrassing message to a number of recipients, with a view to cause harm to someone (generally a colleague of same or lower rank), and then pretends that this happened by accident (an excuse that few will consider true…) This happens quite often indeed with the "Forward" button, rather than with the "Send" one, thus reinforcing the fact that one should think twice before sending a message. Checking the content of all messages to be sent and the list of recipients is a must.

- **Silence:** *Silence* occurs when someone deliberately doesn't answer to someone else's communications (e.g. e-mail), with a view to give offence, or simply to ignore the (justified) requests from the other person. The offence is even worse when the individual who didn't answer to the messages later claims that he/she actually *did* reply.

- **Thank You:** *Thank You* e-mail messages are generally sent to all incoming message recipients, and have generally the same effects as *reply to all*. Unfortunately, a number of e-mail users still consider necessary to send such messages, thus adding to overload.

- **Weekend Catch up:** Using weekends to catch up on the week e-mail overload has become usual business practice for many employees (and managers). Some organizations are nevertheless making attempts to discourage such behavior.

Table 2 provides an overview of the various unethical behaviors described in this section and of their potential consequences.

a. Misusing shared common resources: refers to an abuse of the available resources, such as bandwidth, memory, or the network and the infrastructure in general.

b. Misusing e-mail (for emotional resources): denotes a poor usage of a communication channel for communications that should be run over another medium or face-to-face.

c. Generating mail overload: multiplying inappropriately the volume of e-mail, thus creating e-mail and information overload.

d. Lacking mutual respect: misbehaving through the usage of e-mail, thus demonstrating a lack of respect towards the others.

e. Generating time pressure: voluntary or accidental behavior putting too much pressure on the mail recipients by misusing this channel.

f. Creating misunderstanding: all types of behaviors that can lead to misunderstanding.

Unethical Behavior or Inexperience?

As noted by Langford, misuse of e-mail can be explained by inexperience. Originally, e-mail was not developed for the general public and few users take *Netiquette* or e-mail ethics lessons before clicking the "send" button. As previously noted, behavior is typically only viewed as unethical if it is voluntary. Therefore, one might argue that most behaviors listed in the previous section are a result of poor training, or poor understanding of computer-mediated communications. While this might be true for personal communications and the general public, this is no longer acceptable in the professional environment. Business management must ensure that appropriate training is provided to their staff, and their employees must make all efforts to acquire proper knowledge. By not doing so, both managers and employees expose themselves to unethical behaviors.

Inexperience: within the list of behaviors described in the previous sections, broadcast, cc:, click and rush, debate, reply to all, and thank you mails denote a lack of understanding of the nature of e-mail as a communication channel and the inexperience of the user.

Cultural Ignorance: Behaviors described as ambiguity, grammar, gossip, humor, interpretation, or sarcasm can be particularly disturbing in multicultural environments and denote the inexperience of the user as well.

Harmful Voluntary Behavior: Actions such as address disclosure, delay, forgery, masquerade, misquote, sent by mistake, denote tense relationships in the working environment. These might the result from the management style, of corporate policies such as internal competition, and must be addressed by management: are corporate cultures fostering unethical behavior?

Managerial Aggressive E-mail Usage: Attitudes such as power flaunting, omission or sarcasm are generally the result of management style and corporate culture. They may affect staff morale, motivation and performance, and must be addressed properly.

Interpersonal Aggressive E-mail Usage: Behaviors such as angry discussions, lengthy debates, and flaming between peers and colleagues can be resolved at the direct supervisor level. Common sense and management authority generally suffice to stop escalation.

Overload: Finally, management should take careful action regarding the risks of mail overload and of its impact on staff performance, staff morale, as well as on home/work balance.

The above-mentioned behaviors, actions or situations might be addressed by a combination of technical training and of management guidelines. For example, on-site technical training or e-learning modules could help employees use e-mail more efficiently and avoid behaviors resulting from senders' inexperience. Internal policies should aim at addressing behaviors and attitudes such harmful or aggressive e-mail usage by indicating clearly what is acceptable and more specifically what is considered unacceptable behavior in the organization. Also, in organizations where some staff may have privileged access to private, confidential and or business-critical data (such as IT personnel), policies must be developed to ensure no misuse of this date, by example through the development of codes of conduct.

Multicultural organizations should foster interpersonal communications and develop an internal culture of mutual respect and of cultural awareness. Finally, managerial guidelines should address issues such as work/life balance and help avoiding mail overload.

FUTURE RESEARCH DIRECTIONS

This paper is essentially based on existing literature, available information about core business practices, and industry experience. It develops several hypotheses regarding the links between unethical behaviors and e-mail specific characteristics. It doesn't provide broad empirical and quantitative evidence of these hypotheses. Future research may concentrate in getting appropriate quantitative data. This paper focuses on entities having a thorough knowledge and practice of electronic communications, in particular e-mail. Future research may also address other types of entities, followers rather than early adopters, which are not so familiar with ICTs, and may present other business and staff profiles. Future research should also tackle business segments more sensitive to misunderstanding in communications, for instance in the legal or diplomatic field.

CONCLUSION

Electronic communications have provided new tools to support and develop e-businesses. These communication channels also raise new ethical challenges. Companies who will address these challenges properly will get the trust and respect of their stakeholders – customers, employees, managers and owners, as well as the general public and policy makers. Their brand and reputation will be reinforced, a must in e-business, where viral and customer-lead marketing is commonplace. They will also gain a competitive advantage, with enhanced staff morale, dedication and productivity.

Ethics find its roots at the top: managing by example, carefully assessing the internal communication scheme, developing and enforcing proper e-mail policies, are steps that modern e-businesses must engage in, if they want to achieve their corporate responsibility goals as well as their commercial objectives.

REFERENCES

Akrich, M., Méadel, C., & Paravel, V. (2000). Le temps du mail: écrit instantané ou oral médiat. *Sociologie et Sociétés*, *32*(2), 154–171.

Aristote (1994). *Ethique à Eudème* (Translation by P. Maréchaux). Paris: Editions Payot & Rivage

Aristote (2008). *Sur la justice. Ethique à Nicomaque.* (Translation by R. Bodéüs). Paris: Editions Flammarion.

Bennett, M. (1998). *Basic Concepts of Intercultural Communication*. Yarmouth, ME: Intercultural Press.

Byron, K. (2008). Carrying too heavy a load? The communication and miscommunication of emotions by e-mail. *Academy of Management Review*, *33*(2), 309–327.

Carlile, P. (2002). A Pragmatic View of Knowledge and Boundaries: Boundary Objects in New Product Development. *Organization Science*, *13*(4), 442–455. doi:10.1287/orsc.13.4.442.2953

Carnevale, P. J., & Probst, T. M. (1997). Conflict on the Internet. In Kiesler, S. (Ed.), *Culture of the Internet*. Hillsdale, NJ: Lawrence Erlbaum.

Danna, A., & Gandy, O. H. Jr. (2002). All That Glitters is Not Gold: Digging Beneath the Surface of Data Mining. *Journal of Business Ethics*, *40*, 373–386. doi:10.1023/A:1020845814009

Davenport, T. H., & Harris, J. G. (2007). The Dark Side of Customer Analytics. *Harvard Business Review*, May

Dawley, D. D., & Anthony, W. P. (2003). User perceptions of E-mail at Work. *Journal of Business and Technical Communication*, *17*, 170–200. doi:10.1177/1050651902250947

Dimension Data and Datamonitor. (2007). *Dimension Data Unified Communications Study*. Retrieved December 10, 2008, from http://www.dimensiondata.com/uc/

Edmunds, A., & Morris, A. (2000). The problem of information overload in business organizations: a review of the literature. *International Journal of Information Management*, *20*, 17–28. doi:10.1016/S0268-4012(99)00051-1

Enquist, H., & Makrygiannis, N. (1998). Understanding misunderstandings. In *Proceedings of the 31st Hawaii International Conference on System Science*, 83-92.

Fadden, R. B. (2008). *BlackBerry blackout. Message from the Deputy Minister to all CIC employees*. Retrieved February 15, 2009, from http://www.canada.com/ottawacitizen/story.html?id=00e90468-cdb5-4181-abc1-42b245334df2

Falzon, P. (1989). *L'ergonomie cognitive du dialogue*. Grenoble, France: PUG.

Ferry, L. (2006). *Apprendre à vivre: Traité de philosophie à l'usage des jeunes générations*. Paris, France: Plon.

Friedman, R. A., & Currall, S. C. (2003). Conflict escalation: Dispute exacerbating elements of e-mail communication. *Human Relations*, *56*, 1325–1347. doi:10.1177/00187267035611003

Garton, L. E., & Wellman, B. (1993). *Social impacts of electronic mail in organizations: A review of the research literature.* Toronto, Canada: University of Toronto, Centre for Urban and Community Studies and the Ontario Telepresence Project.

Giddens, A. (1990). *The Consequences of Modernity.* Cambridge, UK: Polity Press.

Hardin, G. (1968). The tragedy of the Commons. *Science, 162*(3859), 1243–1248. doi:10.1126/science.162.3859.1243

House, J. (2003). Misunderstanding in intercultural university encounters. In House, J., Kasper, G., & Ross, S. (Eds.), *Misunderstanding in Social Life.* London: Longman.

Huberman, B. A., & Lukose, R. M. (1997). *Social Dilemmas and Internet Congestions.* Retrieved from http://ssrn.com/abstract=41207

Humphrey-Jones, C. (1986). Cited. In House, J., Kasper, G., & Ross, S. (Eds.), *Misunderstanding in Social Life.* London: Longman.

Internet Engineering Task Force. (2008). *Request for Comment 5322.* Retrieved from http://www.rfc-editor.org/rfc/rfc5322.txt

Isaac, H., Campoy, E., & Kalika, M. (2007). Surcharge informationnelle, urgence et TIC. L'effet temporal des technologies de l'information. *Management & Avenir, 12,* 153–172.

Karsenty, L. (2008). *L'incompréhension dans la communication.* Paris: Presses Universitaires de France.

Karsenty, L., & Lacoste, M. (2004). Communication et travail. In Falzon, P. (Ed.), *Traité d'ergonomie.* Paris: Presses Universitaires de France.

Langford, D. (1996). Ethics and the Internet: Appropriate Behavior in Electronic communication. *Ethics & Behavior, 6*(2), 91–106. doi:10.1207/s15327019eb0602_2

Lea, M., & Spears, R. (1991). Computer-mediated communication, deindividuation and group decision-making. *International Journal of Man-Machine Studies, 34,* 283–301. doi:10.1016/0020-7373(91)90045-9

Loch, K. D., Conger, S., & Oz, E. (1998). Ownership, Privacy and Monitoring in the Workplace: A Debate on Technology and Ethics. *Journal of Business Ethics, 17,* 653–663.

McGregor, J. (2008, May 19). Can't It Wait Till Monday? *Business Week,* 54.

McLuhan, M. (1962). *The Gutenberg Galaxy: The Making of Typographic Man.* Toronto, Canada: University of Toronto Press.

Miller, S., & Weckert, J. (2000). Privacy, the Workplace and the Internet. *Journal of Business Ethics, 28,* 255–265. doi:10.1023/A:1006232417265

Monster Press Room. (2009). *Plus de la moitié des salariés européens envoient des courriels personnels ou surfent sur le Net durant leurs heures de travail.* Retrieved May 9, 2009, from http://presse.monster.fr/13162_fr_p1.asp

Moor, J. H. (1990). The Ethics of Privacy Protection. *Library Trends, 39*(1 & 2), 69–82.

Nussbaum, K., & duRivage, V. (1986). Computer Monitoring: Mismanagement by Remote Control. *Business and Management Review,* 56.

Palmer, D. E. (2005). Pop-Ups, Cookies, and Spam: Toward a Deeper Analysis of the Ethical Significance of Internet Marketing Practices. *Journal of Business Ethics, 58,* 271–280. doi:10.1007/s10551-005-1421-8

Peyrelong, M.-F., & Accart, J. P. (2002). Du système d'information personnel au système d'information collectif réalités et mirages du partage d'information en entreprise. In *Conference Proceedings, Canadian Association for Information Science,* 135-149.

Rogers, C., & Kinget, G. M. (1966). *Psychothérapie et relations humaines.* Louvain, Belgique: Publications Universitaires.

Roman, S. (2007). The Ethics of Online Retailing: A Scale Development and Validation from the Consumers' Perspective. *Journal of Business Ethics, 72,* 131–148. doi:10.1007/s10551-006-9161-y

Sarbaugh-Thompson, M., & Feldman, M. S. (1998). Electronic mail and organizational communication: Does saying "hi" really matter? *Organization Science, 9,* 685–698. doi:10.1287/orsc.9.6.685

Sproull, L. O., & Kiesler, S. (1986). Reducing social context cues: Electronic mail in organizational communication. *Management Science, 32,* 1492–1512. doi:10.1287/mnsc.32.11.1492

Stead, B. A., & Gilbert, J. (2001). Ethical Issues in Electronic Commerce. *Journal of Business Ethics, 34,* 75–85. doi:10.1023/A:1012266020988

Taylor, H., Fieldman, G., & Altman, Y. (2008). E-mail at work: A cause of concern? The implications of the new technologies for health, wellbeing and productivity at work. *Journal of Organizational Transformation and Social Change, 5*(2), 159–173. doi:10.1386/jots.5.2.159_1

Warren, S. D., & Brandeis, L. D. (1890). The right to privacy. *Harvard Law Review, 4*(5). doi:10.2307/1321160

Weigand, E. (1999). Misunderstanding: The standard case. *Journal of Pragmatics, 31,* 763–785. doi:10.1016/S0378-2166(98)00068-X

Womack, L., Braswell, D. E., & Harmon, W. K. (2004). Email Policy Enforcement: Techniques and Legality. *Journal of Accounting and Finance Research, Summer*(2), 102-108.

Compilation of References

A&M Records v. Napster, No.00-16401, D.C. No. CV-99-05183-MHP (2001).

Abelson, H. (2004, November). Universities, the Internet, and the Intellectual Commons. Paper presented at the Conference on the Intellectual Commons, Orono, ME.

Acquisti, A. (2002). Privacy and security of personal information: Economic incentives and technological solutions. Presented at the 1st SIMS Workshop on Economics and Information Security. Berkley, CA

Acquisti, A., & Grossklag, S. J. (2004). Privacy attitudes and privacy behavior: Losses, gains, and hyperbolic discounting. In Camp, J., & Lewis, R. (Eds.), The Economics of Information Security. Boston: Kluwer.

Aitken, J., Childerhouse, P., Christopher, M., & Towill, D. (2005). Designing and managing multiple pipelines. Journal of Business Logistics, 26(2), 73–95.

Akerlof, G. (1970). The Market for "Lemons": Quality Uncertainty and the Market Mechanism. The Quarterly Journal of Economics, 84(3), 488–500. doi:10.2307/1879431

Akrich, M., Méadel, C., & Paravel, V. (2000). Le temps du mail: écrit instantané ou oral médiat. Sociologie et Sociétés, 32(2), 154–171.

Alderman, E., & Kennedy, C. (1995). The right to privacy. New York: Knopf.

AllAboutJazz.com. (2004). Maria Schneider - 4 Grammy nods and NOT ONE RETAIL SALE! Retrieved from http://www.allaboutjazz.com/php/news.php?id=4785

Allen, A. L. (1988). Uneasy Access. Privacy for women in a free society. Totowa, NJ: Rowman and Littlefield.

Ardagna, C., Cremonini, M., Damiani, E., De Capitani di Vimercate, S., & Samarati, P. (2008). Privacy-Enhanced Location Services Information. In Acquisti, A., Gritzalis, S., Lambrinoudakis, C., & De Capitani di Vimercati, S. (Eds.), Digital Privacy: Theory, Technologies, and Practices. Boca Raton, FL: Auerbach Publications.

Aristote (1994). Ethique à Eudème (Translation by P. Maréchaux). Paris: Editions Payot & Rivage

Aristote (2008). Sur la justice. Ethique à Nicomaque. (Translation by R. Bodéüs). Paris: Editions Flammarion.

Aristotle. (Trans. 1999). Nicomachean Ethics, Second Edition. (T. Irwin, Trans.) Indianapolis, IN: Hackett Publishing.

Arthur, C. (2009, July 8). Google's Marissa Mayer on the importance of real-time search. Retrieved September 7, 2009, from http://www.guardian.co.uk/technology/2009/jul/08/google-search-marissa-mayer

Assemblée nationale française. (n.d.). Law on the Intellectual Property Code (Legislative Part) (No. 92-597 of July 1, 1992, as last amended by Law No. 97-283 of March 27, 1997). Retrieved from http://www.wipo.int/clea/en/details.jsp?id=1610

AT&T revises privacy policy. (2006, June 22). Retrieved September 11, 2009, from The New York Times web site: http://query.nytimes.com/gst/fullpage.html?res=9F00E0DB1F31F931A15755C0A9609C8B63

Austin, M. J., & Reed, M. L. (1999). Targeting children online: Internet advertising ethics issues. Journal of Consumer Marketing, 16(6), 590–602. doi:10.1108/07363769910297579

Barclay, L., & Markel, K. (2009). Ethical Fairness and Human Rights: The Treatment of Employees with Psychiatric Disabilities. Journal of Business Ethics, 85(3), 333–345. doi:10.1007/s10551-008-9773-5

Barkhuus, L., & Dey, A. (2003). Is Context-Aware Computing Taking Control away from the User? Three Levels of Interactivity Examined. In Proceedings of UBIComp 2003: Ubiquitous Computing. Berlin: Springer Publishing.

Bartlett, A., & Preston, D. (2000). Can ethical behaviour really exist in business? Journal of Business Ethics, 23(2), 199–209.Boone L. E., &. Kurtz, D. L. (1995). Contemporary marketing, 8. Ed., Fort Worth, TX: The Dryden Press.

Beacon. (2009). Retrieved May 29, 2009, from Facebook Web Site: http://www.facebook.com/business/?beacon

Beard, F. (2007). The ethicality of in-text advertising. Journal of Mass Media Ethics, 22(4), 356–359.

Bel de Tienne, K., & Lewis, L. W. (2005). The pragmatic and ethical barriers to corporate social responsibility disclosure: The Nike case. Journal of Business Ethics, 60(4), 359–376. doi:10.1007/s10551-005-0869-x

Bell, T. (2002). Indelicate Imbalancing in Copyright and Patent Law, In A. Thierer and C.W. Crews Jr. (Eds.), Copy Fights: the Future of Intellectual Property in the Information Age (pp. 1-17). Washington, DC: Cato Institute.

Bennett, A. (2008). Consumers are watching you: Ignore the role of the consumer in corporate governance at your own risk. Advertising Age, 19.

Bennett, C. J. (1992). Regulating privacy. Ithica, NY: Cornell University Press.

Bennett, M. (1998). Basic Concepts of Intercultural Communication. Yarmouth, ME: Intercultural Press.

Berk, J., & Hughson, E. N. (2006). Can Bounded Rational Agents Make Optimal Decisions? A Natural Experiment. (Robert Day School Working Paper No. 2008-7). Retrieved November 18, 2008, from http://ssrn.com/abstract=1281150.

Berle, A. A. Jr, & Means, G. C. (1933). The Modern Corporation and Private Property. New York: Macmillan.

Blair, M. M., & Stout, L. A. (1999). A Team Production Theory of Corporate Law. Virginia Law Review, 85, 247–328. doi:10.2307/1073662

Blois, K. (2003). Is it Commercially Irresponsible to Trust? Journal of Business Ethics, 45(3), 183–193. doi:10.1023/A:1024115727737

Boatright, J. (1994). Fiduciary Duties and the Shareholder-Management Relation: Or, What's So Special about Shareholders? Business Ethics Quarterly, 4, 423–429. doi:10.2307/3857339

Bolton, G. E., & Ockenfels, A. (2008, February). The Limits of Trust in Economic Transactions - Investigations of Perfect Reputation Systems (CESifo Working Paper Series No. 2216). Retrieved July 24, 2009, from http://ssrn.com/abstract=1092394

Bolton, G. E., Loebbecke, C., & Ockenfels, A. (2008, April). How Social Reputation Networks Interact with Competition in Anonymous Online Trading: An Experimental Study. (CESifo Working Paper Series No. 2270). Retrieved July 19, 2009, from http://ssrn.com/abstract=1114755

Borden, N. H. (1964). The concept of the marketing mix. Journal of Advertising Research, 24(4), 7–12.

Boucher, D. (1994). Introduction . In Boucher, D. (Ed.), The Social Contract from Hobbes to Rawls. New York: Routledge. doi:10.4324/9780203392928

Bower, J. L., & Christensen, C. M. (1995). Disruptive technologies: Catching the wave. Harvard Business Review, 73, 43–53.

Bowie, N. E. (2002). A Kantian approach to business ethics . In Frederick, R. E. (Ed.), A companion to business ethics (pp. 3–16). Malden, MA: Blackwell Publishing.

Brenkert, G. (2009). Marketing and the Vulnerable . In Beauchamp, T., Bowie, N., & Arnold, D. (Eds.), Ethical Theory and Business (8th ed., pp. 297–306). Upper Saddle River, NJ: Pearson/Prentice Hall.

Brin, D. (1998). The transparent society. Reading, MA: Perseus Books.

Brink, D. O. (2006). Some Forms and Limits of Consequentialism . In Copp, D. (Ed.), The Oxford Handbook of Ethical Theory. New York: Oxford University Press.

Brodie, R., Winklhofer, H., Coviello, N., & Johnston, W. (2007). Is e-marketing coming of age? An examination of the penetration of e-marketing and firm performance. Journal of Interactive Marketing, 21(1), 2–21. doi:10.1002/dir.20071

Brohan, M. (2007). Internet Retailer Survey: Form and function. Retrieved September 7, 2009, from http://www.internetretailer.com/article.asp?id=23262

Brudney, V. (1997). Contract and Fiduciary Duty in Corporate Law. Boston College Law Review. Boston College. Law School, 38, 595–665.

Burrows, A. (2002). We Do This at Common Law But That in Equity. Oxford Journal of Legal Studies, 22, 1–16. doi:10.1093/ojls/22.1.1

Bush, V. D., Venable, B. T., & Bush, A. J. (2000). Ethics and marketing on the internet: practitioners' perceptions of societal, industry and company concerns. Journal of Business Ethics, 23(3), 237–248. doi:10.1023/A:1006202107464

Bynum, T. W. (2001). Ethics and the Information Revolution . In Spinello, R., & Taviani, H. (Eds.), Readings in Cyberethics. Sudbury, MA: Jones and Bartlett.

Byron, K. (2008). Carrying too heavy a load? The communication and miscommunication of emotions by e-mail. Academy of Management Review, 33(2), 309–327.

Callaway, S. K., & Hamilton, R. D. III. (2008). Managing Disruptive Technology — Internet Banking Ventures For Traditional Banks. International Journal of Innovation & Technology Management, 5, 55–80. doi:10.1142/S0219877008001242

Camenisch, P. F. (1991). Marketing ethics: Some dimensions of the challenge. Journal of Business Ethics, 10(4), 245–248. doi:10.1007/BF00382961

Carlile, P. (2002). A Pragmatic View of Knowledge and Boundaries: Boundary Objects in New Product Development. Organization Science, 13(4), 442–455. doi:10.1287/orsc.13.4.442.2953

Carnevale, P. J., & Probst, T. M. (1997). Conflict on the Internet . In Kiesler, S. (Ed.), Culture of the Internet. Hillsdale, NJ: Lawrence Erlbaum.

Carson, T. L. (2002). Ethical Issues in Selling and Advertising . In Bowie, N. E. (Ed.), The Blackwell Guide to Business Ethics (pp. 186–205). Malden, MA: Blackwell Publishing.

Carter, M. (2007). Internet Advertising: Video/ethics: Online ads must clean up their act. The Guardian Supplement, 6.

Castelfranchi, C. (Ed.) & Tan, Y. H. (Ed.). (2001). Trust and deception in virtual societies. New York: Springer.

Caudill, E. M., & Murphy, P. E. (2000). Consumer online privacy: Legal and ethical issues. Journal of Public Policy & Marketing, 19(1), 7–19. doi:10.1509/jppm.19.1.7.16951

Cavusgil, S. T. (2002, March/April). Extending the reach of e-business. Marketing Management, pp. 24-29.

Chafkin, M. (2007). Reverse auctions a supplier's survival guide. Inc., 29(5), 27–30.

Charters, D. (2002). Electronic monitoring and privacy issues in business-marketing: The ethics of the doubleclick experience. Journal of Business Ethics, 35(4), 243–254. doi:10.1023/A:1013824909970

Chaudhury, A., Mallick, D. N., & Rao, H. R. (2001). Web channels in e-commerce. Communications of the ACM, 44(1), 99–104. doi:10.1145/357489.357515

Chen, C. Y. (2006). The comparison of structure differences between internet marketing and traditional marketing. International Journal of Management and Enterprise Development, 3(4), 397–417.

Chen, G., & Kotz, D. (2000). A Survey of Context Aware Mobile Computing Research (Dartmouth Computer Science Technical Report TR2000-381). Retrieved June 1, 2009, from http://www.cs.dartmouth.edu/~dfk/papers/chen-survey-tr.pdf

Chen, P., Dhanasobhon, S., & Smith, M. D. (2008, May) All Reviews are Not Created Equal: The Disaggregate Impact of Reviews and Reviewers at Amazon. Com. Retrieved July 23, 2009, from http://ssrn.com/abstract=918083

Chonko, L. B. (1995). Ethical decision making in marketing. Thousand Oaks, CA: Sage Publications.

Christensen, C., & Raynor, M. (2003). The innovator's solution: Creating and sustaining successful growth. Boston, MA: Harvard Business School Press.

Christman, J. (1994). The Myth of Property: Toward an Egalitarian Theory of Ownership. New York: Oxford University Press.

Chua, C. E. H., & Wareham, J. (2004). Fighting Internet Auction Fraud: An Assessment and Proposal. Computer, 37(10), 31–37. doi:10.1109/MC.2004.165

Cisco Systems, Inc. (2009). Online Privacy Statement. Retrieved June 13, 2009, from http://www.cisco.com/web/siteassets/legal/privacy.html

Coase, R. (1960). The problem of social cost. The Journal of Law & Economics, 3(1), 1–44. doi:10.1086/466560

Coles, A., & Harris, L. (2006) Ethical Consumers and E-Commerce: The Emergence and Growth of Fair Trade in the UK. Journal of Research for Consumers, 10.

Coltman, T., Devinney, T., Latukefu, A., & Midgley, D. (2001). E-Business: Revolution, evolution, or hype? California Management Review, 44(1), 57–86.

Constantinides, E. (2002). The 4S web-marketing mix model. Electronic Commerce Research and Applications, 1(1), 57–76. doi:10.1016/S1567-4223(02)00006-6

Consumers Paying More Bills Online. (2008, September 1). Point for Credit Union Research and Advice. Credit Union National Association, Inc., p. 15.

Cook, D. L., & Coupey, E. (1998). Consumer behavior and unresolved regulatory issues in electronic marketing. Journal of Business Research, 41(3), 231–238. doi:10.1016/S0148-2963(97)00066-0

Copp, D. (2005). Introduction: Metaethics and Normative Ethics . In Copp, D. (Ed.), The Oxford Handbook of Ethical Theory. New York: Oxford University Press. doi:10.1093/0195147790.001.0001

COPPA - Children's Online Privacy Protection. (n.d.). Retrieved September 11, 2009, from http://www.coppa.org/comply.htm

Corp, Q. v. Heitkamp, 504 U.S. 298 (1992). Retrieved from http://laws.findlaw.com/us/504/298.html

Crisp, R. (1987). Persuasive Advertising, Autonomy, and the Creation of Desire. Journal of Business Ethics, 6(5), 413–418. doi:10.1007/BF00382898

Cullen, A. J., & Webster, M. (2007). A model of B2B e-commerce, based on connectivity and purpose. International Journal of Operations & Production Management, 27(2), 205–225. doi:10.1108/01443570710720621

Cummings, B. (2008). It's a 24/7 world. Wearables Business, 12(1), 57–60.

Dai, Q., & Kauffman, R. J. (2002). Business models for internet-based B2B electronic markets. International Journal of Electronic Commerce, 6(4), 41.

Danna, A., & Gandy, O. H. Jr. (2002). All That Glitters is Not Gold: Digging Beneath the Surface of Data Mining. Journal of Business Ethics, 40, 373–386. doi:10.1023/A:1020845814009

Das, T. K., & Teng, B. (2004). The Risk-based View of Trust: A Conceptual Framework. Journal of Business and Psychology, 19(1), 85–116. doi:10.1023/B:JOBU.0000040274.23551.1b

Dash, E. (2007, January 19). Data Breach could affect millions of TJX shoppers. Retrieved September 12, 2009, from The New York Times web site. http://www.nytimes.com/2007/01/19/business/19data.html

Davenport, T. H., & Harris, J. G. (2007). The Dark Side of Customer Analytics. Harvard Business Review, May

Davidson, H. R. (1969). The legend of lady godiva. Folklore, 80(2), 107–122.

Davies, P. L., & Prentice, D. D. (1997). Gower's Principles of Modern Company Law (6th ed.). London: Sweet & Maxwell.

Davis, C. H., & Vladica, F. (2007). The value of Internet technologies and e-business solutions to micro-enterprises in Atlantic Canada . In Barnes, S. (Ed.), E-commerce and v-business (2nd ed., pp. 125–156). Amsterdam: Elsevier. doi:10.1016/B978-0-7506-6493-6.50009-5

Dawley, D. D., & Anthony, W. P. (2003). User perceptions of E-mail at Work. Journal of Business and Technical Communication, 17, 170–200. doi:10.1177/1050651902250947

Dayal, S., Landesberg, H., & Zeisser, M. (1999). How to build trust online. MM, Fall, 64–69.

De Angeli, A., & Brahnam, S. (2008). I hate you! Disinhibition with virtual partners. Interacting with Computers, 20(3), 302–310. doi:10.1016/j.intcom.2008.02.004

De George, R. T. (2003). The Ethics of Information Technology and Business. Malden, MA: Blackwell Publishing. doi:10.1002/9780470774144

Dean, J. (2008). Communicative Capitalism: Circulation and Foreclosure of Politics . In Boler, M. (Ed.), Digital democracy and media: Tactics in hard times (pp. 101–121). Cambridge, MA: The MIT Press.

DeCrew, J. (1997). In pursuit of privacy. Ithica, NY: Cornell University Press.

Dedrick, J., Kraemer, L. K., King, L. J., & Lyytinen, K. (2006). The United States: adaptive integration versus the Silicon valley model . In Kraemer, K. L., Dedrick, J., Melville, N. P., & Zhu, K. (Eds.), Global e-Commerce: Impacts of national environment and policy (pp. 62–107). Cambridge, UK: Cambridge University Press. doi:10.1017/CBO9780511488603.003

DeGeorge, R. T. (2000). Business ethics and the challenge of the information age. Business Ethics Quarterly, 10(1), 63–72. doi:10.2307/3857695

DeGeorge, R. T. (2002). Ethical issues in information technology . In Bowie, N. E. (Ed.), The Blackwell guide to business ethics (pp. 267–288). Malden, MA: Blackwell Publishing.

Dellarocas, C. N. (2001, October). Building Trust On-Line: The Design of Reliable Reputation Reporting: Mechanisms for Online Trading Communities (MIT Sloan Working Paper No. 4180-01). Retrieved July 24, 2009, from http://ssrn.com/abstract=289967

Dellarocas, C. N. (2003, March). The Digitization of Word-of-Mouth: Promise and Challenges of Online Feedback Mechanisms. (MIT Sloan Working Paper No. 4296-03). Retrieved July 19, 2009, from http://ssrn.com/abstract=393042

Deutscher Bundestag. Law on Copyright and Neighboring Rights (Copyright Law) of September 9, 1965, as last amended by the Law of July22,1997. Retrieved from http://www.wipo.int/clea/en/details.jsp?id=1032

Dey, A. K. (2001). Understanding and using Context. Personal and Ubiquitous Computing, 5(1), 4–7. doi:10.1007/s007790170019

Dibona, C., Cooper, D., Stone, M. (2005, October). Open Sources 2.0: the continuing evolution. Sebastopol, California: O'Reilly Press.

Dimension Data and Datamonitor. (2007). Dimension Data Unified Communications Study. Retrieved December 10, 2008, from http://www.dimensiondata.com/uc/

Dobosz, B., Green, K., & Sisler, G. (2006). Behavioral Marketing: Security and Privacy Issues. Journal of Information Privacy & Security, 2(4), 45–59.

Dominici, G. (2009). From marketing mix to e-marketing mix: A literature overview and classification. International Journal of Business and Management, 4(9), 17–24.

Doolin, B., Dillon, S., Thompson, F., & Corner, J. L. (2008). Perceived Risk, the Internet Shopping Experience, and Online Purchasing Behavior . In Becker, A. (Ed.), Electronic Commerce: Concepts, Methodologies, Tools and Applications (Vol. 1, pp. 324–345). Hershey, PA: Information Science Reference.

Dreyfus, H. L. (2009). On the internet (2nd ed.). New York: Routledge.

Easterbrook, F. H. (1995). Cyberspace and the Law of the Horse. The University of Chicago Legal Forum, 207–216.

Edition, M. (2008). GE loses consumers' personal records. Retrieved September 12, 2009, from http://www.npr.org/templates/story/story.php?storyId=18212217

Edmunds, A., & Morris, A. (2000). The problem of information overload in business organizations: a review of the literature. International Journal of Information Management, 20, 17–28. doi:10.1016/S0268-4012(99)00051-1

Eldred v. Ashcroft. 1 U.S. 537 (2003).

Elgesem, D. (1999). The structure of rights in directive 95/46/EC. Ethics and Information Technology, 1, 283–293. doi:10.1023/A:1010076422893

Enquist, H., & Makrygiannis, N. (1998). Understanding misunderstandings. In Proceedings of the 31st Hawaii International Conference on System Science, 83-92.

Ergin, E. A., & Özdemir, H. (2007). Advertising ethics: A field study on Turkish consumers. The Journal of Applied Business Research, 23(4), 17–26.

Evan, W., & Freeman, R. E. (1988). A Stakeholder Theory of the Modern Corporation: Kantian Capitalism . In Beauchamp, T., & Bowie, B. (Eds.), Ethical Theory and Business (3rd ed.). Englewood Cliffs, NJ: Prentice Hall.

Facebook. (2009). Retreived May 28, 2009, from Facebook web site: http://www.facebook.com/about.php

Fadden, R. B. (2008). BlackBerry blackout. Message from the Deputy Minister to all CIC employees. Retrieved February 15, 2009, from http://www.canada.com/ottawacitizen/story.html?id=00e90468-cdb5-4181-abc1-42b245334df2

Falzon, P. (1989). L'ergonomie cognitive du dialogue. Grenoble, France: PUG.

Federal Convention of 1787. (1787). The Constitution of the United States of America. Retrieved from http://www.law.cornell.edu/constitution/constitution.overview.html

Feist Publications, Inc. v. Rural Telephone Service Co. 499 U.S. 340 (1991).

Fernando, A. (2007). Transparency under attack. Communication World, 24(2), 9–11.

Ferre, F. (1995). Philosophy of technology. Athens, GA: The University of Georgia Press.

Ferrell, O. C., & Gresham, L. (1985). A contingency framework for understanding ethical decision making in marketing. Journal of Marketing, 49(3), 87–96. doi:10.2307/1251618

Ferry, L. (2006). Apprendre à vivre: Traité de philosophie à l'usage des jeunes générations. Paris, France: Plon.

Finn, P. D. (1989). The Fiduciary Principle . In Youdan, T. G. (Ed.), Equity, Fiduciaries and Trusts (pp. 1–56). Toronto, ON: De Boo.

Finn, P. D. (1992). Fiduciary Law and the Modern Commercial World . In McKendrick, E. (Ed.), Commercial Aspects of Trust and Fiduciary Obligations (pp. 7–42). Oxford, UK: Clarendon.

Fisher, M. J. (1997, September). Moldovascam.com: A Complicated Case of Electronic and Telephone Fraud Suggests Just How Vulnerable Internet Users May Be. Atlantic Monthly, 280(3), 19-22. Retrieved April 26, 2009, from http://www.theatlantic.com/issues/97sep/moldova.htm

Fitzgibbon, S. (1999). Fiduciary Relationships are not Contracts. Marquette Law Review, 82, 303–353.

Flannigan, R. (2004). The Boundaries of Fiduciary Accountability. The Canadian Bar Review, 83, 35–90.

Flannigan, R. (2006). The Strict Character of Fiduciary Liability. New Zealand Law Review, 209-242.

Foucoult, M. (1977). Discipline and punish: The birth of the prison . In Rabinow, P. (Ed.), The Foucoult reader. New York: Pantheon Books.

Fox Film Corp. v. Doyal, 286 U.S. 123 (1932).

Fox, S. (2000). Trust and privacy online: Why Americans want to rewrite the rules. The Pew Internet and American Life Project. Retrieved June 13, 2009, from http://www.pewinternet.org

Foxman, E. R., & Kilcoyne, P. (1993). Information technology, marketing practice, and consumer privacy: Ethical issues. *Journal of Public Policy & Marketing*, 12(1), 106–119.

Frankel, T. (1983). Fiduciary Law. *California Law Review*, 71, 795–836. doi:10.2307/3480303

Frankel, T. (2001). Trusting and Non-Trusting on the Internet. *Boston University Law Review*. Boston University. School of Law, 81, 457–478.

Franzak, F., Pitta, D., & Fritsche, S. (2001). Online relationships and the consumer's right to privacy. *Journal of Consumer Marketing*, 18(7), 631–641. doi:10.1108/EUM0000000006256

Free Software Foundation. (2002). Selling Free Software. Retrieved May 6, 2009, from http://www.gnu.org/philosophy/selling.html

Freeman, R. E. (2008). Managing for Stakeholders . In Zakhem, A., Palmer, D., & Stoll, M. (Eds.), *Stakeholder Management Theory: Essential Readings in Ethical Leadership and Management*. New York: Prometheus Books.

Freeman, R. E., & Gilbert, D. R. (1988). *Corporate strategy and the search for ethics*. Englewood Cliffs, NJ: Prentice Hall.

Fried, C. (1984). Privacy . In Schoeman, F. D. (Ed.), *Philosophical dimensions of privacy: An anthology*. Cambridge, UK: Cambridge University Press.

Friedman, R. A., & Currall, S. C. (2003). Conflict escalation: Dispute exacerbating elements of e-mail communication. *Human Relations*, 56, 1325–1347. doi:10.1177/00187267035611003

Friedman, T. (2005). *The World is Flat: a brief history of the globalized world in the twenty –first century*. London: Allen Lane.

Fudge, R. S., & Schlacter, J. L. (1999). Motivating employees to act ethically: An expectancy theory approach. *Journal of Business Ethics*, 18(3), 295–304. doi:10.1023/A:1005801022353

Garfunkel, S. L. (2008). Wikipedia and the meaning of truth: Why the online encyclopedia's epistemology should worry those who care about traditional notions of accuracy. *Technology Review*, 111(6), 84–86.

Garton, L. E., & Wellman, B. (1993). *Social impacts of electronic mail in organizations: A review of the research literature*. Toronto, Canada: University of Toronto, Centre for Urban and Community Studies and the Ontario Telepresence Project.

Gaski, J. F. (1999). Does marketing ethics really have anything to say? – A critical inventory of the literature. *Journal of Business Ethics*, 18(3), 315–334. doi:10.1023/A:1017190829683

Gattiker, T. F., Huang, X., & Schwarz, J. L. (2007). Negotiation, email, and internet reverse auctions: How sourcing mechanisms deployed by buyers affect suppliers' trust. *Journal of Operations Management*, 25(1), 184–202. doi:10.1016/j.jom.2006.02.007

Gauthier, D. (1986). *Morals by Agreement*. New York: Oxford University Press.

Gauthier, D. (2007). Why Contractarianism? In Theory, E. (Ed.), *Russ Shafer-Landau* (pp. 620–630). Malden, MA: Blackwell Publishing.

Gauzente, C., & Ranchhod, A. (2001). Ethical marketing for competitive advantage on the internet. *Academy of Marketing Science Review*, 5(4), 1–7.

Gefen, D., Karahanna, E., & Straub, D. W. (2003). Trust and TAM in Online Shopping: An Integrated Model. *Management Information Systems Quarterly*, 27(1), 51–90.

Genge, W. (1985). Ads stimulate the economy. *Business and Society Review*, 1(55), 58–59.

Gewirtz, P. (1996). On "I Know It When I See It." . *The Yale Law Journal*, 105, 1023–1047. doi:10.2307/797245

Gibbs, P. (2004). Marketing and the Notion of Well-Being. *Business Ethics*. European Review (Chichester, England), 13(1), 5–13. doi:10.1111/j.1467-8608.2004.00344.x

Giddens, A. (1990). *The Consequences of Modernity*. Cambridge, UK: Polity Press.

Godiva, L. (1950, January). Lady Godiva . Western Folklore, 77–78.

Goldsmith, J. (1998). Regulation of the Internet: Three Persistent Fallacies. Chicago-Kent Law Review, 73, 1119–1131.

Goodin, R. (1985). Protecting the Vulnerable. Chicago: University of Chicago Press.

Goodman, B., & Rushkoff, D. (Writers), Goodman, B., & Dretzin, R. (Directors). (2003). The Persuaders [Television series episode]. In D. Fanning (Producer), PBS Frontline. Boston: WGBH.

Google. (n.d.). Google Analytics - Case Studies - American Cancer Society. Retrieved June 13, 2009, from http://www.google.com/analytics/case_study_acs.html

Google. (n.d.). Google Analytics. Retrieved September 7, 2009, from http://www.google.com/analytics/index.html

Grabner-Kraeuter, S. (2002). The Role of Consumers' Trust in Online Shopping. Journal of Business Ethics, 39(1/2), 43–50. doi:10.1023/A:1016323815802

Green, K. (2008, March/April). TR10: Reality Mining. MIT Technology Review.com. Retrieved June 1, 2009, from http://www.technologyreview.com/Infotech/20247/

Greenberg, A. (2008, April 23). What your Cell Phone Knows about You. Forbes.com. Retrieved June 1, 2009, from http://www.forbes.com/2008/05/22/reality-mining-cellphone-tech-wire-cx_ag_0523reality.html

Greene, J. D., Nystrom, L. E., Engell, A. D., Darley, J. M., & Cohen, J. D. (2004). The Neural Bases of Cognitive Conflict and Control in Moral Judgment. Neuron, 44, 389–400. doi:10.1016/j.neuron.2004.09.027

Greene, J. D., Sommerville, R. B., Nystrom, L. E., Darley, J. M., & Cohen, J. D. (2001). An fMRI investigation of Emotional Engagement in Moral Judgment. Science, 293, 2105–2108. doi:10.1126/science.1062872

Greene, J., & Haidt, J. (2002). How (and where) Does Moral Judgment Work? Trends in Cognitive Sciences, 6(12), 517–523. doi:10.1016/S1364-6613(02)02011-9

Greenfield, K. (2005). New Principles for Corporate Law. Hastings Business Law Journal, 1, 87–118.

Griffith, D. A., & Krampf, R. F. (1998). An examination of the web-based strategies of the top 100 U.S. retailers. Journal of Marketing Theory and Practice, 6(3), 12–23.

Grogan, D. (2009, April 23). Minnesota Internet Sales Tax Bills Rolled Into Omnibus Tax Bills Bookselling This Week, American Booksellers Association. Retrieved from: http://news.bookweb.org/news/6761.html

Guest, S. (2005). Integrity, Equality, and Justice. Revue Internationale de Philosophie, 59(3), 335–362.

Hafner, K. (2006). Researchers yearn to use AOL logs but they hesitate. Retrieved September 12, 2009, from The New York Times web site: http://www.nytimes.com/2006/08/23/technology/23search.html

Hair, N., & Clark, M. (2007). The ethical dilemmas and challenges of ethnographic research in electronic communities. International Journal of Market Research, 49(6), 781–800.

Handfield, R. B., Straight, S. L., & Sterling, W. A. (2002). Reverse auctions: How do supply managers really feel about them? Inside Supply Management, 13(11), 56-61.

Hansell, S. (2008, May 1). Amazon Sues New York State to Void Tax Rules. The New York Times. Retrieved from http://bits.blogs.nytimes.com/2008/05/01/amazon-sues-new-york-state-to-void-sales-tax-rules/

Hansmann, H., & Kraakman, R. (2001). The End of History for Corporate Law. The Georgetown Law Journal, 89, 439–468.

Hardin, G. (1968). The tragedy of the Commons. Science, 162(3859), 1243–1248. doi:10.1126/science.162.3859.1243

Harmon, R., & Daim, T. (2008). Assessing the Future of Location-Based Services: Technologies, Applications, and Strategies . In Unhelkar, B. (Ed.), Handbook of Research in Mobile Business: Technical, Methodological, and Social Perspectives (2nd ed.). Hershey, PA: IGI Global.

Harridge-March, S. (2004). Electronic marketing, the new kid on the block. Marketing Intelligence & Planning, 22(3), 297–309. doi:10.1108/02634500410536885

Hartland, E. (1890). Peeping Tom and Lady Godiva. Folklore, 1(2), 207–226.

Hasnas, J. (1998). The Normative Theories of Business Ethics: A Guide for the Perplexed. Business Ethics Quarterly, 8, 19–43. doi:10.2307/3857520

Heath, J. (2005). Liberal Autonomy and Consumer Sovereignty. In, John Christman (ed.), Autonomy and the Challenges to Liberalism: New Essays (204-225). Cambridge, UK.Cambridge University Press, Hobbes, T. [1651] (1991). Leviathan. R. Tuck (Ed.) New York: Cambridge University Press.

Held, V. (1984). Advertising and program content. Business & Professional Ethics Journal, 3(3/4), 61–76.

Helft, M. (2007). Google zooms in too close for some. Retrieved September 12, 2009 from http://www.nytimes.com/2007/06/01/technology/01private.html

Hermalin, B. E., & Katz, M. L. (2006). Privacy, property rights, and efficiency: The economics of privacy as secrecy. Quantitative Marketing and Economics, 4(3), 209–239. doi:10.1007/s11129-005-9004-7

Herschel, R. I., & Andrews, P. H. (1997). Ethical implications of technological advances on business communication. Journal of Business Communication, 34(2), 160–170. doi:10.1177/002194369703400203

Hobbes, T. (1996). Leviathan (Gaskin, J. C. A., Ed.). New York: Oxford University Press.

Hoffman, D. L., Novak, T. P., & Peralta, M. (1999). Building consumer trust online. Communications of the ACM, 41(4), 80–85. doi:10.1145/299157.299175

Hollis, M. (1982). Dirty Hands. British Journal of Political Science, 12, 385–398. doi:10.1017/S0007123400003033

Hollis, M. (2002). The Philosophy of Social Science: An Introduction (Revised and Updated). Cambridge, UK: Cambridge University Press.

Holsapple, C. W., & Singh, M. (2000). Toward a unified view of electronic commerce, electronic business, and collaborative commerce: A knowledge management approach. Knowledge and Process Management, 7(3), 151–164. doi:10.1002/1099-1441(200007/09)7:3<151::AID-KPM83>3.0.CO;2-U

Holson, L. M. (2007a). Verizon letter on privacy stirs debate. Retrieved September 12, 2009, from http://www.nytimes.com/2007/10/16/business/16phone.html.

Holson, L. M. (2007b). Privacy lost: These phones can find you. Retrieved September 12, 2009, web site: http://www.nytimes.com/2007/10/23/technology/23mobile.html.

Hospers, J. [1974] (2005). What Libertarianism Is. In James Sterba (Ed.), Morality in Practice. Belmont, CA: Thomson-Wadsworth.

House, J. (2003). Misunderstanding in intercultural university encounters . In House, J., Kasper, G., & Ross, S. (Eds.), Misunderstanding in Social Life. London: Longman.

Howard, K. (2007, March 14). The damaging effects of identity theft. Retrieved July 15, 2009, from http://www.ezinearticles.com/?The-Damaging-Effects-of-Identity-Theft&id=488838

Hsu, P, & Kraemer, L., K., & Dunkle, D. (2006). Determinants of e-business use in U.S. firms. International Journal of Electronic Commerce, 10(4), 9–45. doi:10.2753/JEC1086-4415100401

Huberman, B. A., & Lukose, R. M. (1997). Social Dilemmas and Internet Congestions. Retrieved from http://ssrn.com/abstract=41207

Hull, G. (2009). Clearing the Rubbish: Locke, the Waste Proviso, and the Moral Justification of Intellectual Property. Public Affairs Quarterly, 23(1), 67–93. Retrieved from http://papers.ssrn.com/sol3/papers.cfm?abstract_id=1082597.

Hume, D. (1978). A Treatise of Human Nature. New York: Oxford University Press.

Humphrey-Jones, C. (1986). Cited . In House, J., Kasper, G., & Ross, S. (Eds.), Misunderstanding in Social Life. London: Longman.

Hunt, S. D., & Parraga, A. Z. V. (1993). Organizational consequences, marketing ethics and sales force supervision. JMR, Journal of Marketing Research, 30(1), 78–90. doi:10.2307/3172515

Hunt, S. D., & Vitell, S. J. (1986). A general theory of marketing ethics. Journal of Macromarketing, 6(1), 5–16. doi:10.1177/027614678600600103

Hunter, L. M., Kasouf, C. J., Celuch, K. G., & Curry, K. A. (2004). A classification of business-to-business buying decisions: risk importance and probability as a framework for e-business benefits. Industrial Marketing Management, 33(2), 145–154. doi:10.1016/S0019-8501(03)00058-0

Hur, D., Mabert, V. A., & Hartley, J. L. (2005). Getting the most out of e-auction investment. Omega, 35, 403–416. doi:10.1016/j.omega.2005.08.003

Husted, B. (1998). The Ethical Limitations of Trust in Business Relations. Business Ethics Quarterly, 8(2), 233–248. doi:10.2307/3857327

Hyman, M. R., Tansey, R., & Clark, J. W. (1994). Research on advertising ethics: Past, present, and future. Journal of Advertising, 23(3), 5–15.

Ian, J. (2005). The Internet Debacle—An Alternate View. Retrieved May 6, 2009, from http://www.janisian.com/article-internet_debacle.html

Infonetics Research. (2008). Mobile Subscribers to Hit 5.2B in 2011. Retrieved June 1, 2009, from http://www.infonetics.com/pr/2008/ms08.sub.nr.asp

Internet Engineering Task Force. (2008). Request for Comment 5322. Retrieved from http://www.rfc-editor.org/rfc/rfc5322.txt

Internetretailer.com. (2008, May 29). Multivariate testing produces winning product pages for ShopNBC.com & InternetRetailer.com, Daily News. Retrieved May 28, 2009, from http://www.internetretailer.com/dailyNews.asp?id=26545

Internetretailer.com. (2009, March 17). Marketers plan to spend more online and aim to measure results better. Retrieved September 7, 2009, from http://www.internetretailer.com/dailyNews.asp?id=29781

Introna, L. D., & Petrakaki, D. (2007). Defining the virtual organization . In Barnes, S. (Ed.), E-commerce and v-business (2nd ed., pp. 181–191). Amsterdam: Elsevier. doi:10.1016/B978-0-7506-6493-6.50011-3

Isaac, H., Campoy, E., & Kalika, M. (2007). Surcharge informationnelle, urgence et TIC. L'effet temporal des technologies de l'information. Management & Avenir, 12, 153–172.

Jacobs, R., & Stone, B. (2008). Successful Direct Marketing Methods: Interactive, Database, and Customer Marketing for the Multichannel Communications Age. New York: McGraw-Hill.

James, W. (1931). Pragmatism: A new name for some old ways of thinking. New York: Longmans, Green, and Co.

Jap, S. D. (2003). An exploratory study of the introduction of on-line reverse auctions. Journal of Marketing, 67, 96–107. doi:10.1509/jmkg.67.3.96.18651

Jap, S. D. (2007). The impact of Online Reverse Auction design on Buyer-Supplier Relationships. Journal of Marketing, 71, 146–159. doi:10.1509/jmkg.71.1.146

Ja-Shen Chen, & Ching, R. K. H. (2002). A proposed framework for transitioning to an E-business model. Quarterly Journal of Electronic Commerce, 3(4), 375.

Jefferson, T. (1813). Thomas Jefferson To Isaac McPherson, Monticello, August 13th, 1813. Retrieved from http://www.temple.edu/lawschool/dpost/mcphersonletter.html

Jesdanun, A. (2007). Facebook retreat shows ad-targeting risk. The Associated Press Retrieved September 12, 2009, from http://www.washingtonpost.com/wp-dyn/content/article/2007/11/30/AR2007113001668_pf.html

Jhally, S. (Director). (1998). Advertising and the End of the World [Documentary]. USA: Media Education Foundation.

Jiang, X. (2002). Safeguard privacy in ubiquitous computing with decentralized information spaces: Bridging the technical and the social. Presented at the 4th International Conference on Ubiquitous Computing (UBICOMP 2002), Gotenborg, Sweden.

Jiang, X., & Landay, J. A. (2002). Modeling privacy control in context-aware systems. Pervasive Computing, IEEE, 1(3), 59–63. doi:10.1109/MPRV.2002.1037723

Jiang, X., Hong, J. L., & Landay, J. A. (2002). Approximate information flows: Socially based modeling of privacy in Ubiquitous Computing. Presented at the 4th International Conference on Ubiquitous Computing (UBICOMP 2002), Gotenborg, Sweden

Johnson, D. (2001). Is the Global Information Infrastructure a Democratic Technology? In Spinello, R., & Taviani, H. (Eds.), Readings in Cyberethics. Sudbury, MA: Jones and Bartlett.

Johnson, D. J. (2001). Computer ethics. Upper Saddle River, NJ: Prentice Hall.

Johnston, D. (1988). The Roman Law of Trusts. Oxford, UK: Clarendon.

Johnston, D. R., & Post, D. G. (1997). And How Shall the Net be Governed? A Meditation on the Relative Virtues of Decentralized, Emergent Law . In Kahn, B., & Keller, J. H. (Eds.), Coordinating the Internet (pp. 62–91). Cambridge, MA: MIT Press.

Jones, T. M., Wicks, A. C., & Freeman, R. E. (2002). Stakeholder theory: The state of the art . In Bowie, N. E. (Ed.), The Blackwell guide to business ethics (pp. 19–37). Malden, MA: Blackwell Publishing.

Kahn, R. E., & Cerf, V. G. (1999). What Is The Internet (And What Makes it Work). Retrieved April 29, 2009, from http://www.policyscience.net/cerf.pdf

Kalyanam, K., & McIntyre, S. (2002). The E-marketing mix: A contribution of the e-tailing wars. Academy of Marketing Science Journal, 30(4), 487–499. doi:10.1177/009207002236924

Kang, J. (1998). Information Privacy in Cyberspace Transactions. Stanford Law Review, 50, 1193–1294. doi:10.2307/1229286

Kant, I. (1993). Grounding for the Metaphysics of Morals With on a Supposed Right to Lie Because of Philanthropic Concerns. Indianapolis, IN: Hackett Pub Co Inc.

Kant, I. (1996a). Groundwork of the Metaphysics of Morals . In Gregor, M. (Ed.), Practical Philosophy (Cambridge Edition of the Works of Immanuel Kant) (pp. 37–108). Cambridge, UK: Cambridge University Press.

Kant, I. (1996b). The Metaphysics of Morals . In Gregor, M. (Ed.), Practical Philosophy (Cambridge Edition of the Works of Immanuel Kant) (pp. 353–604). Cambridge, UK: Cambridge University Press.

Kant, I. (1996c). On the Wrongfulness of Unauthorized Publication of Books . In Gregor, M. (Ed.), Practical Philosophy (Cambridge Edition of the Works of Immanuel Kant) (pp. 23–36). Cambridge, UK: Cambridge University Press.

Kapelus, P. (2002). Mining, corporate social responsibility, and the community: The case of Rio Tinto, Richard's Bay Minerals, & Mbonambi. Journal of Business Ethics, 39(3), 275–296. doi:10.1023/A:1016570929359

Karsenty, L. (2008). L'incompréhension dans la communication. Paris: Presses Universitaires de France.

Karsenty, L., & Lacoste, M. (2004). Communication et travail . In Falzon, P. (Ed.), Traité d'ergonomie. Paris: Presses Universitaires de France.

Keeton, G. W. (1965). An Introduction to Equity (6th ed.). London: Pitman.

Kehoe, W. J. (1985). Ethics, price fixing, and the management of price strategy . In Laczniak, G. R., & Murphy, P. E. (Eds.), Marketing ethics - Guidelines for manager (pp. 71–83). Lexington, KY: Lexington Books.

Kelly, E. P., & Rowland, H. C. (2000). Ethical and online privacy issues in electronic commerce. Business Horizons, 43(3), 3–12. doi:10.1016/S0007-6813(00)89195-8

Kickul, J. (2001). When Organizations Break Their Promises: Employee Reactions to Unfair Processes and Treatment. Journal of Business Ethics, 29(4), 289–307. doi:10.1023/A:1010734616208

Kiesler, S., & Sproull, L. (1992). Group Decision Making and Communication Technology. Organizational Behavior and Human Decision Processes, 52, 96–123. doi:10.1016/0749-5978(92)90047-B

Kiesler, S., Siegel, J., & McGuire, T. W. (1984). Social psychological aspects of computer-mediated communication. The American Psychologist, 39(10), 1123–1134. doi:10.1037/0003-066X.39.10.1123

Kim, E. Y., & Kim, Y. K. (2004). Predicting online purchase intentions for clothing products. European Journal of Marketing, 38(7), 883–897. doi:10.1108/03090560410539302

Kımıloğlu, H. (2004). The "E-Literature": A framework for understanding the accumulated knowledge about internet marketing. Academy of Marketing Science Review, 6, 1-36.Kotler, P., & Armstrong, G. (1991). Principles of marketing, 5th Ed., Upper Saddle River, NJ: Prentice-Hall.

Korsgaard, C. (1996). Creating the Kingdom of Ends. Cambridge, UK: Cambridge University Press.

Kracher, B., & Corritore, C. L. (2004). Is There a Sepcial E-Commerce Ethics? Business Ethics Quarterly, 14(1), 71–94.

Kracher, B., & Corritore, C. L. (2004). Is there a special e-commerce ethics? Business Ethics Quarterly, 14(1), 71–94.

Kraemer, L. K., Dedrick, J., & Melville, N. P. (2006). Globalization and national diversity: e-commerce diffusion and impacts across nations. In In K. L. Kraemer, J. Dedrick, N. P. Melville, and K. Zhu (Eds.), Global e-Commerce: Impacts of national environment and policy (pp. 13-61). Cambridge, UK: Cambridge University Press.

Kraeuter, S. G. (2002). The role of consumers' trust in online-shopping. Journal of Business Ethics, 39(1/2), 43–50. doi:10.1023/A:1016323815802

Krishnamurthy, S. (2006). Introducing e-markplan: A practical methodology to plan e-marketing activities. Business Horizons, 49(1), 51–60. doi:10.1016/j.bushor.2005.05.008

Kukathas, C., & Pettit, P. (1990). Rawls: A Theory of Justice and its Critics. Stanford, CA: Stanford University Press.

Kymlica, W. (2002). Contemporary Political Philosophy: An Introduction (2nd ed.). New York: Oxford University Press.

Laczniak, G. R., & Murphy, P. E. (1993). Ethical marketing decisions: The Higher Road. Boston: Allyn and Bacon.

Laczniak, G. R., & Murphy, P. E. (2006). Marketing, consumers and technology: Perspectives for enhancing ethical transactions. Business Ethics Quarterly, 16(3), 313–321.

Laertius, D. (1985). The Lives and Opinions of Eminent Philosophers. In C. D. Yonge (Eds.), Retrieved from http://classicpersuasion.org/pw/diogenes/

Landes, W., & Posner, R. (2003). The Economic Structure of Intellectual Property Law. Cambridge, UK: Belknap.

Langford, D. (1996). Ethics and the Internet: Appropriate Behavior in Electronic communication. Ethics & Behavior, 6(2), 91–106. doi:10.1207/s15327019eb0602_2

Lea, M., & Spears, R. (1991). Computer-mediated communication, deindividuation and group decision-making. International Journal of Man-Machine Studies, 34, 283–301. doi:10.1016/0020-7373(91)90045-9

Lee, C. S. (2001). An Analytical framework for evaluating e-commerce business models and strategies. Internet Research: Electronic Networking Applications and Policy, 11(4), 349–359. doi:10.1108/10662240110402803

Lessig, L. (1995). The Path of Cyberlaw. The Yale Law Journal, 104, 1743–1755. doi:10.2307/797030

Lessig, L. (1998). The architecture of privacy. Retrieved September 12, 2009, from http://cyber.law.harvard.edu/works/lessig/architecture_priv.pdf

Lessig, L. (1999). Code and Other Laws of Cyberspace. New York: Basic Books.

Lessig, L. (1999). The Law of the Horse: What Cyberlaw Might Teach. Harvard Law Review, 113, 501–546. doi:10.2307/1342331

Lessig, L. (2006). Code 2.0. New York: Basic Books. Retrieved from http://codev2.cc/download+remix/

Liedtke, M. (2007). Facebook lets users block marketing tool. The Associated Press. Retrieved September 12, 2009, from SFGate.com

Limaye, M. R., & Victor, D. A. (1991). Cross-cultural business communication research: State of the art and hypotheses for the 1990s. Journal of Business Communication, 28(3), 277–299. doi:10.1177/002194369102800306

Lippke, R. L. (1989). Advertising and the Social Conditions of Autonomy. Business & Professional Ethics Journal, 8(4), 35–58.

Loch, K. D., Conger, S., & Oz, E. (1998). Ownership, Privacy and Monitoring in the Workplace: A Debate on Technology and Ethics. Journal of Business Ethics, 17, 653–663.

Locke, J. (1980). Second Treatise of Government (Macpherson, C. B., Ed.). Indianapolis, IN: Hackett Publishing Company.

Locke, J. (2005). Two Treatises of Government. Project Gutenberg. Retrieved from http://www.gutenberg.org/ebooks/7370

Locke, J. [1690] (2008). Cambridge Texts in the History of Political Thought: Two Treatises of Government. Edited, introduction, and notes by P. Laslett (Student Edition). New York: Cambridge University Press.

Lohr, S. (2007). As its stock tops $600, Google cases growing risks. Retrieved September 12, 2009, from http://www.nytimes.com/2007/10/13/technology/13google.html

Lucas, A. (2000, Feb. 2). Safeguarding Internet Tax Fairness. Center for Trade Policy Studies. CATO Institute. Retrieved from: http://www.freetrade.org/node/350

Lund, D. B. (2000). An emprical examination of marketing professional's ethical behavior in differing situations. Journal of Business Ethics, 24(4), 331–342. doi:10.1023/A:1006005823045

MacDonald, C. (2008). Green, Inc.: An environmental insider reveals how a good cause has gone bad. Guilford. CT: The Lyons Press.

Mahadevan, B. (2000). Business models for internet-based E-commerce: An anatomy. California Management Review, 42(4), 55–69.

Marcoux, A. (2003). A Fiduciary Argument against Stakeholder Theory. Business Ethics Quarterly, 13(1), 1–24.

Marcoux, A. M. (2003). Snipers, stalkers and nibblers: Online auction business ethics. Journal of Business Ethics, 46, 163–173. doi:10.1023/A:1025001823321

Masclet, D., & Pénard, T. (2008, January). Is the eBay Feedback System Really Efficient? An Experimental Study. Retrieved July 24, 2009, from http://ssrn.com/abstract=1086377

Mattsson, J., & Rendtorff, J. D. (2006). E-marketing ethics: A theory of value priorities. International Journal of Internet Marketing and Advertising, 3(1), 35–47.

Maury, M. D., & Kleiner, D. S. (2002). E-commerce, ethical commerce? Journal of Business Ethics, 36(1/2), 21–31. doi:10.1023/A:1014274301815

McCarthy, E. J. (1964). Basic marketing: A managerial approach (2nd ed.). Boston: Irwin.

McChesney. R. W. (2008). The political economy of media: Enduring issues, emerging dilemmas. New York: Monthly Review of Foundation.

McGregor, J. (2008, May 19). Can't It Wait Till Monday? Business Week, 54.

McLuhan, M. (1962). The Gutenberg Galaxy: The Making of Typographic Man. Toronto, Canada: University of Toronto Press.

McQuivey, J., & DeMaulin, G. (2000). States Lose Half a Billion in Taxes to Web Retail. New York: Forrester Research, Inc.

Meel, M., & Saat, M. (2002). Ethical life cycle of an innovation. *Journal of Business Ethics*, 39(1/2), 21–27. doi:10.1023/A:1016319714894

Megarry, R. Hon. Sir R. E. (1991). Historical Development. In Fiduciary Duties, Law Society of Upper Canada Special Lectures, 1990 (pp. 1-14). Toronto, ON: De Boo.

Menezes, M. A. J. (1993). Ethical issues in product policy . In Smith, N. C., & Quelch, J. A. (Eds.), Ethics in marketing (pp. 283–301). Boston: Irwin.

Meyers, D. (n.d.). The Secret to Happiness, YES! Magazine. Retrieved June 9, 2009, from http://www.yesmagazine.org/article.asp?ID=866

Mill, J. S. (2004). Utilitarianism. Boston: Public Domain Books. doi:10.1522/cla.mij.uti

Miller, S., & Weckert, J. (2000). Privacy, the Workplace and the Internet. *Journal of Business Ethics*, 28, 255–265. doi:10.1023/A:1006232417265

Millett, Hon. P. J. (1998). Equity's Place in the Law of Commerce. *The Law Quarterly Review*, 114, 214–227.

Milne, G. R. (2000). Privacy and ethical issues in database/interactive marketing and public policy: A research framework and overview of the special issue. *Journal of Public Policy & Marketing*, 19(1), 1–6. doi:10.1509/jppm.19.1.1.16934

MIT OpenCourseWare. (2007). MIT OpenCourse-Ware. Retrieved September 4, 2007, from http://ocw.mit.edu/index.html Archived at http://web.archive.org/web/20070904051417/http://ocw.mit.edu/index.html

Mitchell, D. (2007). Online ads vs. privacy. Retrieved September 12, 2009, from http://www.nytimes.com/2007/05/12/technology/12online.html

Monster Press Room. (2009). Plus de la moitié des salariés européens envoient des courriels personnels ou surfent sur le Net durant leurs heures de travail. Retrieved May 9, 2009, from http://presse.monster.fr/13162_fr_p1.asp

Montecino, V. (1996). Copyright and the Internet. Retrieved September 11, 2009, from http://mason.gmu.edu/~montecin/copyright-internet.htm

Moor, J. (1997, September). Towards a theory of privacy in the information age. *Computers & Society*, 27(3), 27–32. doi:10.1145/270858.270866

Moor, J. H. (1990). The Ethics of Privacy Protection. *Library Trends*, 39(1 & 2), 69–82.

Morash, E. A. (2001). Supply chain strategies, capabilities, and performance. *Transportation Journal*, 41(1), 37.

Morgan, F. W. (1993). Incorporating a consumer safety perspective into the product development process . In Smith, N. C., & Quelch, J. A. (Eds.), Ethics in marketing (pp. 350–358). Boston: Irwin.

Motion Picture Association of America. (2002). Study shows copyright industries as largest contributor to the U.S. Economy. Retrieved December 3, 2005, from http://mpaa.org/copyright/2002_04_22.htm Archived at http://web.archive.org/web/20051203235151/http://www.mpaa.org/copyright/2002_04_22.htm

Motion Picture Association of America. (2005). Internet Piracy. Retrieved May 6, 2009, from http://www.mpaa.org/piracy_internet.asp

Muscatello, J., & Emens, S. (2007). Do Reverse Auctions Violate Professional Standards and Codes of Conduct? In Parente, D. H. (Ed.), Best Practices in Online Procurement Auctions (1st ed.). Hershey, PA: IGI Global.

Myskja, B. K. (2008). The categorical imperative and the ethics of trust. *Ethics and Information Technology*, 10(4), 213–220. doi:10.1007/s10676-008-9173-7

Nadel, M. (2004). How Current Copyright Law Discourages Creative Output: The Overlooked Impact of Marketing. *Berkeley Technology Law Journal*, 19, 785-856. Available at SSRN: http://ssrn.com/abstract=489762

Nairn, A., & Dew, A. (2007). Pop-ups, pop-unders, banners and buttons: The ethics of online advertising to primary school children. *Journal of Direct . Data and Digital Marketing Practice*, 9(1), 30–46. doi:10.1057/palgrave.dddmp.4350076

Nakashima, E. (2007). Feeling betrayed, Facebook users force site to hone their privacy. Retrieved September 12, 2009, from http://www.washingtonpost.com/wp-dyn/content/article/2007/11/29/AR2007112902503.html

Neil, P. (1996). Electronic Commerce: Taxation without Clarification. New York: Klynveld, Peat, Marwick, and Gordeler (KPMG).

Newton, L. (2002). A passport for the corporate code: From Borg Warner to the Caux Principles . In Frederick, R. E. (Ed.), A companion to business ethics (pp. 374–385). Malden, MA: Blackwell Publishing.

Nissenbaum, H. (1995). Should I Copy My Neighbor's Software . In Johnson, D., & Nissenbaum, H. (Eds.), Computers, Ethics, and Social Values (pp. 200–212). Upper Saddle River, NJ: Prentice Hall.

Noddings, N. (1984). Caring: A Feminine Approach to Ethics and Moral Education. Berkeley, CA: University of California Press.

Nokia Technological Insights. (2009). Location, Context, and Mobile Services: The Context. Retrieved June 1, 2009, from http://research.nokia.com/files/insight/NTI_Location_&_Context-Jan_2009.pdf

Noll, J. (2001). The Importance of Confidence for Success in E-Commerce. Retrieved July 19, 2009. from http://ssrn.com/abstract=288940

Nozick, R. (1974). Anarchy, State, Utopia. New York: Basic Books.

Nussbaum, K., & duRivage, V. (1986). Computer Monitoring: Mismanagement by Remote Control. Business and Management Review, 56.

Olson, D. L. (2008). Ethical aspects of web log data mining. International Journal of Information Technology and Management, 7(2), 190–200. doi:10.1504/IJITM.2008.016605

Oosterhof, N. N., & Todorov, A. (2008). The Functional Basis of Face Evaluation. Proceedings of the National Academy of Sciences of the United States of America, 105, 11087–11092. doi:10.1073/pnas.0805664105

Oreilly, T. (2007). What is Web 2.0: Design Patterns and Business Models for the Next Generation of Software. Communications & Strategies, 1(1), 17.

Osterwalder, A., & Pigneur, Y. (2002, June 17-19). An e-Business Model Ontology for Modeling e-Business. Presented at the 15th Bled Electronic Commerce Conference e-Reality: Constructing the e-Economy. Bled, Slovenia.

Ozanne, J. L., & Saatcioglu, B. (2008). Participatory Action Research. The Journal of Consumer Research, 35(3), 423–439. doi:10.1086/586911

Palmer, D. (2005). Pop-Ups, Cookies, and Spam: Toward a Deeper Analysis of the Ethics Significance of Internet Marketing Practices. Journal of Business Ethics 58, 271-280.

Palmer, D. E. (2005). Pop-Ups, Cookies, and Spam: Toward a Deeper Analysis of the Ethical Significance of Internet Marketing Practices. Journal of Business Ethics, 58, 271–280. doi:10.1007/s10551-005-1421-8

Palmer, D. E., Stoll, M. L., & Zakhem, A. (2008). Introduction . In Palmer, D. E., Stoll, M. L., & Zakhem, A. (Eds.), Stakeholder theory: Essential readings in ethical leadership and management (pp. 15–25). Amherst, NY: Prometheus Books.

Paradigm - On Definition. (n.d.). Criticism Of Kuhn's Paradigms, Revolutions, Leaps Of Faith, Criticism Of Kuhn's Relativism. Retrieved August 4, 2009, from http://science.jrank.org/pages/7948/ Paradigm.html#ixzz0P1OrmeJJ

Parental Control Software. (n.d.). Internet Pornography Statistics, Retrieved September 27, 2009, from http://www.parental-control-software-top5.com/internet-statistics.html

Parliament of England. (1710). The Statute of Anne. Retrieved from http://www.copyrighthistory.com/anne.html

Pateli, A. G., & Giaglis, G. M. (2005). Technology innovation-induced business model change: A contingency approach. Journal of Organizational Change Management, 18(2), 167–183. doi:10.1108/09534810510589589

Peace, G., Weber, J., Hartzel, K., & Nightingale, J. (2002). Ethical Issues in eBusiness: A Proposal for Creating the eBusiness Principles. [from Business Source Complete database.]. Business and Society Review, 107(1), 41. Retrieved April 20, 2009. doi:10.1111/0045-3609.00126

Pedrasa, J. R., Perera, E., & Seneviratne, A. (2008). Context Aware Mobility Management . In Unhelkar, B. (Ed.), In Handbook of Research in Mobile Business: Technical, Methodological, and Social Perspectives (2nd ed.). Hershey, PA: IGI Global.

Perelman, M. (1991). Information, Social Relations and the Economics of High Technology. New York: St. Martin's.

Perelman, M. (2002). Steal this Idea: Intellectual Property Rights and the Corporate Confiscation of Creativity. New York: Palgrave.

Perez, J. C. (2007a). Facebook admits ad service tracks logged-off users. Retrieved September 12, 2009, from http://www.pcworld.com/article/140225/facebook_admits_ad_service_tracks_loggedoff_users.html

Perez, J. C. (2007b). Facebook Tweaks beacon again; CEO apologizes. Retrieved September 12, 2009, http://www.pcworld.com/article/140322/facebook_tweaks_beacon_again_ceo_apologizes.html

Perez, J. C. (2007c). Facebook doesn't budge on Beacon's broad user tracking. Retrieved September 12, 2009, from http://www.pcworld.com/article/140385/facebook_doesnt_budge_on_beacons_broad_user_tracking.html

Perez, J. C., & Gohring, N. (2007). Facebook partners quiet on beacon fallout. Retrieved September 12, 2009, from http://www.pcworld.com/businesscenter/article/140450/facebook_partners_quiet_on_beacon_fallout.htm

Peyrelong, M.-F., & Accart, J. P. (2002). Du système d'information personnel au système d'information collectif réalités et mirages du partage d'information en entreprise. In Conference Proceedings, Canadian Association for Information Science, 135-149.

Phillips, M. (1994). The Inconclusive Ethical Case Against Manipulative Advertising. Business & Professional Ethics Journal, 13(4), 31–64.

Posner, R. (1981). The economics of privacy. The American Economic Review, 71(2), 405–409.

Posner, R. A. (2002). Economic Analysis of Law (6th ed.). New York: Aspen.

Postman, N., & Powers, S. (1992). How to Watch TV News. Boston: Penguin (Non-Classics).

Powell, D. C. (2000). Internet Taxation and U.S. Intergovernmental Relations: From Quill to the Present. Publius, 30(1), 39–51. doi:10.2307/3331120

Prosser, W. D. (1960). Privacy. California Law Review, 48, 383–396. doi:10.2307/3478805

Quinn, M. (2005). Ethics for the Information Age. Boston, MA: Pearson Addison Wesley Inc.

Rachels, J. (1975). Why is privacy important? Philosophy & Public Affairs, 4(4), 323–333.

Rallapalli, K. C., Vitell, S. J., & Szeinbach, S. (2000). Marketers' norms and personal values: An empirical study of marketing professionals. Journal of Business Ethics, 24(1), 65–75. doi:10.1023/A:1006068130157

Rampell, C. (2008). What Facebook knows that you don't. Retrieved September 12, 2009, from http://www.washingtonpost.com/wp-dyn/content/article/2008/02/22/AR2008022202630.html

Rao, S., & Quester, P. (2006). Ethical marketing in the internet era: A research agenda . International Journal of Internet Marketing and Advertising, 3(1), 19–34.

Rawls, J. (1971). A Theory of Justice. Cambridge, MA: Harvard University Press.

Reiman, J. H. (1975). Privacy, intimacy, and personhood. Philosophy & Public Affairs, 6(1), 26–44.

Reiman, J. H. (1995). Driving to the panopticon. Santa Clara Computer and High-Technology Law Journal, 11(1), 27–44.

Ribstein, L. E. (2001). Law v. Trust. Boston University Law Review. Boston University. School of Law, 81, 553–590.

Robins, F. (2000). The E-marketing mix. The Marketing Review, 1(2), 249–274. doi:10.1362/1469347002529134

Robischom, N. (1997). Rx for Medical Privacy. Netly News, Sept 3.

Roche, M. (2004, July 26). Scientific Web Site Optimization using AB Split Testing, Multi Variable Testing, and The Taguchi Method | WebProNews. Retrieved June 12, 2009, from http://www.webpronews.com/topnews/2004/07/26/scientific-web-site-optimization-using-ab-split-testing-multi-variable-testing-and-the-taguchi-method

Rodin, T. (2001). The privacy paradox: E-commerce and personal information on the internet. Business & Professional Ethics Journal, 20(3-4), 145–170.

Rogers, B. (2008). Contract sales organisations: Making the transition from tactical resource to strategic partnering. Journal of Medical Marketing, 8(1), 39–47. doi:10.1057/palgrave.jmm.5050119

Rogers, C., & Kinget, G. M. (1966). Psychothérapie et relations humaines. Louvain, Belgique: Publications Universitaires.

Roman, S. (2007). The ethics of online retailing: A scale development and validation from the consumers' perspective. Journal of Business Ethics, 72(2), 131–148. doi:10.1007/s10551-006-9161-y

Roman, S. (2007). The Ethics of Online Retailing: A Scale Development and Validation from the Consumers' Perspective. Journal of Business Ethics, 72, 131–148. doi:10.1007/s10551-006-9161-y

Roman, S., & Cuestas, P. J. (2008). The perceptions of consumers regarding online retailers' ethics and their relationship with consumers' general internet expertise and word of mouth: A preliminary analysis. Journal of Business Ethics, 83(4), 641–656. doi:10.1007/s10551-007-9645-4

Rosen, J. (2001). The unwanted gaze: The destruction of privacy in America. New York: Vintage Books.

Rotman, L. I. (1996). Fiduciary Doctrine: A Concept in Need of Understanding. Alberta Law Review, 34, 821–852.

Rotman, L. I. (1996). The Vulnerable Position of Fiduciary Doctrine in the Supreme Court of Canada. Manitoba Law Journal, 24, 60–91.

Rotman, L. I. (2005). Fiduciary Law. Toronto, ON: Thomson.

Rousseau, J.-J. (1987). The Basic Political Writings. Indianapolis, IN: Hackett Publishing Company.

Sama, L. M., & Shoaf, V. (2002). Ethics on the web: Applying moral decision-making to the new media. Journal of Business Ethics, 36(1/2), 93–103. doi:10.1023/A:1014296128397

Samarajiva, R. (1997). Interactivity as though privacy mattered . In Agre, P. E., & Rogenberg, M. (Eds.), Technology and privacy: The new landscape. Cambridge, MA: MIT Press.

Samavi, R., Yu, E., & Topaloglou, T. (2009). Strategic reasoning about business models: A conceptual modeling approach. Information Systems & e-Business Management, 7(2), 171-198.

Samuelson, P. (2000). Privacy as intellectual property? Stanford Law Review, 52(5), 1125–1173. doi:10.2307/1229511

Sandholm, T., Levine, D., Concordia, M., Martyn, P., Hughes, R., & Jacobs, J. (2006). Changing the game in strategic sourcing at procter & gamble: Expressive competition enabled by optimization. Interfaces, 36(1), 55–68. doi:10.1287/inte.1050.0185

Sarbaugh-Thompson, M., & Feldman, M. S. (1998). Electronic mail and organizational communication: Does saying "hi" really matter? Organization Science, 9, 685–698. doi:10.1287/orsc.9.6.685

Saunders, Ben. (2008). The Equality of Lotteries. Philosophy: Journal of the Royal Institute of Philosophy, 83(325), 359-372.

Schenk, D. (1996). Taxation. In New York University School of Law (Ed.), Fundamentals of American Law. New York: Oxford University Press.

Schlegelmilch, B. B. (1998). Marketing ethics: An international perspective. London: International Thomson Business Press.

Schoell, W. F., Dessler, G., & Reinecke, J. A. (1993). Introduction to Business. Boston: Alyn and Bacon.

Schoeman, F. (1984). Privacy: Philosophical dimensions of the literature . In Schoeman, F. D. (Ed.), Philosophical dimensions of privacy: An anthology. Cambridge, UK: Cambridge University Press. doi:10.1017/CBO9780511625138.002

Schwartz, B. (2004). The Paradox of Choice: why more is less. New York: Ecco.

Schweizer, L. (2005). Concept and evolution of business models. Journal of General Management, 31(2), 37–56.

Scolove, D. J. (2006). A taxonomy of privacy. University of Pennsylvania Law Review, 154(3), 477.

Scolove, D. J. (2007). 'I've got nothing to hide,' and other misunderstandings of privacy. The San Diego Law Review, 44.

Search Engine Marketing (SEM) Glossary. (n.d.). Retrieved September 7, 2009, from http://www.anvilmediainc.com/search-engine-marketing-glossary.html sitespect.com (n.d.). Online Businesses Doing More With Less In An Uncertain Economy Via Web Optimization Technology. Retrieved May 20, 2009, from http://www.sitespect.com/news-online-business-121608.shtml

Simmel, G., & Wolff, K. H. (1950). The sociology of georg simmel. Glencoe, IL: Free Press.

Simmons, J. A. (1989). Locke's State of Nature. Political Theory, 17(3), 449–470.

Singleton, S. (1998). Privacy as censorship. Policy Analysis, 295.

Sirgy, J., & Lee, D. (2008). Well-being Marketing: An Ethical Business Philosophy for Consumer Goods Firms. Journal of Business Ethics, 77(4), 377–403. doi:10.1007/s10551-007-9363-y

Sirgy, J., & Su, C. (2000). The Ethics of Consumer Sovereignty in the Age of High Tech. Journal of Business Ethics, 28, 1–14. doi:10.1023/A:1006285701103

Slywotzky, A. J. (1996). Value Migration: How to think several moves ahead of the competition. Boston, MA: Harvard Business School Press.

Smith, A. (1981). An Inquiry Into the Wealth of Nations. Indianapolis, IN: Liberty Fund.

Smith, A. (1982). The Theory of Moral Sentiments. Indianapolis, IN: Liberty Fund.

Smith, N. C. (1993). Ethics and the marketing manager . In Smith, N. C., & Quelch, J. A. (Eds.), Ethics in marketing (pp. 3–34). Boston: Irwin.

Smith, R. E. Privacy: How to protect what's left of it. Garden City, NJ: Doubleday Books.

Sneddon, A. (2001). Advertising and Deep Autonomy. Journal of Business Ethics, 33(1), 15–28. doi:10.1023/A:1011929725518

Solomon, R. C. (1992). Ethics and excellence: Cooperation and integrity in business. Oxford, UK: Oxford University Press.

Spence, E. (Ed.), & Van Heekeren, B. (Ed.). (2005). Advertising ethics. Upper Saddle, NJ: Prentice Hall.

Spence, M. (1973). Job Market Signaling. The Quarterly Journal of Economics, 87(3), 355–374. doi:10.2307/1882010

Spinello, R. A. (2006). Cyberethics: Morality and Law in Cyberspace (3rd ed.). Boston, MA: Jones and Bartlett Publishers.

Sproull, L. O., & Kiesler, S. (1986). Reducing social context cues: Electronic mail in organizational communication. Management Science, 32, 1492–1512. doi:10.1287/mnsc.32.11.1492

Stallman, R. (2004). Why Software Should Not Have Owners. The GNU Project and the Free Software Foundation. Retrieved from http://www.gnu.org/philosophy/why-free.html

Stead, B. A., & Gilbert, J. (2001). Ethical issues in electronic commerce. Journal of Business Ethics, 34(2), 75–85. doi:10.1023/A:1012266020988

Stiglitz, J. E. (1975). Information and Economic Analysis . In Parkin, J. M., & Nobay, A. R. (Eds.), Current Economic Problems (pp. 27–52). Cambridge, UK: Cambridge University Press.

Stoll, K. (2004, August 18). Color your hunger. Retrieved September 13, 2009, from http://findarticles.com/p/articles/mi_qn4179/is_20040818/ai_n11815804/

Stoll, M. L. (2002). The ethics of marketing good corporate conduct. Journal of Business Ethics, 41(1), 121–129. doi:10.1023/A:1021306407656

Stoll, M. L. (2008). Backlash hits business ethics: Finding effective strategies for communicating the importance of corporate social responsibility. Journal of Business Ethics, 78(1-2), 17–24. doi:10.1007/s10551-006-9311-2

Sultan, F., Urban, G. L., Shankar, V., & Bart, Y. Y. (2002, December). Determinants and Role of Trust in E-Business: A Large Scale Empirical Study (MIT Sloan Working Paper No. 4282-02). Retrieved July 19, 2009, from http://ssrn.com/abstract=380404

Tassabehji, R., Taylor, W. A., Beach, R., & Wood, A. (2006). Reverse e-auctions and supplier-buyer relationships: An exploratory study. International Journal of Operations & Production Management, 26(2), 166–184. doi:10.1108/01443570610641657

Tavani, H. T. (2005). Locke, Intellectual Property Rights, and the Information Commons. Ethics and Information Technology, 7, 87–97. doi:10.1007/s10676-005-4584-1

Taylor, H., Fieldman, G., & Altman, Y. (2008). E-mail at work: A cause of concern? The implications of the new technologies for health, wellbeing and productivity at work. Journal of Organizational Transformation and Social Change, 5(2), 159–173. doi:10.1386/jots.5.2.159_1

Technology Weekly. (2007, May 30). Quant is King. Retrieved June 13, 2009, from http://technologyweekly.mad.co.uk/Main/Home/Articlex/6f452b7cb4db47a8afc00ff9b8fb9752/Quant-is King.html

Thomson, J. J. (1975)... Philosophy & Public Affairs, 4(4), 295–322.

Today, U. S. A. (2007). Fired Wal-Mart worker reveals covert operations. Retrieved September 12, 2009, from http://www.usatoday.com/money/industries/retail/2007-04-04-walmart-spying_N.htm

Travani, H. T., & Moor, J. H. (2001, March). Privacy protection, control of information, and privacy –enhancing technologies. Computers & Society, 31(1), 6–11. doi:10.1145/572277.572278

Tsalikis, J., & Fritzsche, D. (1989). Business ethics: A literature review with a focus on marketing ethics. Journal of Business Ethics, 8(9), 695–743. doi:10.1007/BF00384207

U.S. Census Bureau. (2009, May). E-Stats. Retrieved July 9, 2009, from U.S. Census http://www.census.gov/estats

U.S. Department of Justice. (1998). The No Electronic Theft ("NET") Act: Relevant portions of 17 U.S.C. and 18 U.S.C. as amended (redlined). Retrieved from http://www.usdoj.gov/criminal/cybercrime/17-18red.htm

United States Access Board. (n.d.). Retrieved September 11, 2009, from http://www.access-board.gov/

Unites States Census Bureau. (2008). E-Stats: E-Commerce 2006. Retrieved April 20, 2009, from http://www.census.gov/eos/www/2006/2006reportfinal.pdf

Urban, G. L. (2003, March). The Trust Imperative (MIT Sloan Working Paper No. 4302-03). Retrieved July 19, 2009, from http://ssrn.com/abstract=400421

Urban, G. L., Sultan, F., & Qualls, W. J. (2000). Placing trust at the center of your internet strategy. Sloan Management Review, 42(1), 39–48.

Vaidhyanathan, S. (2003). Copyrights and Copywrongs: The Rise of Intellectual Property and How It Threatens Creativity. New York: New York University Press.

Vinter, E. A. (1955). A Treatise on the History and Law of Fiduciary Relationships and Resulting Trust. Cambridge, UK. W.: Heffer.

Vitell, S. J., Dickerson, E. B., & Festervand, T. A. (2000). Ethical problems, conflicts and beliefs of small business professionals. Journal of Business Ethics, 28(1), 15–24. doi:10.1023/A:1006217129077

Wallace, K. A. (2008). Online Anonymity. In K. E. Himma. & H. T. Tavani (Eds.), The Handbook of Information and Computer Ethics (pp. 165-189). Hoboken, NJ: John Wiley & Sons, Inc.

Wallace, P. M. (1999). The psychology of the Internet. Cambridge, UK: Cambridge University Press.

Warren, S. D., & Brandeis, L. D. (1890). The right to privacy. Harvard Law Review, 4(5), 193. doi:10.2307/1321160

Warren, S. D., & Brandeis, L. D. (1890). The right to privacy. Harvard Law Review, 4(5). doi:10.2307/1321160

Wasserstrom, R. A. (1978). Privacy: some arguments and assumptions . In Schoeman, F. D. (Ed.), Philosophical dimensions of privacy: An anthology. Cambridge, UK: Cambridge University Press.

Weigand, E. (1999). Misunderstanding: The standard case. Journal of Pragmatics, 31, 763–785. doi:10.1016/S0378-2166(98)00068-X

Weinrib, E. J. (1975). The Fiduciary Obligation. The University of Toronto Law Journal, 25, 1–22. doi:10.2307/824874

Werhane, P. H. (2002). Business ethics and the origins of contemporary capitalism: economics and ethics in the work of Adam Smith and Herbert Spencer . In Frederick, R. E. (Ed.), A companion to business ethics (pp. 325–341). Malden, MA: Blackwell Publishing.

Westin, A. (1967). Privacy and freedom. New York: Atheneum.

Westin, A. F. (1966). Science, privacy, and freedom: Issues And proposals for the 1970s. Part I-the current impact of surveillance on privacy. Columbia Law Review, 66(6). doi:10.2307/1120997

Whitfield, T., & Wiltshire, T. (1990). Color psychology: a critical review. Genetic, Social, and General Psychology Monographs, 116(4), 385–411.

Williams, D. H. (2006, October 25th). LBS Development – Determining Privacy Requirements. Directions Magazine. Retrieved June 1, 2009, from http://www.directionsmag.com/article.php?article_id=2323

Williams, O. F., & Murphy, P. E. (1990). The ethics of virtue: A moral theory for marketing. Journal of Macromarketing, 10(1), 19–29. doi:10.1177/027614679001000103

Wilson, J. M., Straus, & McEvily, B. (2006). All In due time: the development of trust in computer-mediated and face-to-face teams. Organizational Behavior and Human Decision Processes, 99, 16–33. doi:10.1016/j.obhdp.2005.08.001

Womack, L., Braswell, D. E., & Harmon, W. K. (2004). Email Policy Enforcement: Techniques and Legality. Journal of Accounting and Finance Research, Summer(2), 102-108.

Yudelson, J. (1999). Adapting McCarthy's four P's for the twenty-first century. Journal of Marketing Education, 21(1), 60–67. doi:10.1177/0273475399211008

Zahay, D., Peltier, J., Schultz, D., & Griffin, A. (2004). The role of transactional versus relational data in IMC programs: bringing customer data together. Journal of Advertising Research, 44(1), 3–18. doi:10.1017/S0021849904040188

Zeller, T. (2006). Your life as an open book. Retrieved September 12, 2009, from http://www.nytimes.com/2006/08/12/technology/12privacy.html

Zemer, L. (2006). The Making of a New Copyright Lockean. Harvard Journal of Law & Public Policy, 29(3), 891–947.

Zhu, K. (2004). Information transparency of business-to-business electronic markets: A game-theoretic analysis. Management Science, 50(5), 670–685. doi:10.1287/mnsc.1040.0226

About the Contributors

Daniel E. Palmer is an associate professor of philosophy at Kent State University, Trumbull Campus. His primary research interests are in ethical theory and applied ethics, with particular emphasis upon business ethics and health care ethics. He is a co-editor of the volume Stakeholder Theory: Essential Readings in Ethical Leadership and Management (Prometheus Books, 2008), and his publications on business ethics have appeared in such scholarly journals as the Journal of Business Ethics and Business Ethics Quarterly. Dr. Palmer's current research interests include exploring the ethical issues found in e-business, international business communication, and green marketing practices.

***

James B. Coleman received B.A. and M.A. degrees in philosophy from Kent State University in Ohio, and received a Ph.D. in philosophy Purdue University in 2002. Professor Coleman is currently an Assistant Professor at Central Michigan University in Mount Pleasant, Michigan. He works on issues involving applied and professional ethics, and teaches classes in a variety of areas in philosophy, including philosophy of law, philosophy of religion, business ethics, and ethical theory. Among his current projects is the composition and design of a web-based course in professional ethics. He is also working on Hume's ethical theory.

Susan Emens is an Assistant Professor at Kent State University's Trumbull Campus in the Business Management Technology Program. Her primary research interests are in the areas of marketing for the service and non-profit industry. Previously published work includes a chapter in IGI Global's volume, *Best Practices for Online Procurement Auctions*. Professor Emens has previous industry experience in a variety of marketing positions in the manufacturing and software industries.

Matt Hettche is an Assistant Professor of Marketing at Christopher Newport University (Newport News, Virginia USA). His research interests include topics in consumer behavior, social marketing, and European intellectual history. His most recent research project involves a meta-analysis of how fear and shock as advertising appeals can be used to demarket drug addiction and substance abuse. Dr. Hettche is particularly interested in understanding how the tools and techniques of marketing can be applied to contemporary social problems.

Fernando Lagraña is an adjunct professor of management at the Webster University Geneva, Switzerland. He is also a research fellow at the Grenoble School of Management, France. Fernando is currently in charge of private-public partnerships in the Telecommunication Development Bureau of the International Telecommunication Union and has more than thirty years of experience in the field of telecommunications and of information and communication technologies, both at the national and international level. His current research interests include electronic communications, and, in particular, issues of e-mail and friendly spam, as well as business ethics and interpersonal and organizational behaviors.

Kirsten Martin is an assistant professor at The Catholic University of America where she teaches business ethics, leadership, and strategy. Dr. Martin received her Ph.D. and MBA from the University of Virginia's Darden School of Business and her B.S. Engineering from the University of Michigan. In conjunction to her work at CUA, Dr. Martin is a Business Roundtable Institute for Corporate Ethics fellow conducting research and writing cases on innovation and privacy. Dr. Martin has published several articles in the *Journal of Business Ethics*, has chapters in several ethics academic texts, and co-authored a business ethics textbook. Her work has been reprinted overseas and has been featured in the *Financial Times*. Dr. Martin's research is focused on business ethics, technology, and privacy.

Erkan Özdemir received his PhD in marketing from Uludag University in 2005. He is currently a research assistant in the marketing program of the Department of Business Administration at Uludag University in Bursa, Turkey. His current research and educational interests include marketing ethics, e-marketing, online consumer behaviors, high-tech marketing, and gender based marketing strategies.

Leonard I. Rotman is Professor of Law at the Faculty of Law, University of Windsor. He has authored 5 books and over 80 chapters, essays, and articles in Aboriginal Law, Constitutional Law, Corporate Law and Governance, Equity, Fiduciary Law, Trusts, Remedies, Unincorporated Associations, and Unjust Enrichment/Restitution. His work has been cited by domestic and international commissions and courts, including the Supreme Court of Canada. He has presented papers at law schools and conferences in Canada, the United States, England, France, and Israel, including the American Association of Law Schools, British Association of Canadian Studies, the International Association of Legal Methodology, Boston College School of Law, Boston University School of Law, Duke University School of Law, Southeastern Association of Law Schools, and the 12th Biennial Jerusalem Conference in Canadian Studies. He was a visiting professor at Washington and Lee University, School of Law in the Fall Term, 2008.

Eric M. Rovie is currently a Visiting Instructor in the Philosophy Department at Agnes Scott College in Atlanta. He holds graduate degrees in Philosophy from Georgia State University and Washington University in St Louis, and is co-editor (with Larry May and Steve Viner) of *The Morality of War: Classical and Contemporary Readings* (Prentice-Hall). He works in normative and applied ethics, with particular interests in the problem of dirty hands and virtue ethics. Recent publications include "Reevaluating the History of Double Effect: Anscombe, Aquinas, and the Principle of Side Effects" (*Studies in the History of Ethics*) and "Tortured Knowledge: Epistemological Problems With Ticking Bomb Cases" (forthcoming in the *International Journal of Applied Philosophy*).

J.J. Sylvia IV recently graduated from The University of Southern Mississippi with an M.A. in philosophy. He was also the creator, webmaster, and customer service manager for the e-commerce website for Unfinished Furniture Showcase for almost a decade. One of his major areas of interest includes the ethical and epistemological issues involved with all forms of media. Because of how immersed we are daily in these forms of communication and entertainment, be it radio, television, or the internet, he believes it is important to stop and ask questions about how such media may affect us, both positively and negatively. It is only through asking these questions that we can then make the best use of such media.

Mary Lyn Stoll is currently a member of the Department of Philosophy and Political Science at the University of Southern Indiana in Evansville, Indiana. Dr. Stoll's past research has been devoted to questions concerning the overlap between media ethics and business ethics. She has published articles discussing boycotts, communicating the value of corporate social responsibility, the moral obligations of multimedia conglomerates, and the nature and extent of corporate rights to free speech. She continues to be concerned with issues involving corporate accountability and developing the measures necessary to create socioeconomic conditions conducive to fostering morally acceptable corporate conduct.

Andrew Terjesen is currently a visiting Assistant Professor of Philosophy at Rhodes College in Memphis, TN. He previously held positions at Washington & Lee University and Austin College. His dissertation was on the role of empathy in moral judgment with special emphasis on the work of David Hume and Adam Smith. His work on Adam Smith's moral theory has instilled an interest in the philosophy of economics and problems in business ethics, with particular attention being paid to how the developments in business and economics have diverged from Smith's ideas. In addition, he remains interested in the possibility of developing a modern form of 18th Century moral sentimentalism that does not fall prey to charges that it is committed to moral relativism.

D.E. Wittkower received a Ph.D in Philosophy from Vanderbilt University in 2006. His training concentrated on German philosophy and the history of value theory (ethics, aesthetics, social/political philosophy), and his research has concentrated primarily on issues of technology and political philosophy. Current projects focus on the way digital media shape online culture, including questions of how social networks impact our ethical reasoning, how community membership and political action are formulated online, and why online communications have trended towards an unforeseen predominance of the aesthetics of the cute. He also writes philosophy for a broader audience, and is editor of *iPod and Philosophy* (Open Court, 2008), *Monk and Philosophy* (Open Court, forthcoming 2009), and *Facebook and Philosophy* (Open Court, forthcoming 2010).

Abe Zakhem holds a PhD in philosophy and is currently an assistant professor of philosophy at Seton Hall University. Before returning to academia, he worked in private industry as an organizational management consultant and chief operating officer. His areas of specialization are in business ethics, corporate social responsibility, corporate governance, and management theory. Dr. Zakhem's published work covers a range of theoretical and practical ethical issues in stakeholder management theory, organizational communication, and corporate governance.

Index

A

A/B and multivariate testing 91, 92, 95, 96, 103

A/B and multivariate tests 92, 101, 103

A/B and multivariate website optimization 91, 93

A/B split testing 92

Advertising Standards Authority (ASA) 125

Amazon 18, 29, 35, 36, 41, 43, 44, 45, 46, 55, 56, 136

Amazon.com's Marketplace 5, 29, 35, 44, 46

American Cancer Society 93, 98, 103

applied ethics 3, 12

Aristotelian 51, 100

Aristotle 202, 203, 210

A-to-Z Guarantee 35

Australian Advertiser Code of Ethics 121

automatic teller machine (ATM) 60

B

bandwidth 85

Barnes and Noble 17

Beatles, the 157, 158

behavioral marketing 136, 137

Better Business Bureau Code of Advertising 121

Blackberry 209

blind carbon copy 203, 204

blog 80, 127, 130

Boeing 23

brick-and-mortar 173, 183, 189, 191, 192, 194, 196

brick-and-mortar businesses 5, 62

bricks and mortar store 41, 43, 44, 47, 48

British Codes of Advertising Sales Promotion 121

British Petroleum (BP) 122

BSD license 160

business model 15, 16, 17, 18, 19, 20, 21, 22, 23, 24, 25, 26, 152, 153, 167

business practices 1, 2, 3, 5, 6, 7, 9, 11, 12, 172, 179, 184, 185

Business Software Alliance 155

business to business (B2B) 16, 17, 24, 25, 61, 201

business to consumer (B2C) 61, 201

business transactions 1, 6, 8, 9, 10, 188, 191

buyer 41, 42, 44, 45, 46, 47

buyers 15, 17, 18, 19, 20, 21, 22, 24, 25, 28, 172, 173, 174, 175, 178

buyer-supplier relationship 21, 24

C

Canadian Code of Advertising Standards 121

capitalism 97

Children's Online Privacy Protection Act (COPPA) 91, 103

cognitive morality development 108

computer-mediated communications (CMC) 201

consumer to consumer 29

contractarianism 188, 193, 194, 195, 196, 197

cookies 180

Copyright Act of 1790 155

corporate social responsibility (CSR) 120, 121, 122, 123, 124, 126, 127, 128, 129, 130, 131, 132, 133

Craigslist 5, 62, 63

Creative Commons license 153, 155

Breinigsville, PA USA
03 December 2010
250585BV00003B/9/P